Does America Need More Innovators?

D0933446

Lemelson Center Studies in Invention and Innovation

Joyce Bedi, Arthur Daemmrich, and Arthur P. Molella, general editors

Arthur P. Molella and Joyce Bedi, editors, *Inventing for the Environment*

Paul E. Ceruzzi, *Internet Alley: High Technology in Tysons Corner, 1945–2005*

Robert H. Kargon and Arthur P. Molella, *Invented Edens: Techno-Cities of the Twentieth Century*

Kurt W. Beyer, *Grace Hopper and the Invention of the Information Age*

Michael Brian Schiffer, *Power Struggles: Scientific Authority and the Creation of Practical Electricity before Edison*

Regina Lee Blaszczyk, *The Color Revolution*

Sarah Kate Gillespie, *The Early American Daguerreotype: Cross-Currents in Art and Technology*

Matthew Wisnioski, Eric S. Hintz, and Marie Stettler Kleine, editors, *Does America Need More Innovators?*

Does America Need More Innovators?

Matthew Wisnioski, Eric S. Hintz, and Marie Stettler Kleine, editors

The MIT Press
Cambridge, Massachusetts
London, England

This book was set in Stone Serif by Westchester Publishing Services. Printed and bound in the United States of America.

Library of Congress Cataloging-in-Publication Data is available.

Names: Wisnioski, Matthew H., 1978– editor. | Hintz, Eric S., editor. |
    Kleine, Marie Stettler, editor.
Title: Does America need more innovators? / edited by Matthew Wisnioski,
    Eric S. Hintz, and Marie Stettler Kleine.
Description: Cambridge, MA : The MIT Press, [2019] | Series: Lemelson center
    studies in invention and innovation series | Includes bibliographical references
    and index.
Identifiers: LCCN 2018034415 | ISBN 9780262536738 (pbk. : alk. paper)
Subjects: LCSH: Engineering and state—United States. | Technological
    innovations—United States.
Classification: LCC T21 .D637 2019 | DDC 338.973/06—dc23 LC record available
    at https://lccn.loc.gov/2018034415

10 9 8 7 6 5 4 3 2 1

Contents

Series Foreword

Is it possible to go an entire day in our interconnected, technology-mediated world without hearing the word "innovation" at least once? Used for everything from marketing gimmicks to truly game changing breakthroughs, innovation has simultaneously become a goal, a measure of success, and a compulsion for many individuals and corporations. We are urged to be innovative at work and in our personal lives, and to help our children be innovative as well. What are the pros and cons of this mandate to innovate? This volume begins to answer that question by putting champions of innovation in conversation with those who question the imperative to innovate, as well as those who trust in the imperative but see ways to improve it. The result is a richly textured picture of the ubiquity of innovation in American life.

Invention and innovation have long been recognized as transformational forces in American history, not only in technological realms but also in politics, society, and culture, and they are arguably more important than previously thought in other societies as well. Innovation especially has become a universal watchword of the twenty-first century, so much so that nations are banking their futures on its economic and social effects.

Since 1995, the Smithsonian's Lemelson Center has been investigating the history of invention and innovation from broad interdisciplinary perspectives. Books in the Lemelson Center Studies in Invention and Innovation continue this work to enhance public understanding of humanity's inventive impulse. Authors in the series raise new questions about the work of inventors and the technologies they create, while stimulating cross-disciplinary dialogue. By opening channels of communication between the various disciplines and sectors of society concerned with technological innovation, the Lemelson Center Studies advance scholarship in the history of technology, engineering, science, architecture, the arts, and related fields and disseminate it to a general interest audience.

Joyce Bedi, Arthur Daemmrich, and Arthur P. Molella
Series editors, Lemelson Center Studies in Invention and Innovation

Acknowledgments

Engaging in dialogue with those from different communities and differing points of view requires commitment and generosity. We are grateful to the many contributors who gave us both as they submitted their ideas about innovation for testing in this multidisciplinary project, especially as debates about innovation's relative merits grow increasingly heated.

The volume emerged from a workshop jointly held at Virginia Tech's Arlington Research Center and the Lemelson Center for the Study of Invention and Innovation at the Smithsonian's National Museum of American History. For financial support, we are thankful to the National Science Foundation (SES 1354121); the Lemelson Center; Virginia Tech's Institute for Creativity, Arts, and Technology; the Virginia Tech Department of Science, Technology, and Society; and Virginia Tech's College of Liberal Arts and Human Sciences. The arguments in the volume were sharpened greatly by conference participants including Janet Abbate, David Kirsch, Benjamin Knapp, Doreen Lorenzo, Sonja Schmid, Timothy Sands, Monica Smith, Walter Valdivia, and Kari Zacharias.

The volume would not have been possible without the hard work of a large supporting cast. This has included Audra Wolfe, Doris Shelor, Karen Snider, Chris Gauthier, Sanjay Raman, Paula Byron, Joyce Bedi, and Katie Helke. Finally, we thank Cindy Rosenbaum, Emma Hintz, and Kaleb Kleine for tolerating this ongoing dialogue over many weeks and months.

1 The Innovator Imperative

Matthew Wisnioski

"Our nation knows what it takes to innovate," the prestigious American Academy of Arts and Sciences (AAAS) declared in its 2015 report "Innovation: An American Imperative." To be a "global innovation leader" requires federal support, tax incentives, the pursuit of emerging technologies, a welcome environment for talent, better STEM education, and a meritocratic culture. But, warns "Innovation" and its five hundred signatories from Google to the American Dairy Science Association, "now is not the time to rest on past success." While competitors have adopted our playbook, the United States has stagnated, putting the American dream at risk.[1]

Variations on the AAAS's manifesto dominate visions of the future of the United States. Corporate executives, government leaders, and local schoolboards agree that Americans must innovate. The imperative is remarkably capacious. Innovation today describes everything from the commercialization of new technology to economic policy, design, artistic imagination, and grassroots community renewal.

The demand for innovation is as much a call for *new kinds of people* as it is for national investment. Implicit in the AAAS's plan is an imperative to create *innovators*, the citizens who will make new discoveries, disrupt old ways, solve once intractable social problems, create wealth, and ensure national supremacy. These innovators include not only engineers and scientists but also entrepreneurs, inventors, designers, and civic leaders with the mindsets and tools of "change makers."

The movement to cultivate a new generation of innovators has fueled the rise of *innovation experts*. These champions of innovation lead initiatives to make innovators at all career stages. Business gurus sell how-to books, while universities such as Stanford and Arizona State offer models for producing entrepreneurs, start-up companies, and regional growth.[2] Innovator initiatives are far ranging: the National Science Foundation (NSF), traditionally viewed as a funder of basic science, now teaches midcareer biologists to translate scientific discoveries into marketable products through its Innovation Corps (I-Corps). On the outskirts of Moscow, Singapore, and Abu Dhabi, the Massachusetts Institute of Technology (MIT) partners with other nations to export its blueprint for making innovators around the world.[3] Finally, innovation experts urge parents to mold their children into innovators through creative play, facilitated by invention camps and coding programs.[4]

But what makes someone an innovator? Are these programs actually effective? What purposes and whose ends do they serve? Does America really need more innovators?

As innovation initiatives proliferate, critics question these programs' goals and outcomes and identify their shortcomings. Until recently, academics and activists have been the only serious challengers of innovation. They are now joined by journalists who document fallacies in the mantra of "disruption" and by popular television shows such as HBO's *Silicon Valley*, which skewers the tech industry with portrayals of sexist and self-absorbed innovators.[5] Many of these observers, who are innovation experts in their own right, point out flaws in innovator initiatives in order to improve them. An increasingly prominent group of critics, however, considers the valorization of innovation to be delusional and destructive.

Yet another group of reform-minded experts work from inside the innovation enterprise to critique and improve the training and practices of innovators. Their initiatives include nonprofit organizations that build more welcoming cultures for women and underrepresented minorities in the tech industry, enrichment programs for children that emphasize self-discovery over marketable skills, entrepreneurship education that engages with history to cultivate more effective innovators, and laboratories that pair scientists with humanists to alter the innovation process in action.

This volume provides a critical survey of the "American imperative" for innovation by bringing together leading champions, critics, and reformers in dialogue. While numerous prior works have investigated *innovation*, this volume emphasizes *innovators* and how they are made. The focus on innovators is especially valuable because it is through the initiatives documented in this volume that the motivations, values, and best practices of innovation are crafted, adopted, and spread. Despite otherwise divergent views, the contributors assembled here agree that the widespread effort to educate and train new innovators has become a dominant imperative of our time, one that is increasingly on trial.

In what follows, policymakers, design executives, and educators explore the imperative alongside historians, ethnographers, and social critics. Contributors ask themselves and one another: Why did programs for making innovators emerge? How have they evolved? What is their track record? What are their collective assumptions and shortcomings? How might they be improved? What is their future?

Championing Innovation

From Thomas Edison's laboratory in Menlo Park, New Jersey, to Facebook's headquarters in Menlo Park, California, stories abound of technological wizards whose very force of personality drives breakthroughs and generates fortunes.[6] These young, gritty, and creative men (in such tales they are almost always men) overcome failure and naysayers to create products that remake the world. With varying shades of plausibility, their biographical accounts offer the prospect that you, too, can follow in their footsteps to become the next great innovator. But what characterizes an innovator? Why, over the past five decades, have experts claimed that such individuals are vital to national progress? And who has sought to make them?

The first systematic attempt to understand the characteristics of innovators emerged in tandem with new kinds of expertise for producing technological innovation. In the 1950s and 1960s, innovation became deeply linked with scientific, technological, and economic progress. The United States emerged victorious in World War II thanks to cutting-edge military innovations such as radar and the atomic bomb. After the war, new federal agencies, including the NSF and the National Aeronautics and Space Administration (NASA), were tasked with accelerating the flow of research and development. Meanwhile, economists such as Robert Solow, Richard S. Nelson, and Kenneth Arrow made innovation synonymous with "technological change" as the driver of economic growth.[7] Numerous experts in fields ranging from anthropology to engineering, history, management, and sociology likewise sought to understand and accelerate how new ideas and inventions spread. Some turned to *innovators*, the human agents of innovation, as key drivers of change.

Rural sociologist Everett Rogers emerged as one of the most influential theorists to explore the traits of innovators. Rogers showed that personal reactions to innovation followed a similar pattern across communities as diverse as elementary schools and Native American tribes.[8] He described innovators as having a "propensity for venturesomeness," for the "hazardous, the rash, the avant-garde, and the risky." These "agents that promote change," according to Rogers, had six qualities in common. They were young, high in social status, drawn to "impersonal" information, cosmopolitan, thought leaders, and frequently viewed as "deviant."[9] His conclusion: innovators are curious and intelligent mavericks who can be found anywhere.

Programs designed to cultivate innovators emerged and grew in the United States during the 1960s and 1970s. They blossomed out of professional networks of corporate technology managers, entrepreneurs, venture capitalists, and social scientists who described a global economy in which older forms of invention and discovery were no longer adequate.[10] But the main agent for promoting this agenda was the federal government. The Department of Commerce and the NSF created public-private "incubators," such as the State Technical Services program, that looked for ways to transfer the fruits of basic research and weapons development to the domestic economy. To remake scientists and engineers as "innovators," they also created college entrepreneurship programs that combined science, technology, and small business development.[11]

In the 1980s and 1990s, programs for making innovators expanded in scope and scale. New organizations for research and development, such as the NSF's Engineering Research Centers, elevated interdisciplinarity as a key feature of successful innovators in the global struggle for economic competitiveness.[12] Meanwhile, feminist innovation experts called attention to the importance of diversity, interpersonal relationships, and empathy in successful innovation.[13] Additionally, innovators became synonymous with a "creative class" of designers, artists, and technologists who would spark urban renewal across the United States.[14] During the 1990s, as the United States faced increasing competition from Europe, Japan, and China, private foundations saw innovators as the solution to a nation at risk. Their reports described a dysfunctional government and a nation of youth who worshiped athletes and entertainers over scientists and entrepreneurs. For example, the Lemelson Foundation, created by the inventor Jerome Lemelson, funded programs at MIT, the Smithsonian, and beyond; similarly, the Kauffman Foundation shifted its mission from anti–drug abuse initiatives to programs that educate and cultivate entrepreneurs.[15]

Today's efforts to create innovators build on this legacy and have diversified our ideas about the characteristics of "agents that promote change." National policymakers' focus on global competitiveness has raised questions about where innovation happens and what role immigrants play in national growth. Programs once targeted to technology executives now shape approaches to elementary education. Last, but hardly least, the rise of personal computing and the internet has spawned visions of college dropouts turned billionaires.

Across the board, these innovation experts share an optimistic faith that technology can be used to improve society. They encourage Americans to tap the country's legacy of invention to keep pace with rapid technological advances in the face of growing inequality and increasingly complex problems.[16] By unlocking our "creative confidence," they suggest, each of us must learn to thrive in a knowledge economy that rewards entrepreneurship and ingenuity.[17] Innovators are made, not born, they conclude, and we are not doing enough to cultivate this national resource.

Challenging Innovation

Much of the rhetoric about innovation portrays it as a natural and unquestioned engine of economic and social progress. Innovation's aura of societal benefit via insurgent but noble champions obscures uncomfortable truths about the innovator imperative. However, with the growth of so many pro-innovation initiatives, critics are beginning to ask: To what end?

Contemporary challenges to innovation are based in decades of research that explores technology in its political and social context. From Karl Marx's *Capital* to contemporary analyses of a "fourth industrial revolution," scholars have described the de-skilling of factory jobs, the degradation of workers, and the scourge of technological unemployment wrought by innovations in mechanization, robotics, and artificial intelligence.[18] Other critics have shown how risky and dangerous technologies, such as nuclear power, can make citizens feel imperiled by innovation.[19] Historians also chronicle that from the late nineteenth century to the present, the scientific and technological professions have predominantly served corporate and military prerogatives.[20]

Most critiques of what we now call innovation were previously directed at *science* and *technology*. In the late 1960s and early 1970s, a combination of critical scholars and activist practitioners coalesced in the new interdisciplinary field of science and technology studies (STS) to interrogate the belief that science and technology inevitably lead to social progress. "Innovation" initially had an ambivalent place in these critiques. On the one hand, it was a term rooted in visions of progress through technology; on the other, many theorists saw innovation as an explicitly "socio-technical" process that took into account the values, politics, and social consequences of technology.[21]

In the last decade, as *innovation* has become synonymous with science and technology as a dominant social category, critics now attack it directly. Recent critiques have focused primarily on innovation as an ideology and as an economic and technological process. As such, scholars utilize historical and sociological analyses of innovation in the aggregate. But their insights have direct bearing on initiatives to train innovators.

Many challengers to the innovator imperative recognize innovation's benefits but marshal convincing evidence that its outcomes are not sufficiently accessible or equitably distributed. For example, scholars have long decried the historical exclusion of women and minorities from the technical professions, a pattern that is especially stark in fields most closely aligned with innovation.[22] A 2016 study, *The Demographics of Innovation in the United States*, supports these claims, finding that the median innovator in the United States remains a white man in his late forties with an advanced degree; women represent only 12 percent of innovators, and minorities born in the United States make up only 8 percent, with African Americans numbering less than 0.5 percent.[23]

An emerging field of "critical innovation studies" further interrogates the social value of innovation.[24] These scholars denounce the gadget-centric "solutionism" of innovation's champions and the fallacy that technology alone can solve most problems. They point to the unequal distribution of innovation's burdens and rewards and how the corporate appropriation of innovation matches a larger pattern of economic neoliberalism.[25] They assert that the very ideal of the innovator as a technocratic hero reinforces

structures of gender and racial inequality.[26] They claim that many prominent innovation experts are selling little more than snake oil, and that the
lessons proffered by innovation experts are difficult, if not impossible, to
reproduce.[27] Importantly, these critics also promote alternative values such
as stewardship, care, and maintenance that have been overshadowed by the
focus on disruptive innovation.[28] A key feature of many of the questions
raised by critical innovation studies is the interrogation of the definition of
"innovation" itself, an increasingly expansive concept that is in danger of
losing all meaning.[29]

Reforming Innovation

While many critics diagnose innovation's shortcomings at a distance, there
is a reformist tradition among innovation experts who integrate these
insights directly into innovator initiatives. They train and cultivate innovators but recognize the flaws and trade-offs of the competing imperatives
that guide their efforts.

The rise of government innovation initiatives in the 1960s resulted in
part from efforts to redirect the uses of science and technology to better
serve social needs. Theories of innovation came into being as much in
response to technology's critics as for the desire to create new economic
markets. Across the presidential administrations of John F. Kennedy, Lyndon
Johnson, and Richard Nixon, a coalition of bureaucrats, business consultants, and science advisors created policies designed to bring the successes
of microelectronics, weapons development, and the space program to the
civilian economy.[30]

Attempts to use innovation as a tool of progressive reform overlapped
with a political awakening of Cold War scientists and engineers. The same
physicists who built the first nuclear weapons founded the Federation
of Atomic Scientists to advocate for disarmament, and activist engineers
encouraged colleagues to rethink pesticides, napalm, and the military-
industrial complex that employed them.[31] Some dissenting technologists
turned to "innovation" and the identity of the innovator to assert technology's capacity for human creativity and to address overlooked societal
problems.[32]

In the 1980s and 1990s, feminist sociologists marshaled analyses of
gender and power to expand access to careers in innovation by remaking

institutional structures and realigning corporate values. Rosabeth Moss Kanter, for example, a one-time observer of utopian communes, identified the struggles of underrepresented groups inside corporations and sought to teach institutions and individuals to distribute power so that new innovators would flourish.[33] Critics of innovation's narrow participation have since worked to diversify the images of innovators and replace behaviors such as hypercompetitiveness with ideals of collaborative creativity through play and "making" detached from business concerns.[34]

Sophisticated theoretical models also now accompany efforts to make socially conscious innovators. Since 2000, reform-minded STS scholars have adopted an approach known as critical participation to enhance the impact of their findings among scientists, engineers, and inventors.[35] Advocates of critical participation argue that scholars must go beyond critique and diagnosis to reflectively engage and shape the STEM communities they study.[36] Congress endorsed this brand of reform when it mandated that at least 5 percent of research funding for the Human Genome Project and the National Nanotechnology Initiative be earmarked for research on the ethical, legal, and social implications of those emerging technologies.[37] The field of "responsible innovation" is a growing branch of critical participation bolstered by new international journals and university centers. Responsible innovation's proponents apply theories of reflective practice toward changing innovation policy, redirecting bench-level research, and reimagining the training of future innovators.[38]

A Dialogue on Innovation

This volume convenes champions, critics, and reformers of innovation for three purposes. First, it provides a multifaceted survey of past and present innovation initiatives. Such a perspective is valuable for those engaged in training innovators, who can understand the historical and political contexts of their programs and learn best practices from leading programs, but it is equally valuable for innovation's critics to learn firsthand how and why leading practitioners go about their work. Second, the volume contributes to critical studies of innovation by making emerging scholarship accessible to innovation's practitioners and reformers. These critical insights, we believe, should push innovators, and those who train them, to pursue their work with a greater sense of reflection and moral responsibility. Finally, by initiating a dialogue on equal footing, the volume explores the potential for remaking the innovator imperative.[39]

The volume builds on our own efforts as critical humanists to develop and communicate insights about innovation. It extends our ongoing inquiry into the culture and ideology of invention and innovation from the nineteenth century to the present. But our goals are also practical. As critical scholars working in pro-innovation institutions, we participate in the innovator imperative as we analyze its historical and contemporary implications.[40] We have seen firsthand the aspirations that drive young people to want to become innovators. We also regularly collaborate in interdisciplinary teams on open-ended problems, and we recognize the need for challenging entrenched and often unquestioned routines. At the same time, in our historical research and in our everyday lives, we witness the inequities perpetuated in innovation's name. We also encounter an almost willful avoidance of critique among many of the practitioners with whom we work. We have found it possible, however, to create environments for reflective engagement and mutual understanding.[41]

To achieve such a dialogue requires good-faith participation on equally unsettled territory. We asked architects of innovator initiatives to speak frankly and personally about where their programs come from and what makes them tick. We asked our colleagues in economics, history, and STS to engage innovation's champions with as little academic jargon as possible. We encouraged reformers to describe their motives and explore the challenges of their ambiguous roles. At the best moments, participants in this volume achieve a constructive dialogue; at other points, readers will find that the ideological gulf between contributors is too wide to cross.

The volume is organized into three parts according to contributors' practices and commitments. To establish a common understanding of what drives different perspectives on innovation, each part begins with a brief essay that introduces and analyzes the shared assumptions, strengths, and limitations of that part's contributors. Part I, "Champions," is a tour of innovator training today. It explores the antecedents, motivations, and philosophies of programs that produce innovators across contexts from private industry to universities and governments. Part II, "Critics," offers a primer on critical innovation studies. It includes essays that historicize, contextualize, and problematize the imperative to cultivate innovators. Part III, "Reformers," is an introduction to initiatives that seek to reshape what it means to be an innovator, from programs that support self-discovery among children to organizations that target discrimination in high technology industries. The volume may be straightforwardly read from front to back, though readers interested in particular themes, such as access and inclusion, may find it helpful to follow those threads across the book. The volume concludes with a call for reconsidering America's demand for more innovators.

"Our nation" may know "what it takes to innovate," but why, for what, and by whom? Contributors to this volume demonstrate that the answers are neither simple nor uniform. Those who proffer solutions, moreover, often do so with different assumptions and even different languages. But ideas and tools—whether designed to increase shareholder value or to assert alternative societal values—are only successful if they are taken up, modified, and shared. This volume creates a forum for such an exchange.

Notes

1. "Innovation: An American Imperative," American Academy of Arts and Sciences, 23 June 2015, http://www.amacad.org/content/innovationimperative/.

2. For an example of a how-to book, see Steven Johnson, ed., *The Innovator's Cookbook: Essentials for Inventing What Is Next* (New York: Riverhead Books, 2011). On universities as incubators of innovation, see Michael M. Crow and William B. Dabars, *Designing the New American University* (Baltimore: Johns Hopkins University Press, 2015).

3. On NSF's I-Corps, see Arkilic (chapter 5), and on MIT and its imitators, see Pfotenhauer (chapter 11), both in this volume.

4. On efforts to make innovative children through creative play, see Tony Wagner and Robert A. Compton, *Creating Innovators: The Making of Young People Who Will Change the World* (New York: Scribner, 2012), and Rusk (chapter 15) in this volume.

5. Karl Ulrich, "The Fallacy of 'Disruptive Innovation,'" *Wall Street Journal*, 6 November 2014, https://blogs.wsj.com/experts/2014/11/06/the-fallacy-of-disruptive -innovation/; Andrew Marantz, "How 'Silicon Valley' Nails Silicon Valley," *New Yorker*, 9 June 2016, https://www.newyorker.com/culture/culture-desk/how-silicon -valley-nails-silicon-valley.

6. Paul Israel, *Edison: A Life of Invention* (New York: Wiley, 1998); Ben Mezrich, *The Accidental Billionaires: The Founding of Facebook, a Tale of Sex, Money, Genius, and Betrayal* (New York: Doubleday, 2009); Walter Isaacson, *The Innovators: How a Group of Hackers, Geniuses, and Geeks Created the Digital Revolution* (New York: Simon & Schuster, 2014).

7. Benoît Godin, *Models of Innovation* (Cambridge, MA: MIT Press, 2017).

8. Everett M. Rogers, *The Diffusion of Innovations* (New York: The Free Press, 1962).

9. Everett M. Rogers, "What Are Innovators Like?" *Theory into Practice* 2 (1963): 252–256.

10. Matthew Wisnioski, "How the Industrial Scientist Got His Groove," in *Groovy Science: Knowledge, Innovation, and American Counterculture*, ed. David Kaiser and W. Patrick McCray (Chicago: University of Chicago Press, 2016), 337–365.

11. National Science Foundation, "Incubators for Entrepreneurs," *Mosaic* 9, no. 4 (July/August 1978): 11–16.

12. Syl McNinch, "National Science Foundation Engineering Research Centers (ERC): How They Happened, Their Purpose, and Comments on Related Programs," National Science Foundation, 1984.

13. A recent variation on this argument can be found in Sylvia Ann Hewlett, Melinda Marshall, and Laura Sherbin, "How Diversity Can Drive Innovation," *Harvard Business Review*, December 2013, https://hbr.org/2013/12/how-diversity-can-drive-innovation.

14. Richard Florida, *The Rise of the Creative Class: And How It's Transforming Work, Leisure, Community, and Everyday Life* (New York: Basic Books, 2002).

15. "Congress Knows. The American People Know. Our Competitors Know," *New York Times*, 9 March 1994, D3; Anne Morgan, *Prescription for Success: The Life and Values of Ewing Marion Kauffman* (Kansas City, MO: Andrews and McMeel, 1995), 345–360.

16. "Remarks by President Obama in Mission Innovation Announcement," 30 November 2015, accessed 10 September 2016, https://obamawhitehouse.archives.gov/the-press-office/2015/11/30/remarks-president-obama-mission-innovation-announcement.

17. Tom Kelley and David Kelley, *Creative Confidence: Unleashing the Creative Potential within Us All* (New York: Crown Business, 2013).

18. Karl Marx, *Capital*, vol. 1, trans. Ben Fowkes (London: Penguin Books, 1990 [1867]); Harry Braverman, *Labor and Monopoly Capital: The Degradation of Work in the Twentieth Century*, 25th anniversary ed. (New York: Monthly Review Press, 1998 [1974]); David F. Noble, *Forces of Production: A Social History of Industrial Automation* (New York: Knopf, 1984); Amy Sue Bix, *Inventing Ourselves Out of Jobs? America's Debate over Technological Unemployment, 1929–1981* (Baltimore: Johns Hopkins University Press, 2000); Arthur Daemmrich, "Invention, Innovation Systems, and the Fourth Industrial Revolution," *Technology and Innovation* 18, no. 4 (2017): 257–265.

19. Langdon Winner, *Autonomous Technology: Technics-out-of Control as a Theme in Political Thought* (Cambridge, MA: MIT Press, 1977).

20. Edwin T. Layton, *The Revolt of the Engineers: Social Responsibility and the American Engineering Profession* (Baltimore: Johns Hopkins University Press, 1971); David F. Noble, *America by Design: Science, Technology, and the Rise of Corporate Capitalism* (New York: Knopf, 1977).

21. Matthew Wisnioski, *Engineers for Change: Competing Visions of Technology in 1960s America* (Cambridge, MA: MIT Press, 2012), 148–160.

22. For example, Ruth Oldenziel, *Making Technology Masculine: Men, Women, and Modern Machines in America, 1870–1945* (Amsterdam: Amsterdam University Press, 1999); Bruce Sinclair, ed., *Technology and the African-American Experience: Needs and Opportunities for Study* (Cambridge, MA: MIT Press, 2004); Amy E. Slaton, *Race, Rigor, and Selectivity in U.S. Engineering: The History of an Occupational Color Line* (Cambridge, MA: Harvard University Press, 2010).

23. Adams Nager, David M. Hart, Stephen Ezell, and Robert D. Atkinson, *The Demographics of Innovation in the United States*, Information Technology and Innovation Foundation, 24 February 2016, https://itif.org/publications/2016/02/24/demographics-innovation-united-states.

24. Benoît Godin and Dominique Vinck, eds., *Critical Studies of Innovation: Alternative Approaches to the Pro-Innovation Bias* (Cheltenham, UK: Edward Elgar Publishing, 2017).

25. David Harvey, *A Brief History of Neoliberalism* (Oxford: Oxford University Press, 2007). For connections of neoliberalism to innovation, see Philip Mirowski, *Science-Mart: Privatizing Modern Science* (Cambridge, MA: Harvard University Press, 2011); Lee Vinsel, "95 Theses on Innovation," 12 November 2015, accessed 10 September 2018, http://leevinsel.com/blog/2015/11/12/95-theses-on-innovation.

26. While recent critiques of innovation call for increased access to places of innovation, they often ignore more radical and intersectional scholarship, which finds diversity for diversity's sake insufficient. For example, see Sara Ahmed, *On Being Included: Racism and Diversity in Institutional Life* (Durham, NC: Duke University Press, 2012); Donna Riley, Amy E. Slaton, and Alice L. Pawley, "Social Justice and Inclusion: Women and Minorities in Engineering," in *Cambridge Handbook of Engineering Education Research*, ed. Johri Aditya and Barbara M. Olds (Cambridge: Cambridge University Press, 2014).

27. Jill Lepore, "The Disruption Machine: What the Gospel of Innovation Gets Wrong," *New Yorker*, 23 June 2014, http://www.newyorker.com/magazine/2014/06/23/the-disruption-machine.

28. David Edgerton, *The Shock of the Old: Technology and Global History since 1900* (Oxford: Oxford University Press, 2007).

29. Benoît Godin, *Innovation Contested: The Idea of Innovation over the Centuries* (London: Routledge, 2015). See also Godin (chapter 9) in this volume.

30. Elizabeth Popp Berman, *Creating the Market University: How Academic Science Became an Economic Engine* (Princeton, NJ: Princeton University Press, 2012); Fred Block and Matthew R. Keller, eds., *State of Innovation: The U.S. Government's Role in Technology Development* (Boulder, CO: Paradigm Publishers, 2011).

31. Sarah Bridger, *Scientists at War: The Ethics of Cold War Weapons Research* (Cambridge, MA: Harvard University Press, 2015).

32. Matthew Wisnioski, *Engineers for Change: Competing Visions of Technology in 1960s America* (Cambridge, MA: MIT Press, 2012).

33. Rosabeth Moss Kanter, *The Change Masters: Innovation for Productivity in the American Corporation* (New York: Simon & Schuster, 1983).

34. Monica M. Smith, "Playful Invention, Inventive Play," *International Journal of Play* 5, no. 3 (2016): 244–261.

35. For example, see Teun Zuiderent-Jerak and Casper Bruin Jensen, "Unpacking 'Intervention' in Science and Technology Studies," *Science as Culture* 16, no. 3 (2007): 227–235; Gary Downey, "What Is Engineering Studies For? Dominant Practices and Scalable Scholarship," *Engineering Studies* 1, no. 1 (2009): 55–76.

36. This approach has roots in the 1970s and 1980s; for example, the educational philosopher Donald Schön's 1983 classic, *The Reflective Practitioner*, encouraged professionals in fields such as engineering and urban planning to engage in a continuous feedback loop of work experience, critical reflection, and reform. Schön, *The Reflective Practitioner: How Professionals Think in Action* (New York: Basic Books, 1983). Schön, who had worked as a corporate consultant and federal innovation expert in the 1960s and 1970s, especially encouraged collaboration across disparate disciplines.

37. Jean E. McEwen, Joy T. Boyer, Kathie Y. Sun, Karen H. Rothenberg, Nicole C. Lockhart, and Mark S. Guyer, "The Ethical, Legal, and Social Implications Program of the National Human Genome Research Institute: Reflections on an Ongoing Experiment," *Annual Review of Genomics and Human Genetics* 15 (August 2014): 481–505; Erik Fisher, "Lessons Learned from the Ethical, Legal, and Social Implications Program (ELSI): Planning Societal Implications Research for the National Nanotechnology Program," *Technology in Society* 27 (2005): 321–328.

38. Richard Owen, Jack Stilgoe, Phil Macnaghten, Mike Gorman, Erik Fisher, and Dave Guston, "A Framework for Responsible Innovation," in *Responsible Innovation: Managing the Responsible Emergence of Science and Innovation in Society*, ed. Richard Owen, John Bessant, and Maggy Heintz (Chichester, UK: John Wiley & Sons, 2013), 27–50.

39. The volume's multiperspectival approach is motivated by methodological developments in the field of STS. Our understanding of critical participation is

shaped especially by conversations with our colleague Gary Downey. A pioneer in this field, Downey argues that critical participation can influence and ultimately change the way that science and engineering are practiced. Gary Lee Downey and Teun Zuiderent-Jerak, "Making and Doing: Engagement and Reflexive Learning in STS," in *The Handbook of Science and Technology Studies, 4th ed.*, ed. Ulrike Felt, Rayvon Fouché, Clark A. Miller, and Laurel Smith-Doerr (Cambridge, MA: MIT Press, 2017), 223–252. Other recent advances include Jason Chilvers and Matthew Kearnes, eds., *Remaking Participation: Science, Environment, and Emergent Publics* (London: Routledge, 2015); and Javier Lezaun, Noortje Marres, and Manuel Tironi, "Experiments in Participation," in *The Handbook of Science and Technology Studies, 4th ed.*, ed. Ulrike Felt, Rayvon Fouché, Clark A. Miller, and Laurel Smith-Doerr (Cambridge, MA: MIT Press, 2017), 195–222.

40. Wisnioski is an associate professor and Kleine is a PhD candidate in the Department of Science, Technology, and Society at Virginia Tech, where the university's branding tagline is "Invent the Future." Hintz is a historian at the Smithsonian Institution's Lemelson Center for the Study of Invention and Innovation; the center's vision statement imagines "a world in which everyone is inventive and inspired to contribute to innovation."

41. Our interventions have included courses and museum exhibits that put innovation in context by staging debates among advocates and critics, but also by reflectively introducing students and the public to the tools of innovation experts. This project extends our effort to expert communities.

1 Champions

Innovation experts often use a problem-setting tool known as the "how might we" statement to imagine social change.[1] This simple heuristic is designed to spark solutions to a market opportunity, technical problem, or societal concern. "*How might we*... meet the customer's need and solve their problem?"[2] "*How might we*... use tech to create a culture of civic engagement?"[3] "*How might we*... help corporations, universities, and societies to accelerate innovation in ways that keep pace with these challenges?"[4]

In this section, unapologetic champions of innovation describe the *how might we* mentality that guides nationally prominent initiatives for making innovators. The contributors are a more heterogeneous group than one might imagine. They work variously in academia, government, and the private sector. Not surprisingly, three of the authors reside in Silicon Valley, but the others work in Pittsburgh, Washington, DC, and Chapel Hill, North Carolina. They pursue different objectives that include fostering innovative mindsets among students, delivering solutions for clients, stimulating regional economic growth, turning academic scientists into entrepreneurs, and engaging technically inclined citizens through open innovation.

Presented in expanding scale from individuals to nationwide collaborations, these innovation experts provide first-person accounts of the origins and outcomes of their programs. They invite readers to learn from their stories, to emulate their methods, and to join their causes. Viewed in context, their accounts offer important insights about the strengths, limitations, and consequences of a *how might we* mentality.

Advocates consider innovation to be a requisite skill set for the twenty-first century because of its association with highly valued competencies such as creativity, critical thinking, and problem-solving. They claim that these skills are especially important since today's college graduates face an uncertain future in which they will have multiple careers. Yet they assert that universities remain organized around disciplinary "silos" that train graduates for well-defined jobs.[5] They ask: *How might we* give young people the tools to thrive in a future of continuous change?

In chapter 3, "An Innovators' Movement," Humera Fasihuddin and Leticia Britos Cavagnaro describe how their University Innovation Fellows (UIF) program simultaneously seeks to prepare a generation of innovators and to reform higher education. Based at Stanford University's d.school, the UIF network includes students from more than one hundred colleges, in fields ranging from mechanical engineering to the creative arts. Fasihuddin and Britos Cavagnaro explain how UIF's Silicon Valley training equips young people from across the country with "empathetic" design methods and the confidence to "ignite" social change. As student fellows organize hackathons and set up maker spaces, UIF encourages them to improve the innovation ecosystem on their local campuses.

Fasihuddin and Britos Cavagnaro argue that collective change begins with personal empowerment. They define innovation as a mindset for questioning the status quo and a set of transferable skills for enacting change. For students, UIF is a safe place for experimentation; when a student succeeds at organizing a TED talk, she can acquire leadership skills and improve her resume, but the stakes are low if the college try does not pan out. For universities, a local UIF chapter signals that the institution is taking positive steps to contribute to the national innovator imperative. UIF's critics, however, argue that its design thinking approach gives students false confidence in quick fixes to societal challenges (Russell and Vinsel, chapter 13) and downplays the value of traditional education for cultivating creative students (Carlson, chapter 16).[6]

While UIF aims to inspire individuals, other innovator initiatives stress that collaboration among multiple experts is necessary to solve the twenty-first century's "wicked problems."[7] These champions argue that a revolution in information technology has complicated the already difficult task of creating social change within existing infrastructures, political and economic constraints, and stakeholder demands. Since the 1980s, human-centered design firms such as Frog Design, IDEO, and MAYA have defined innovators as interdisciplinary collaborators who integrate technical and social approaches in order to "tame complexity."[8] These firms ask *how might we* foster collaborative creativity to address sociotechnical problems?

In chapter 4, "Building High-Performance Teams for Collaborative Innovation," Mickey McManus and Dutch MacDonald share lessons learned at their firm MAYA to explain why high-performance, multidisciplinary teams prevail while "heroic" individual approaches to innovation typically fail. Their company works with clients as diverse as Whirlpool, the Pentagon, and public school systems. While each client's challenges differ, MAYA's interdisciplinary model of innovation is consistent: assemble the right people, with the right mix of skills, in the right work environment, with the right set of methods. Utilizing this approach, MAYA has spawned an innovation training company, the LUMA Institute, and a popular handbook, *Innovating for People.*[9]

McManus and MacDonald argue that innovation is first and foremost a problem of human interaction. Their LUMA Institute declares that "the need for more people to be more innovative…is a global, social and economic imperative" that requires tested methods for unlocking collaborative creativity.[10] The authors are self-reflective in codifying their firms' creative process—and they have certainly delivered for clients. But their for-profit model prefigures who gets to deploy innovation expertise and who benefits from it. For example, women and minorities rarely are equal participants on high-performing teams in the innovation economy (Sanders and Ashcraft, chapter 17), and many communities and civic organizations cannot afford expert consulting services.

Private companies such as MAYA play a crucial role in the innovation economy, but since World War II the federal government has been the primary funder of scientific innovation in the United States. Agencies such as the National Science Foundation (NSF) have persistently faced pressure to provide a return on taxpayers' investment by translating government-funded research into commercial technologies. Policymakers and scholars typically view technology transfer as a problem of institutional structure and economic incentives.[11] Academic scientists, however, are not typically trained to translate their research into usable applications. So bureaucratic innovators ask: *How might we* change the attitudes of scientists to reap the social and economic rewards of new discoveries?

In chapter 5, "Raising the NSF Innovation Corps," venture capitalist Errol Arkilic describes how he and his colleagues built a federal innovation "boot camp" that teaches teams of university professors and their students to engage in technology transfer. Innovation Corps (I-Corps) is built on the belief that the innovation process can be studied scientifically and then codified, taught, and learned. In 2011, NSF made its first I-Corps awards consisting of small grants that provided intensive instruction on developing prototypes, interviewing potential customers, and writing business plans. Over a thousand teams have since participated in the program.

By teaching NSF grantees how to commercialize their discoveries, I-Corps has launched over one hundred start-ups. However, I-Corps revives a longstanding debate on NSF's proper role: Should the agency fund "basic" science as a public good and leave commercialization to the private sector?[12] Critics argue that university entrepreneurship and commercialization enhance the corrupting influence of capitalism in science.[13] Furthermore, I-Corps' standardized process and emphasis on economic gain may diminish the values of exploration and self-discovery associated with innovation (Rusk, chapter 15) and overshadow deliberation and social responsibility as core values of science and engineering (Fisher, Guston, and Trinidad, chapter 18).

Innovation experts have long emphasized the role of *place* in the gathering and training of innovators.[14] Silicon Valley, Boston, and North Carolina's Research Triangle Park are hotbeds for innovative activity in part because they provide the right institutions, amenities, and culture for innovators to thrive. After decades of efforts to build the next Silicon Valley have failed, however, experts recognize the perils of a cookbook approach. They ask: *How might we* cultivate successful innovative regions that honor and build on local strengths?

In chapter 6, "Making Innovators, Building Regions," economic geographer Maryann Feldman surveys how "local champions" work in concert with universities, firms, and other institutions to build high-tech regional clusters. Like many economists, Feldman defines innovation as the commercialization of new knowledge and the primary driver of economic growth. She synthesizes insights of a career spent studying high-tech regions and measuring their outcomes. Drawing on examples from places as diverse as Kansas City, Missouri, and Greenwood, Mississippi, she argues that there are common ingredients in making innovative regions but no single recipe. Each innovative region has its own unique blend of institutions, regional capabilities, and social configurations.

Feldman documents how as innovators build their own companies they also build local institutions and shared resources that produce even more innovators. But Feldman only shows us the dynamic processes underlying *successful* innovative places. She does not address regions such as Dallas or Albany, New York, where local champions have assembled all the necessary ingredients yet ultimately failed to ignite high-tech clusters.[15] As we will see in part II, innovative communities can be difficult to replicate and sustain (Hintz, chapter 10). Also, regional efforts to build new innovation hubs often become entangled with national priorities that complicate what counts as success (Pfotenhauer, chapter 11).

Despite insisting that innovators are made not born, the programs described so far portray innovation as an elite activity beyond the reach of most citizens. The innovators they hope to make are PhD scientists and driven entrepreneurs, generally in high-technology regions. However, there is also a longstanding tradition in the United States of amateur scientists and do-it-yourself tinkerers.[16] In the past decade, moreover, experts such as Eric von Hippel and Henry Chesbrough have argued that organizations benefit when they bring end users and outsiders into the innovation process.[17] As government agencies respond to market pressures and the inequities in the innovation economy, policymakers ask: *How might we* democratize innovation to harness the contributions of all Americans?

In the section's last chapter, "Innovation for Every American," Jenn Gustetic, a federal innovation expert, contends that all Americans can contribute to innovation. She recounts how, under President Obama, the White House's Office of Science and Technology Policy (OSTP) encouraged citizens to participate in scientific discovery and technology design through crowd-sourcing and citizen science initiatives. Gustetic recounts how these innovation programs were as much a project for reforming government as a strategy for harnessing open innovation; government bureaucrats were forced to leave their comfort zones, work across departments, and partner with new kinds of innovators to solve their agencies' challenges. She argues that future presidential administrations must attend not only to who produces innovations but also to who owns the results.

The Obama administration's open innovation policies recruited students, retirees, and ordinary Americans to become innovators. The OSTP elevated the innovator imperative to a national goal and perpetuated the belief that innovation is an inherent social good (Godin, chapter 9). But "inclusive" techniques such as crowdsourcing and offering incentive prizes shift many of innovation's risks and costs from the government onto its citizens.[18] Citizen-innovators risk their own money to develop solutions, but the government pays only for those that meet the prize criteria. Volunteer citizen scientists, meanwhile, generally go unpaid. Also, while federal open innovation efforts target participation from underrepresented groups, they do not confront the structural inequities that prevent deeper participation in the innovation economy (Cook, chapter 12).

Collectively, the experts profiled in this section believe that innovation leads to social progress and national prosperity. Their initiatives share the premise that the failures of the status quo and existing bureaucracies can be overcome; that no matter the life stage, everyone can work to better themselves; that innovative skills can be learned; and that large-scale interventions are required to support them. All of these contributors also draw upon a network of institutional support from the government, corporations, and universities. Does America need more innovators? The answer for these practitioners is a resounding "Yes...and let me show you how it's done."

But there are significant differences in the goals of these initiatives. Some programs equip students with new skill sets; others hope to maximize the return on taxpayers' dollars; still others are driven by the potential for profits. These different motivations, in turn, result in programs for different target audiences, tactics, and messages. For example, UIF's methods explicitly focus on empathy and self-actualization, while NSF's I-Corps teaches senior academics to become competitors in an unforgiving market environment.

Finally, innovation's advocates rarely question the necessity or the potentially negative consequences of their work. As Part II will address, where creators of innovator initiatives describe empowerment, detractors find boosterism and false promises. Where these champions promote novel twenty-first-century methods, historians recognize well-trodden patterns with a mixed record. Where this section's contributors describe beneficial collaborations among industry, government, and academia, critics detect the privatization of public goods. And where advocates of innovation training see avenues for personal growth, critics see the redistribution of risk and anxiety from institutions onto individuals. Exploring competing interpretations of the nation's innovator imperative requires first understanding its champions.

Notes

1. Warren Berger, "The Secret Phrase Top Innovators Use," *Harvard Business Review*, 17 September 2012, https://hbr.org/2012/09/the-secret-phrase-top-innovato.

2. Andrew B. Williams, "A Recipe for Unleashing Creative Design Thinking in Your Teams," *In Search of Innovation*, 31 December 2014, accessed 15 May 2017, https://drandrewspeaks.wordpress.com/category/design-thinking/.

3. "How Might We Use Tech to Build a Culture of Civic Engagement?" *CHORUS*, 22 March 2017, accessed 23 May 2017, https://www.jointhechorus.org/blog/2017/4/4/how-might-we-use-tech-to-build-a-culture-of-civic-engagement.

4. Larry Leifer and Christoph Meinel, "Design Thinking for the Twenty-First Century Organization," in *Design Thinking Research: Taking Breakthrough Innovation Home*, ed. Hasso Plattner, Christoph Meinel, Larry Leifer (Cham, Switzerland: Springer International Publishing, 2016), 3.

5. Carolin Kreber, *The University and Its Disciplines: Teaching and Learning within and beyond Disciplinary Boundaries* (New York: Routledge, 2010).

6. Additionally, see Natasha Iskander, "Design Thinking Is Fundamentally Conservative and Preserves the Status Quo," *Harvard Business Review*, 5 September 2018, https://hbr.org/2018/09/design-thinking-is-fundamentally-conservative-and-preserves-the-status-quo.

7. The term "wicked problems," popular among innovation advocates, actually is a twentieth century one. C. West Churchman, "Wicked Problems," *Management Science* 14 (December 1967): B141–B142.

8. Barry M. Katz, *Make It New: The History of Silicon Valley Design* (Cambridge, MA: MIT Press, 2015); Tom Kelley, Jonathan Littman, and Tom Peters, *The Art of Innovation: Lessons in Creativity from IDEO, America's Leading Design Firm* (New York: Currency, 2001); Peter Lucas, Joe Ballay, and Mickey McManus, *Trillions: Thriving in the Emerging Information Ecology* (New York: Wiley, 2012).

9. LUMA stands for Looking, Understanding, Making, Advancing. *Innovating for People: Handbook of Human-Centered Design Methods* (Pittsburgh: LUMA Institute, 2012).

10. "Our Beliefs," LUMA Institute, accessed 9 April 2018, https://www.luma-institute.com/why-luma/our-beliefs/.

11. David Mowery, Richard Nelson, Bhaven Sampat, and Arvids Ziedonis, *Ivory Tower and Industrial Innovation University-Industry Technology Transfer before and after the Bayh-Dole Act* (Redwood City, CA: Stanford University Press, 2015); Henry Etzkowitz, *The Triple Helix: University-Industry-Government Innovation in Action* (London: Routledge, 2008).

12. On debates over basic versus applied research at NSF, see Daniel L. Kleinman, *Politics on the Endless Frontier: Postwar Research Policy in the United States* (Durham, NC: Duke University Press, 1995), and Dian Olson Belanger, *Enabling American Innovation: Engineering and the National Science Foundation* (West Lafayette, IN: Purdue University Press, 1998).

13. For example, see Philip Mirowski, *Science-Mart: Privatizing Modern Science* (Cambridge, MA: Harvard University Press, 2011).

14. Arthur P. Molella and Anna Karvellas, eds., *Places of Invention* (Washington, DC: Smithsonian Institution Scholarly Press, 2015); Margaret Pugh O'Mara, *Cities of Knowledge: Cold War Science and the Search for the Next Silicon Valley* (Princeton, NJ: Princeton University Press, 2005); AnnaLee Saxenian, *Regional Advantage: Culture and Competition in Silicon Valley and Route 128* (Cambridge, MA: Harvard University Press, 1994); Bruce Katz and Julie Wagner, *The Rise of Innovation Districts: A New Geography of Innovation in America,* Washington, DC, Metropolitan Policy Program at Brookings, May 2014, accessed 22 May 2017, https://c24215cec6c97b637db6-9c08 95f07c3474f6636f95b6bf3db172.ssl.cf1.rackcdn.com/content/metro-innovation -districts/~/media/programs/metro/images/innovation/innovationdistricts1.pdf.

15. Stuart W. Leslie and Robert H. Kargon, "Selling Silicon Valley: Frederick Terman's Model for Regional Advantage," *Business History Review* 70, no. 4 (1996): 435–472; Stuart W. Leslie, "Regional Disadvantage: Replicating Silicon Valley in New York's Capital Region," *Technology and Culture* 42, no. 2 (2001): 236–264.

16. Jack Hitt, *Bunch of Amateurs: Inside America's Hidden World of Inventors, Tinkerers, and Job Creators* (New York: Broadway Books, 2013).

17. Eric von Hippel, *Democratizing Innovation* (Cambridge, MA: MIT Press, 2006); Henry Chesbrough, *Open Innovation: The New Imperative for Creating and Profiting from Technology* (Boston: Harvard Business School Press, 2007).

18. Jacob Silverman, "The Crowdsourcing Scam: Why Do You Deceive Yourself?" *The Baffler*, October 2014, https://thebaffler.com/salvos/crowdsourcing-scam.

3   An Innovators' Movement

Humera Fasihuddin and Leticia Britos Cavagnaro[1]

The way we live and work is fundamentally changing because of a convergence of technological advances and socioeconomic forces. Many of the things we used to take for granted, like a lifelong career at one company, are yielding to a future where change is the only constant. Because of this, the number of colleges and universities offering programs in innovation and entrepreneurship (I&E) has increased significantly.[2]

Adapting and flourishing in this new economy requires a new set of skills. Harvard professor Clayton Christensen and his colleagues have identified five core skills that distinguish innovators and entrepreneurs:

- observing and noticing what others don't;
- questioning the status quo and common wisdom;
- associating, or the ability to connect seemingly unrelated data and ideas from different fields;
- experimenting with different approaches; and
- networking with others who have diverse perspectives.[3]

These skills are not only useful in discovering new business opportunities; they are essential to empowering every person to be a successful contributor to the advancement of humanity. Ensuring that these competencies are part of the education of all students is a matter of great importance and urgency.[4] However, the education system is slow to change.

Students can and should be key protagonists in bringing about a revolution in higher education, as they can accelerate the pace of change. This chapter describes the University Innovation Fellows (UIF), a program that empowers university students as change agents uniquely positioned to reimagine education in the twenty-first century. When students are provided the right tools and opportunities, change happens faster and has lasting effects on institutions and the education system at large. By describing why and how this program was created, we hope to inspire other programs to leverage the engagement of young people in building the future.

The Context That Gave Rise to a Movement

In 2010, the National Science Foundation (NSF) published a solicitation for the creation of a science, technology, engineering, and mathematics (STEM) Talent Expansion Program Center. The solicitation, which focused on the role of innovation and entrepreneurship in solving the grand challenges of the twenty-first century and driving the economy, ended with this question: "How can the next generation of engineers be encouraged to be innovators, entrepreneurs, and leaders who improve the quality of life and establish the industries and jobs of the future in the United States?"[5]

Eager to address that question, Stanford University professors Tom Byers and Kathleen Eisenhardt, directors of the Stanford Technology Ventures Program, and Sheri Sheppard, director of the Designing Education Lab, applied for the grant. They brought on a nonprofit, VentureWell, to reach a wider national network of colleges and universities, and SageFox Consulting Group to provide ongoing assessment and feedback.[6] The combination of these partners' diverse expertise uniquely positioned them to pursue such a vital challenge.

NSF awarded the grant, and the National Center for Engineering Pathways for Innovation, or Epicenter for short, officially launched in July 2011. Epicenter's original mission was to "empower U.S. undergraduate engineering students to bring their ideas to life for the benefit of our economy and society, by combining their ability to develop innovative technology that solve important problems with an entrepreneurial mindset of market opportunity and customer focus."[7]

Epicenter's strategic plan recognized that systemic change in engineering education in the United States has long been a challenge.[8] In part, that challenge results from the diversity of the country's 350 accredited engineering schools, which display a great range of missions, sizes, resources, and regional contexts. The Epicenter team knew that it would be impossible to find a one-size-fits-all solution to integrate I&E into undergraduate engineering education. The center could not be rooted in a single pedagogy, methodology, or ideology if it wanted to effectively engage those diverse stakeholders.

Epicenter intended to engage all the stakeholders of the educational system—not only the faculty but also the school administrators and leaders who support them and who determine institutional policies, as well as the students themselves. Initial plans and programs also included government and nonprofit leaders, who shape educational policies and accreditation requirements, as well as representatives from industry, who could speak of the competencies needed by the graduates entering the workforce.

The I&E activities and programs supported by Epicenter evolved significantly during the first two years of the grant, based on evaluation results that challenged some of the initial assumptions and strategies. In 2013, in consultation with NSF, the offerings were reduced to three main thrusts in order to better focus Epicenter's programming: focusing on institutional change from the top down (faculty and administrators), from the bottom up (students), and supporting the change efforts with research findings.

The offering that targeted faculty—the Pathways to Innovation Program—worked with a subset of institutions to initiate new I&E offerings in engineering education. Teams of faculty from these institutions would engage in a long-term strategic planning and doing process supported by Epicenter staff and other experts.

The research initiative, called Fostering Innovative Generations Studies (FIGS), examined I&E program models, assessed the entrepreneurial interests and skills of engineering students, identified ways to infuse I&E into technical engineering classes, and explored how to foster research community connections.[9]

The remainder of this chapter describes Epicenter's student engagement strategy—the UIF program—including its evolution, impact, and future direction.

Students as Change Agents: A Paradigm Shift

The story of our journey starting the UIF program is not unlike other start-up beginnings. Humera Fasihuddin had been asked by VentureWell to form the Student Ambassadors program in 2010 as a means to promote entrepreneurship and to identify potential recipients of VentureWell's own grants. Leticia Britos Cavagnaro had been teaching at Stanford University's

Hasso Plattner Institute of Design, or d.school, and was brought on to help launch the Epicenter in 2011. We joined forces to define Epicenter's student engagement strategy.

Aligning Student Ambassadors with Epicenter's broader mission required enriching the Ambassadors' training to include how to organize and facilitate experiential learning opportunities. In 2012, the first cycle of the program under Epicenter, trained nineteen students from eighteen universities in a four-day workshop of team-building activities, ideation workshops, and information sessions on strategies for fund-raising and team recruitment practices.

Upon returning to their campuses, Student Ambassadors set to work, with various levels of dedication and success, organizing either a basic "how-to" venture creation workshop called Invention to Venture (I2V) or an I&E-themed TEDx event. We soon realized that the program would be more valuable and attractive to students and their faculty if the student-led activities were tailored to each school's local context. While the prescribed events would be a meaningful contribution at a school with limited I&E resources, they would not make much of a difference in a more developed ecosystem. This insight led to a larger one: Why not challenge the students to obtain the knowledge about those ecosystems themselves and then use it to envision new possibilities? It was at this point that the paradigm of "students as change agents" took root as the foundation of the UIF movement.

The idea of students being responsible for lasting institutional change might sound counterintuitive. After all, students spend only a few years in school. However, three inherent characteristics are at the core of students' *potential* to become change agents. First, students, as the primary customers of the education system, understand what works and doesn't work for themselves and for other students. In this sense, University Innovation Fellows (subsequently referred to as Fellows) are the "users" in what innovation scholar Eric von Hippel calls "user innovation."[10] Second, their unfamiliarity with academic bureaucracy gives them a healthy disregard for the impossible. Third, the very fact that they spend only a few years at the institution gives a sense of urgency to those students who want to effect change. A name change from "Student Ambassadors" to "University Innovation Fellows" supported the new paradigm, shifting the perception from one of students who represent a specific organization and strategy, to one of students as strategic thinkers and change agents.

Developing Students as Strategic Thinkers

As part of the program overhaul toward preparing Fellows as change agents, we developed several new training components. Two initiatives, in particular, redefined the role of the student leaders and the impact their efforts had on campus.

The first training component was the creation of a "Landscape Canvas" framework that enables a comprehensive mapping of campus assets related to I&E.[11] The framework organizes I&E campus resources into five "action categories" that map onto a student's learning journey.

- *Discover:* resources that help students become aware of I&E as a set of relevant skill sets and mindsets, including speaker series and courses
- *Learn:* resources for students who, having recognized the relevance of I&E, want to develop their skills, including experiential courses, hackathons, student clubs, competitions, and short-term ideation or venture creation workshops
- *Experiment:* resources to aid students in applying their knowledge of I&E to a specific challenge or opportunity they wish to explore, including infrastructure (e.g., innovation spaces, maker spaces, or labs), engagement with external industry partners, and internship or co-op opportunities
- *Pursue:* resources to support students or student-led teams in committing to an opportunity, licensing a technology, or forming/joining a new venture, including incubators and accelerators, seed-funding sources, technology transfer offices, and grant-writing resources on campus
- *Spin out:* resources designed to help students bring their projects into the real world by launching a new venture or nonprofit initiative, including regional tech parks, local angel investor organizations, and mentor networks

Fellows use this framework to make sense of the range of I&E assets at their schools, to identify and articulate essential gaps in the system, and to benchmark against other ecosystems.

The second training component added exposure to design thinking, an approach to innovation that equips students with the mindsets and skills to identify human-centered problems and to imagine and implement solutions through experimentation.[12] Fellows apply design thinking to their work as change agents, and they have discovered that this approach is a useful way to engage their peers in I&E.

Scaling a National Movement

To create a national movement, we knew we had to think differently about how to identify and cultivate Fellows at a larger scale. We instituted a six-week online training using videoconferencing to help form connections with students located throughout the United States. These virtual meetings expose potential Fellows to custom-designed curricula, national resources, and possible program models. Employing a constructivist approach, candidates are required to apply their learning, produce new knowledge about their campus context, and publish it on the UIF open-source wiki.[13] The online approach enabled the program team to accept many more students while effectively gauging their commitment and potential.[14]

Upon successful completion of the online training, students are officially launched as Fellows. The UIF team contacts faculty and campus communications officers with news of each Fellow's launch, landscape research, and project plans, resulting in hundreds of press mentions annually. Upon launch, Fellows get to work on their plans. They create prototypes of new learning opportunities and get feedback from students, faculty, and administrators without devoting a lot of resources to any one idea before they know it meets a need. Taking action also helps Fellows learn more about other stakeholders from whom they can gain support.

Every spring, newly trained Fellows are invited to a three-day Silicon Valley Meetup at Stanford's d.school and Google to expand their knowledge of innovation ecosystems as well as change and engagement strategies. Meeting hundreds of like-minded young change agents makes Fellows feel part of a national movement.[15] Lastly, each institution's financial sponsorship of its Fellows' enrollment in the program contributes to a stable base of returning schools and strong word of mouth to attract new schools.[16]

The Journey to Become a Fellow

The best way to understand the transformation from student to change agent is to follow one on her journey. Computer science student Angelica Willis was tapped to become a Fellow by faculty and staff at North Carolina A&T State University, a historically black college/university (HBCU) in

Greensboro, North Carolina. A faculty member had heard from colleagues at other institutions about the growth in their own I&E programs, driven by the engagement of student leaders involved in UIF.[17]

Willis was intrigued by the program's mission and inspired by stories of Fellows creating impact at their schools. While the time commitment was substantial in an already jam-packed academic load, Willis was a fan of extracurricular learning opportunities that allowed her to apply her computer science skills to interesting problems. She applied alongside two other students from her campus. We require an institutional endorsement, so her faculty advisor also applied to become her "Faculty Champion." Once her application was submitted, we invited her to a video interview with Nadia Gathers, a current Fellow working on creating a social entrepreneurship major at Converse College. Additionally, Willis met three other candidates from across the nation who shared her passion for creativity, innovation, and entrepreneurship. A few weeks later, she received notice of her acceptance.

The six-week, online training featured hands-on activities, including a design challenge that gave Willis tools and strategies to interview students and faculty at her school to better understand their perspectives on I&E (figure 3.1). She used the Landscape Canvas to uncover the assets present in the I&E ecosystem at North Carolina A&T and identify gaps in the student experience. During weekly videoconference sessions, Willis talked with peers who faced similar challenges and heard about potential solutions that had worked at other campuses.

The final training sessions required Willis to create a YouTube video pitch, along with wiki pages of her research and project plans. She shared these materials at a stakeholders' meeting—a gathering in which Fellows convene the institution's administrative, faculty, and student I&E leaders—to get feedback and garner support.

The launch was exciting for Willis. She and her Faculty Champion received a congratulatory email welcoming them into the Fellows' community. The UIF team issued a press release to the North Carolina A&T communications office, resulting in an article on the campus website.[18] In addition, we sent a "pinning kit" to the Faculty Champion, which contained a small lapel pin for Willis and suggested guidelines for recognizing her new role with a pinning ceremony. The ceremony served as an opportunity for university leaders to interact with Willis and her Faculty Champion and to learn more

Figure 3.1
UIF candidates undergo six weeks of training in an online videoconference environment. Pictured are the UIF team orienting new candidates, including Leticia (bottom
row, right), Humera (middle row, right), Katie (middle row, left), and Laurie (top row,
center). Also pictured is Angelica Willis (top row, right), patching in from North Carolina A&T with her Faculty Champion and other student candidates for the program
in the fall of 2015. Photo: Humera Fasihuddin.

about their plans. Following training, Willis got right to work on her ultimate goal of developing a maker space to nurture interdisciplinary learning
opportunities across campus. Willis started by creating the Aggie Innovation
Network, a social entrepreneurship club, and began convening meetings
with peers in her lab. By opening up their efforts to community members,
including high school students, North Carolina A&T students started working on real issues facing the community. That's where Project Forage, a food
desert mobile app, was born, among other projects. Willis and her team
decided to raise funds for the maker space by entering the HBCU Innovation
Challenge, ultimately winning the top prize of $15,000.

Willis gained even more clarity in her role as Fellow when she attended
the three-day Silicon Valley Meetup a few months later. There she met
nearly three hundred other newly launched Fellows from all over the
country (figure 3.2). The meetup expanded her knowledge of viable strategies to nurture I&E on campus. Willis heard about moonshot thinking
from leaders at Alphabet's X and learned how Google promotes a culture

Figure 3.2
Angelica Willis, a University Innovation Fellow from North Carolina A&T State University, attended her first Silicon Valley Meetup in the spring of 2016. From left: Angelica Willis, Briana Cantos, Kathryn Christopher, Hannah Hund, Asya Sergoyan, LaRissa Lawrie, and Ann Delaney. Photo: Patrick Beaudouin.

of innovation and inclusive practices.[19] She spent the following two days at Stanford's d.school, exploring strategies to design and facilitate engaging learning experiences and ways to evolve plans to effect change on campus. She visited Microsoft and learned about The Garage, a global network of spaces nurturing the entrepreneurial mindset of employees. Willis reflected that the meetup taught her how to ensure the sustainability of her efforts:

> I realized how much more Fellows were able to accomplish with a formalized team, to bring more people into the movement, and to leverage the group as a future selection pool for future A&T Fellows. This helped show our faculty and stakeholders that we were serious about "making" things happen. Additionally, our African American Fellows across the nation created a sort of affinity group within UIF to support each other's efforts within the diversity innovation space. Ultimately the meetup showed me that I, and like-minded students, were in control of the UIF impact on my campus, not my faculty, not my administration, not our funding.[20]

Willis realized her dream of developing a maker space when faculty and administrators matched the $15,000 her team had won with physical space at the campus library. What's more, the meetup helped her realize that it wasn't enough to have an innovation space; what mattered was what people did *inside* the space. What cultural norms, behaviors, and attitudes might she help establish that would draw in participants and keep them coming back? A thriving community grew out of her efforts, and Willis's achievements brought her recognition by President Barack Obama as a White House Champion of Change for Computer Science Education in 2016. When Willis graduated, the Fellows that followed her built on her work and continued to enhance the higher education experience at North Carolina A&T. Willis remains involved in the network, mentoring Fellows and staying connected with her colleagues across the nation.

How Students Effect Change at Their Schools and Beyond

The Fellows' training equips them with the knowledge, skills, and mindsets to be agents of change at their schools. This is the big idea that gives the program its unique impact. In the process of becoming Fellows, students gain *knowledge* about the I&E resources available to students at their campus as well as those at other schools. They also practice the *skills* needed to craft and advance strategic projects, to enrich the landscape of available learning opportunities. But arguably the program's most important contribution is in changing students' *attitudes* (or mindsets). This change is not dependent on how well they can complete a task or apply a given method, but on their beliefs about what they are capable of doing. The program gives these budding change agents permission to act and, in doing so, contributes to building their sense of agency.

Three program elements are particularly effective at building the Fellows' sense of agency. First, Fellows report drawing inspiration from meeting a large number of like-minded students, some of whom face similar challenges and some of whom have already achieved significant impact at their schools (figure 3.3). Second, having access to a repository of concrete resources and strategies allows Fellows to achieve results more quickly. Third, the experience of witnessing program staff continuously improving and scaling the program reinforces the value of an experimental, iterative approach.

Figure 3.3
University Innovation Fellows gathered at Stanford University's d.school for the Silicon Valley Meetup in the spring of 2016. Photo: Ryan Phillips.

As a result of this training, Fellows feel empowered to act. They understand that the institutional challenges they are tackling require learning by doing, as well as courage, persistence, and flexibility. According to entrepreneur and author Peter Sims, rather than planning an entire project or starting with a big idea, organizations should create a set of small experiments designed to inform larger goals. He calls these experiments "little bets."[21] Kettering University Fellow Alan Xia echoed this approach, saying, "The big thing is the rapid execution of ideas. Planning is key, but if you can have many executions, then you subsequently have many learning experiences. Fail fast and often in the early stages of development, and the multitude of learning experiences will help you execute something incredible."[22]

The program also contributes to the mindsets of Fellows by encouraging them to develop empathy. Only by understanding what motivates other students, as well as faculty and administrators, are Fellows able to implement lasting change and create resources that add real value.

Armed with a sense of empathy, agency, and knowledge about campus assets and needs, Fellows embark on their projects. These projects loosely fall into four groups: (1) physical spaces, (2) educational activities, (3) student groups, and (4) outreach to alumni, as well as to the broader community.

Creating Environments That Nurture Student Innovators

Fellows recognize that creating or repurposing physical spaces where students can gather and collaborate is a powerful strategy to achieve lasting change and to attract more students to I&E. The visit to Stanford's d.school and Google during the Silicon Valley Meetup exposes Fellows to environments that have been intentionally designed to promote behaviors associated with innovation, such as collaborating with others across disciplines, sharing ideas, and making. At the meetup, Fellows use these spaces and participate in hands-on activities to understand the design principles that have guided their creation so that they can apply those principles to create and transform spaces at their schools.[23]

In 2014 alone, Fellows were involved in the creation of new spaces at twenty-two schools. Fellow Ryan Phillips, a computer science major with a minor in mathematics at the University of Oklahoma, was instrumental in connecting the efforts of different departments; he raised $5 million from five deans to build a 20,000-square-foot Innovation Hub on campus.

But spaces do not have to be brand-new buildings. In fact, repurposing and redesigning existing, underutilized spaces can have as much impact. When Fellow Alexandra Seda conducted field research to complete the Landscape Canvas for her school, Ohio Northern University, she realized that students did not spend time in the engineering building outside of classes. As a result, the casual encounters that lead to brainstorming and collaboration were not happening. She pitched the idea of a dedicated student space to one of her professors, who in turn shared it with the dean. When the dean provided funding to transform the Freshman Design Studio into a commons, Seda and another student invited all engineering students to join the effort of changing not only the physical environment where they learn but, through it, the school's culture. The group who answered the call became ION (Innovators of Ohio Northern). Over a break, the ION students transformed the space, bringing in comfortable couches for casual conversation and brainstorming, and covering the walls with whiteboards. As Seda put it, "We are more than just students here. We are partners with the University."

Other students are working together to build maker spaces. In November 2015, the Michigan Tech Fellows convened students, faculty, and staff to brainstorm ideas to create the school's first maker space by turning a refurbished bowling alley into a place for anyone in the community to collaborate, design, build, tinker, and bring their ideas to life. Maker spaces like the one at Michigan Tech are being built at universities across the nation, as the value of the maker culture in the education of all students gains acceptance.[24]

At schools where multiple maker and coworking spaces exist, Fellows strive to democratize access to those spaces. At many universities, maker spaces, labs, and machining locations are scattered in several locations across the campus. However, these resources are not universally available for use by undergraduate students, often because a lack of training creates safety concerns. At Kettering University, Fellow Alan Xia came up with the idea to take the kind of resources that were accessible only to students in certain engineering classes and make them accessible to the whole student body and members of the community. For his "Open Lab Days," Xia enlisted the help of other students and instructors who could provide guidance and safety training, and he drew in a full house of students, faculty, community members, and even school-age kids to use the foundry, welding machines, and other equipment for the first time.

Reimagining Learning Experiences

Whether through for-credit courses or cocurricular activities such as hackathons, Fellows design and facilitate experiences that allow students to take ownership of their learning process and expand it beyond the walls of the classroom. While curricula are often slow to change, Fellows use UIF strategies to create opportunities for all students to learn (almost) anything, anywhere, and at any time.

Consider the work of Tanner Wheadon, a Fellow who majored in technology management at Utah Valley University.[25] When he presented university administrators with the idea of opening an innovation space on campus, he was told it would be included for consideration in the five-to-seven year plan. Undaunted, he scraped together what little funding he could, purchased a $500 rolling industrial cart like the one he had seen at the d.school, and filled it with $200 worth of prototyping supplies (figure 3.4). As word

Figure 3.4
Tanner Wheadon, a University Innovation Fellow from Utah Valley University, created a mobile maker space for his school. Photo: Tanner Wheadon.

of his "mobile maker space" spread, Wheadon's "little bet" paid off, and he started receiving invitations to bring the cart to different courses. Based on the success of this project, Utah Valley University expedited the timeline to building an innovation space from five years to just three.[26]

Wheadon also thought about how he might be able to use his mobile maker space to reach all students at Utah Valley University, not only those already engaged in I&E. He discovered that one of the classes that students could take to satisfy the general education requirements, "Understanding Technology" (Tech1010), presented an opportunity to include new content. He designed a pop-up module of design thinking activities that could be included in a two-week section of the course.[27] Wheadon's module had students go out of the classroom to identify problems on campus in need of creative solutions. After teaching a successful pilot using the prototyping cart, Wheadon trained all Tech1010 instructors to teach the design thinking module, reaching 1,500 students annually across all majors.

Fellows often leverage collaborations with other national organizations that program staff regularly spotlight. Jonathan Spiegel, Fellow at the University of New Haven, hosted a 3 Day Startup event, a hands-on workshop used by colleges across the globe to teach entrepreneurial skills.[28] Similarly, Corey Brugh and his team of Fellows helped establish the National Academy of Engineering's Grand Challenge Scholars Program at the Colorado School of Mines. These collaborations allow Fellows to hit the ground running and leverage the resources, know-how, and brand visibility of established programs. In the process of planning such events, Fellows also discover that facilitating learning requires careful planning of the experience, including configuring the physical space and supplying engaging materials that go beyond a slide deck.

Reaching All Students

Many Fellows form new student groups as a vehicle to transform learning at their schools. This was the case of Grand Valley State University Fellows Kathryn Christopher and Leah Bauer. They realized that most students only interacted with other students in their major. To help break down these barriers, they created a group called IDEA (Interdisciplinary Entrepreneurship Alliance) that considers problems from the outside world and encourages students to collaborate with others outside their major, using design thinking to come up with solutions.

In other cases, Fellows join existing groups. Alex Francis, a PhD candidate in mechanical engineering at the University of Wisconsin-Milwaukee, became the president of the UWM chapter of the Collegiate Entrepreneurs Organization (CEO), a national student group traditionally composed of business students. According to Francis's Faculty Champion Ilya Avdeev, having Fellows take standard campus-leadership positions has attracted a more diverse set of students to I&E offerings. Belonging to multiple student groups creates opportunities for cross-pollination and sharing of ideas that benefit everyone.

As strategic thinkers and doers, Fellows build on the work of previous cohorts; their initiatives evolve to address emerging demands. University of Pittsburgh Fellow Karuna Relwani created the student organization Engineers for Sustainable Medical Devices (ESMD) to provide biomedical engineering students with hands-on experience in working with physicians to design medical devices, from surgical mounts to brain stimulators. As the student organization grew and more Fellows joined, ESMD was rebranded and grew into the Design Hub, reaching other majors outside biomedical engineering.

Fellows learn to be mindful of the language they use to design and promote events and activities in order to engage a broader population that might not be attracted by the words "innovation" and "entrepreneurship." As a generation that has lived through the dot-com and real estate bubbles, #occupywallstreet, and ballooning student debt, many students may equate entrepreneurship with business, and business with capitalism and greed. To reach other disciplines, Fellows may instead lead with "creativity" or "making." However, these terms elicit particular associations as well that can limit their appeal to certain groups of students. For instance, innovation may be equated with product design, while creativity and making may be associated with art. Understanding the audience they want to reach and using language that resonates with that audience is part of the empathy-driven strategy Fellows employ to reach all students.

Engaging Community and Alumni

Fellows understand the importance of connecting with the community that surrounds their institutions, including organizations supportive of innovators and entrepreneurs, local industry, and the K–12 education system. Angelica Willis, introduced earlier, is just such an example. The maker space she developed was aimed at drawing in members of the underserved population in Greensboro, North Carolina. As they engage with members of the local community, students tackle real challenges, such as urban food deserts and crime.

Clemson University Fellow Bre Przestrzelski has facilitated several K–12 design thinking experiences, including the Design Discovery (D^2) Program, a semesterlong immersive experience for rising high school juniors and seniors interested in biomedical engineering and health, in partnership with St. Joseph's Catholic High School in Greenville, South Carolina.

Going beyond her individual community outreach effort, Przestrzelski combined forces with Ben Riddle, a Fellow at nearby Furman University, to organize the first UIF Regional Meetup in Greenville in October 2014.[29] Fellows from across the country and students from universities in the area gathered to apply design thinking to a community challenge and learned lean start-up techniques. After that successful pilot, Fellows at the University of Maryland, Kent State University, James Madison University, William

Jewell College, Georgia Tech, the University of Wisconsin-Milwaukee, LaSalle University, Wichita State University, Utah Valley University, University of North Dakota, Universidad Católica del Uruguay, Universidad de Montevideo, Universidad Tecnológica del Uruguay, Universidad CLAEH, and University of Twente, also organized Regional Meetups to showcase their campus assets and attract new students to the movement.

Fellows have also been effective in enlisting members of their school's alumni, many who are successful entrepreneurs, as supporters of I&E initiatives. A team of University of Maryland students co-led by Mackenzie Burnett spearheaded Bitcamp, one of the country's largest hackathons. Fellows produced a video that highlighted Bitcamp as a model for interdisciplinary education that should be available to all students in the classroom. The video set in place a chain reaction that resulted in university leaders inviting the cofounders of Oculus to speak, shortly after the company's $2 billion sale to Facebook. Oculus cofounder Brendan Iribe (a former Maryland computer science undergraduate) cited student leadership and Bitcamp as a significant reason behind a $31 million gift toward a new Center for Computer Science and Innovation.[30]

Bringing the Student Voice to National Conversations about Education and I&E

Fellows not only accelerate change on their campuses, but they also contribute a student perspective to national conversations about education and I&E. This fills an important need, as too many discussions about what students need or want fail to engage the students themselves.

Fellows have brought the student voice to organizations and gatherings that have never before had students as presenters. In April 2014, ten Fellows participated in the "Educating Engineers to Meet the Grand Challenges" workshop held at the National Academy of Engineering. The event brought together leaders from academia, associations, start-ups, service learning organizations, and industry to identify best practices for preparing students to meet global challenges such as poverty and access to water.[31] The Fellows were active contributors and were instrumental in galvanizing the support of the attending engineering deans. Fellows have additionally presented at the White House, the Deshpande Symposium, VentureWell's Open Conference, Epicenter's Research Summit, the Annual Conference of the American Society of Engineering Education, the Global Entrepreneurship Summit,

SXSW Edu, and the Annual Conference of the Association of American Colleges and Universities.

Fellows have also mobilized leaders at their schools to rally behind such national causes as attraction and retention in STEM disciplines. In March 2015, Fellows from ten schools obtained the signatures of their schools' presidents on a joint letter of commitment presented to President Obama on occasion of the White House Science Fair.[32] The campaign, called "#uifresh" (University Innovation Freshmen), has grown to hundreds of schools that are now exposing all incoming freshmen to experiences in design thinking, innovation, and entrepreneurship.

From School to the Real World

When Fellows graduate and join the workforce, they bring valuable skills and mindsets, whether to existing corporations or their own ventures. Ryan Phillips, a Fellow who graduated with an engineering degree from the University of Oklahoma in 2015, accepted a job as a program manager at Microsoft. He soon realized that his employer greatly valued the skills he had learned as part of the UIF program, especially his experience solving problems with unknown solutions and unknown paths. Phillips's core project as a Fellow resulted in a collaborative culture square in the middle of the university's research park—which brought together students from arts and sciences, engineering, and architecture—located right next to world-class research and Fortune 500 companies. Every aspect of the project involved unknowns and collaboration. Phillips discovered that Microsoft values people who can act as "mini-CEOs" of the projects they work on—taking a large problem, breaking it down, finding creative solutions that fit together, and executing to success. These are exactly the skills Fellows gain navigating complex environments, forging collaborations, and creating lasting institutional change.

Nadia Gathers graduated from Converse College in 2015 and began a career at Code2040, a nonprofit working to provide black and Latinx students access to coding careers in Silicon Valley. Gathers's experiences in the Fellows program helped them design communications strategies for Code2040 that successfully attracted hundreds of underserved minorities to summer internships in the Bay Area and encouraged them to seek such career pathways.[33] Gathers's success at Code2040 led to a management level position at the large tech firm Github, where they now design internal communications strategy for the company. As the youngest member of the team (and also one of the youngest in the company), Gathers credits their poise, confidence, and earned credibility to the savvy they obtained from leading Converse College to create the new social entrepreneurship minor that is now available to all students.

Fellows report that their experiences navigating change in academia and persevering in creating an impact rank highly in conversations with potential employers. Hiring managers appreciate talent who can "learn the system" and get things done. Fellows have a saying: "Once a Fellow, always a Fellow." With a passion for innovation, they continue to apply their change agent skills at traditional Fortune 500s such as MasterCard, Lockheed Martin, and Procter & Gamble. Fellows thrive in start-up companies like Handshake, a firm that is reinventing software for university career services departments, or Spira, an innovative venture in the food sector that aims to combat malnutrition and food insecurity. Fellows also lead innovation within nonprofit organizations like St. Jude's Hospital and Capital Area Food Bank. Finally, many Fellows pursue graduate studies or are invited to stay on as staff with their academic institutions, where they continue to help build I&E ecosystems.

A Program Model That Works

The economics of student-led efforts are difficult to sustain. Many of these efforts are funded by student government organizations that compete for limited dollars and/or rely on club dues. Institutional funding is not geared toward student-led initiatives. This has required us to design a business model that minimizes the risk and uncertainty of working with students and yields a return on investment for all involved. A strong value proposition for students, faculty, and institutions supports the appeal of UIF.

Student Value Proposition
Students are at the center of our work. Applicants are attracted to the opportunity of becoming Fellows because they are motivated to make a difference at their schools. They are passionate about what I&E empowers people to achieve. Receiving a nomination from their school inspires candidates, as does the opportunity to participate in the Silicon Valley Meetup. Fellows

report that the affiliation with a national program lends them legitimacy and institutional support.

Above all else, Fellows report that the single most important asset coming out of their participation is the network and the feeling of belonging to a large and growing family of like-minded change agents. Candidates learn from one another and from each other's ecosystems. After training, Fellows gain access to an active online community of peers, where they can share new strategies and pose questions. They remain connected to this network even after they graduate from their institutions.

Faculty Value Proposition

Traditional structures and incentives in academia are almost exclusively centered on disciplines and departments. At many academic institutions, scholarly research comes first, followed by teaching, with service ranked far behind. As a result, faculty who are passionate about and committed to advancing I&E may not experience any direct benefit in terms of promotion and tenure. Faculty who commit to I&E often do so because they see the positive effect on student learning and motivation.[34] This interest often leads them to introduce I&E in the context of their discipline with team-based projects focused on real-world problems, which can result in student innovations.[35] In the absence of an innovation ecosystem, faculty witness too many discontinued student projects that could have resulted in real-world impact. For these faculty, University Innovation Fellows bring the value of a stronger I&E ecosystem that can nurture student projects that have the potential to increase the impact of their research. Faculty Champions, who are themselves change agents, find in Fellows collaborators who not only have been trained to effect change, but who can leverage their student perspective to engage other students.

Institutional Value Proposition

US institutions of higher education are steadily increasing their investment in programs supporting entrepreneurship.[36] Studies show that faculty engagement in institutional change efforts are supported by the positive outcomes and increased motivation seen among participating students. Student engagement is never the sole factor in driving curriculum change, but according to researcher Ruth Graham, "even the more reluctant [faculty] see the difference in the students and the higher levels of motivation."[37]

Consistent with this, institutional leaders report that a Fellow's voice can play a prominent role in accelerating change efforts and informing curricular changes.

In our experience, there is no stronger driver for peer engagement than students themselves. Fellows are magnets for other students, experimenting with ways of framing new offerings that resonate with peers and achieve increased levels of student engagement. Fellows are reaching hundreds of peers on campus, with new spaces, workshops, courses, and clubs.[38] Faculty who regularly sponsor students have confirmed that Fellows are serving as partners, advising and helping ensure sound organizational strategies that positively impact students.

A Growing Movement

UIF serves as a growing platform for student-led change in higher education institutions—one that reimagines learning beyond traditional curricular activities, one that aims to recruit university students of all backgrounds and disciplines, one that connects its fellows with alumni and industry networks to better prepare them to be innovators and entrepreneurs after graduation.

Building such a movement is not without challenges. It requires identifying candidates and Faculty Champions whose motivations are aligned with the UIF mission. Also, the student-faculty partnership toward a common goal of institutional change is different from the traditional relationship within the classroom or the research lab; Faculty Champions need to strike the right balance between support and control, and to empower the Fellows to take initiative. Additionally, at schools where there are many I&E resources, issues of territoriality can arise. Institutional cultures are sometimes difficult for students to navigate successfully. Fellows are encouraged "to lead when leadership is required and exercise humility, when not."[39] These are values identified by Google and other industry leaders as key skills, but they are not necessarily principles employed by traditional management structures in academia.[40]

Despite the obstacles, the program has grown significantly since its start in 2012. With a rapidly increasing number of applicants and a high percentage of schools that have sponsored Fellows over several cycles, it became evident that the program was having a major impact and needed to continue beyond the end of the NSF funding.[41]

With the help of NSF's I-Corps customer discovery training, we identified a variety of new directions to expand UIF.[42] The connection to Stanford and Silicon Valley has been a fundamental piece of the Fellows' experience. In 2016, we transitioned to operate as an initiative of Stanford's d.school, which has always been a center of gravity for the program. We subsequently developed a workshop for faculty to learn design thinking and apply it to creating new learning experiences at the classroom level and beyond. As of November 2018, we have conducted nine workshops and reached over 380 educators.[43] Additionally, we opened the program to higher education institutions outside of the United States, and we presently reach twenty other countries. Through a partnership with Google, we have expanded into India with 306 Fellows from forty-five Indian institutions.

The program continues to cultivate strategic partnerships with industry leaders that are interested in changing how young people are prepared to enter the workforce. We hope that this connection with a growing network of Fellows results in broadening the set of schools from which companies recruit. This is especially important given the growing awareness about the lack of diversity in industry.

As of November 2018, UIF has trained and supports over 1,800 Fellows across 250 institutions worldwide and has engaged over 380 educators in redesigning teaching and learning within their classrooms and more broadly at their institutions. The program's substantial reach makes it well poised to catalyze new ways of thinking from all higher education stakeholders toward collectively reimagining how we prepare students to invent the future.

Notes

1. We would like to thank our dedicated University Innovation Fellows program teammates—Katie Dzugan, Laurie Moore, Stacey Mushenski, and Ghanashyam Shankar—for their contributions in growing this movement. We would also like to thank our colleagues from Epicenter, Stanford Technology Ventures Program, Venture-Well, SageFox, and the d.school. We are incredibly inspired by the students, faculty, and industry partners participating in the program and building a movement to revolutionize higher education. We also wish to acknowledge our families for their support and encouragement through this exciting journey. Contact us at humera@dschool.stanford.edu and leticia@dschool.stanford.edu.

2. Beth McMurtrie, "Now Everyone's an Entrepreneur," *Chronicle of Higher Education*, 20 April 2015, http://chronicle.com/article/Now-Everyone-s-an/229447/; Arnobio

Morelix, "The Evolution of Entrepreneurship on College Campuses," Kauffman Foundation, 29 October 2015, https://www.kauffman.org/currents/2015/10/the-evolution-of-entrepreneurship-on-college-campuses.

3. Jeffrey H. Dyer, Hal B. Gregersen, and Clayton M. Christensen, "The Innovator's DNA," *Harvard Business Review* 87, no. 12 (2009): 60–67.

4. In September 2009, President Barack Obama issued the first-ever innovation strategy for the United States. Updated in 2011 and once more in October 2015, the strategy makes the case for key investments "to ensure America continues to lead as the world's most innovative economy, to develop the industries of the future, and to harness innovation to help address our Nation's most important challenges." One of those key investments, according to the strategy, should be to engage more students in STEM learning and entrepreneurship and ultimately "empowering a nation of innovators." "Fact Sheet: The White House Releases New Strategy for American Innovation, Announces Areas of Opportunity from Self-Driving Cars to Smart Cities," 21 October 2015, https://obamawhitehouse.archives.gov/the-press-office/2015/10/21/fact-sheet-white-house-releases-new-strategy-american-innovation; "A Strategy for American Innovation," National Economic Council and Office of Science and Technology Policy, October 2015, https://obamawhitehouse.archives.gov/sites/default/files/strategy_for_american_innovation_october_2015.pdf.

5. "Science, Technology, Engineering, and Mathematics Talent Expansion Program Centers (STEP Centers) Program Solicitation NSF 10–569," National Science Foundation, accessed 15 December 2015, http://www.nsf.gov/pubs/2010/nsf10569/nsf10569.htm.

6. VentureWell—formerly the National Collegiate Inventors and Innovators Alliance—is a nonprofit founded in 1995 whose mission is to support I&E in higher education.

7. "About Epicenter," Epicenter: National Center for Engineering Pathways to Innovation, accessed 22 February 2016, http://epicenter.stanford.edu/page/about.html.

8. Maura Borrego, Jeffrey E. Froyd, and T. Simin Hall, "Diffusion of Engineering Education Innovations: A Survey of Awareness and Adoption Rates in U.S. Engineering Departments," *Journal of Engineering Education* 99, no. 3 (2010): 185–207; Sheri D. Sheppard, et al., *Educating Engineers: Designing for the Future of the Field* (San Francisco: Jossey-Bass, 2009).

9. The Engineering Majors Survey designed by the FIGS team continues to be deployed as a national longitudinal study to explore engineering students' engineering, innovation, and entrepreneurial interests and experiences over time. For more information, see https://web.stanford.edu/group/design_education/cgi-bin/mediawiki/index.php/Engineering_Majors_Survey.

10. Eric von Hippel, *Democratizing Innovation* (Cambridge, MA: MIT Press, 2005), 4.

11. Humera Fasihuddin, "How to Complete the Landscape Canvas," University Innovation Fellows, last modified 15 January 2016, http://universityinnovation.org/wiki /How_to_complete_the_Landscape_Canvas.

12. Stanford University Institute of Design, "What We Do," accessed 15 December 2015, https://dschool.stanford.edu/about.

13. The wiki, universityinnovation.org, was launched in partnership with the White House Office of Science and Technology Policy in November 2013 to profile innovative universities and colleges across the nation.

14. The external evaluator, SageFox, reports that students spend an average of 6.5 hours per week on these activities, with the vast majority indicating they felt this amount was "about right."

15. Rebecca Zarch et al., "Epicenter University Innovation Fellows Program Annual Meetup Evaluation," SageFox Consulting Group, May 2015, http://universityinno vationfellows.org/wp-content/uploads/2016/01/UIF-Annual-Meetup-2015-Survey -Report.pdf.

16. As of November 2018, the program fee is $4,000 for a team of up to four students, not including travel expenses.

17. "What Fellows Do," University Innovation Fellows, accessed 5 April 2018, http://universityinnovationfellows.org/about-us/what-fellows-do/.

18. "N.C. A&T Junior Joins University Innovation Fellows," North Carolina A&T State University, accessed 18 September 2017, http://www.ncat.edu/news/news-archive /2015/10/innovation-fellows.html.

19. X is a subsidiary of Alphabet, Google's parent company (formerly known as Google X).

20. All quotations from Fellows and Faculty Champions derive from our correspondence and conversations with them as program directors, and they have agreed to be quoted.

21. Peter Sims, *Little Bets: How Breakthrough Ideas Emerge from Small Discoveries* (New York: Simon & Schuster, 2013).

22. Pardeep Toor, "Kettering University Students Participate in University Innovation Fellows Program at Stanford University," Kettering University, 3 June 2015, https://www.kettering.edu/news/kettering-university-students-participate-university -innovation-fellows-program-stanford.

23. Scott Doorley and Scott Witthoft, *Make Space: How to Set the Stage for Creative Collaboration* (Hoboken, NJ: John Wiley & Sons, 2012).

24. In December 2015, the Make Schools Alliance launched with forty-seven institutions (universities, community colleges, art and design schools) committed

to providing students with spaces, projects, and mentors to engage in hands-on making activities. More information is available at http://make.xsead.cmu.edu/.

25. Wheadon has since graduated and is leading an industry innovation team.

26. "University Innovation Fellows Stories: Tanner Wheadon, Utah Valley University," YouTube, 10:34, posted by University Innovation Fellows, 29 October 2015, https://www.youtube.com/watch?v=5N5dcg0cqYY.

27. Pop-up classes are short, noncredit learning opportunities. For more information, see "Pop-Ups Offer Classes on Today's Hot Topics," accessed 5 April 2018, https://www.nytimes.com/2018/04/05/education/learning/pop-ups-offer-classes-on-todays-hot-topics.html.

28. See 3 Day Startup homepage: http://3daystartup.org/.

29. Humera Fasihuddin, "Hypothesis Affirmed! Greenville, SC, Ignited by Southeastern Regional Meetup," University Innovation Fellows, 9 October 2014, http://universityinnovationfellows.org/hypothesis-affirmed-greenville-sc-ignited-by-southeastern-regional-meetup/.

30. See Brendan Iribe Center for Computer Science and Innovation homepage: http://iribe.cs.umd.edu/home.

31. See NAE Grand Challenges for Engineering homepage: http://www.engineeringchallenges.org/.

32. Laurie Moore, "University Innovation Fellows Launch #uifresh Campaign to Attract and Retain First-Year Students to STEM Majors," University Innovation Fellows, 23 March 2015, http://universityinnovationfellows.org/uifresh/.

33. On the dearth of women and underserved minorities in technical careers, see Cook (chapter 12) and Sanders and Ashcraft (chapter 17) in this volume.

34. Ruth Graham, "Achieving Excellence in Engineering Education: The Ingredients of Successful Change," Royal Academy of Engineering, 2012, http://www.raeng.org.uk/publications/reports/achieving-excellence-in-engineering-education.

35. Robert Trebar, "The Influence of the College Environment on the Entrepreneurial Intentions of Students," PhD thesis, University of Toledo, 2014, https://etd.ohiolink.edu/!etd.send_file?accession=toledo1404751268&disposition=inline.

36. Wendy E. F. Torrance et al., "Entrepreneurship Education Comes of Age on Campus: The Challenges and Rewards of Bringing Entrepreneurship to Higher Education," Ewing Marion Kauffman Foundation, August 2013, https://www.kauffman.org/-/media/kauffman_org/research-reports-and-covers/2013/08/eshipedcomesofage_report.pdf.

37. Graham, "Achieving Excellence in Engineering Education," 36.

38. Rebecca Zarch et al., "Epicenter University Innovation Fellows Program Annual Survey Spring 2016," SageFox Consulting Group, July 2016, http://university innovationfellows.org/wp-content/uploads/2016/01/3-UIF-S16-Survey-Report.pdf.

39. Humera Fasihuddin et al., "Manifesto: We Believe Students Can Change the World," University Innovation Fellows, last modified 20 March 2014, http://univer sityinnovationfellows.org/manifesto/.

40. Thomas L. Friedman, "How to Get a Job at Google," *New York Times*, 22 February 2014, http://www.nytimes.com/2014/02/23/opinion/sunday/friedman-how-to-get-a -job-at-google.html; Rebecca Greenfield, "Holawhat? Meet the Alt-Management System Invented by a Programmer and Used by Zappos," *Fast Company*, 30 March 2015, http://www.fastcompany.com/3044352/the-secrets-of-holacracy.

41. While schools contributed a fee for student participation, that revenue did not cover the operating budget, which had been subsidized by the NSF grant that supported the creation of the Epicenter.

42. For more information about I-Corps, see Arkilic (chapter 5) in this volume.

43. More information about the faculty workshop can be found at http://university innovationfellows.org/teachingandlearningstudio/.

4   Building High-Performance Teams for Collaborative Innovation

Mickey McManus and Dutch MacDonald[1]

Innovative people are often described as curious, empathetic, imaginative, collaborative, and fearless. Yet individuals who possess these traits, or teams that possess them collectively, still sometimes fail at solving complex problems. Why? More importantly, what's the solution?

The two of us together have more than twenty years of experience grappling with these questions while tackling complex business problems for clients in virtually every industry, first at MAYA, our Pittsburgh-based design consultancy and innovation lab, and now for the Boston Consulting Group (BCG), which acquired MAYA in 2017. Our interdisciplinary team of strategists, designers, engineers, and human scientists deliver solutions through creative collaboration. Using a human-centered and iterative approach, we architect innovations for clients such as Oreo, Philips, Whirlpool, even the Pentagon. We have found that a diversity of thought, which includes understanding how others think and work, is a critical "superpower" for high-performance teams. Returning to the question, then, of why talented individuals and teams sometimes fail to innovate, we believe there are at least three reasons, all of which are in the control of the individual or, in the case of teams, the creative leader.

The first reason is that many individuals lack an understanding of (and literacy in) the methods successful innovators use to solve different classes of problems. A thwarted innovator may be relying on a prescribed organizational protocol, or perhaps chance, to discover an approach for tackling a particular challenge, rather than making use of specific skills and iterative techniques known to foster innovation.

The second reason is the temptation to take refuge in the popular "hero's journey" narrative, in which the independent rebel innovator can solve any problem through the force of will and genius alone. This mythology sells books and movies and may have even been true in rare cases, but in reality, the solutions to the hardest problems, those at the intersections of disciplines—the problems worth solving—cannot be solved alone. That doesn't mean there isn't room for single-minded creative leadership, but rather that leadership should be expressed at an appropriate place within the life cycle of a team's journey.

The third reason is that the individual or team may be unaware of (or have limited awareness about) the impact of their environment on their ability to innovate. Even a team that does everything right will not innovate in the wrong environment. In short, collaborative work is hard. Different ideas, agendas, and opinions can derail conversations, especially in a working environment that is ill-suited to the task at hand.

In our experiences at MAYA, we have learned to overcome these obstacles by developing a proven set of strategies for building innovative teams. We believe that innovation is about honing individual skills and then applying those skills through the service of interdisciplinary teams. These teams are assembled in a creative environment tuned for their needs, which promotes growth and maturity. We believe this method can be repeatedly applied in a host of circumstances to build and sustain high-performance, innovative teams to solve a host of "wicked" problems.

Increased Complexity and Wicked Problems

Our society has spent the last several centuries discovering and perfecting basic ways to solve problems by making a given *thing* and making it *right*. This approach to problem solving worked well enough when there were far fewer things to make right in the first place and when technological complexity was miniscule compared to what we encounter today—a world saturated with computation and billions (soon to be *trillions*) of connected devices. Today, when we can make almost anything that we can imagine, the most important problems for innovators involve making the *right things* instead of making *things right*.

In the book *Trillions: Thriving in the Emerging Information Ecology*, which one of us (McManus) wrote with our colleagues Peter Lucas and Joe Ballay, we outline MAYA's vision of how the process of innovation is changing.[2]

There are many reasons why tomorrow's innovation will be different from today's. In what follows, we focus on the most critical factor: complexity. Most of the easy problems have been solved, because for much of human history, most people faced relatively little technological complexity. A few people had to cope with what Horst Rittel and later C. West Churchman called "wicked problems," that is, problems that are resistant to resolution because they cannot be easily defined and have contradictory or changing requirements (figure 4.1).[3] Most people, however, could go through their whole lives without coming face-to-face with technological complexity. If you were building the very first factory, or trying to connect the world with lines of communication, or building an airliner, or trying to launch

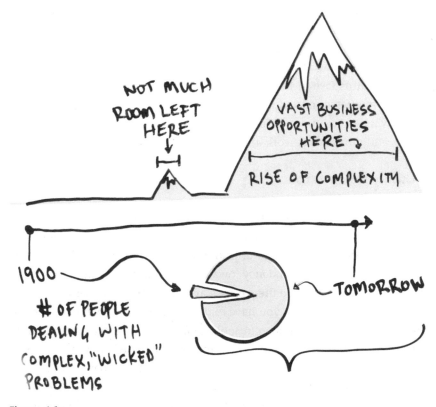

Figure 4.1
The rise of complexity forces an ever increasing number of people to manage complex and "wicked" problems. Courtesy of MAYA.

a rocket ship, or working on solving large-scale social problems, then you had to cope with complexity of one form or another (either due to technological complexity or the complexities of society itself). But the sum total of complexity in the world was comparatively small and localized. Today, however, due largely to the rise of digital connectivity, we are facing an era of *unbounded complexity*.[4]

For context, consider advances in computation. As early as 2007, the world produced more transistors annually—at a lower cost—than grains of rice.[5] In 2010, we produced more microprocessors than there were people on the planet (over ten billion that year alone).[6] If we stopped there, we would still have a relatively manageable level of aggregate complexity in the world. But this supersaturated solution of computation has now been seeded with connectivity. Those billions—and soon trillions—of devices are now getting connected, and each of those trillions of things are sending billions of messages, turning into "bricks" when they get bad updates, or in some cases being co-opted for nefarious reasons. Many of the challenges that innovators face in the coming decades will be focused on addressing this highly interconnected world. To address these wicked problems, innovators will need to improve their innovation literacy, build interdisciplinary teams, and pursue their work in environments optimized for creativity.

A Literacy for Innovation

Why do collaborative teams made up of innovative individuals fail? First, each discipline has its own language, tools, and ways of solving problems. This often leads to poor communication among team members trained in various disciplines and a tendency toward conflict and zero-sum thinking: "If it's not my way, it must be the wrong way." Second, it is harder to navigate through uncertainty when you have to defer judgment, suspend disbelief, or take a leap of faith as part of an unspoken (and necessary) agreement to trust the possible validity of what your teammates have to say. Third, space, time, and a different understanding of the problem tend to separate innovative people who should be working toward a common goal.

Team members need to be on the same page. Yet they often speak a different language derived from their specific discipline's jargon. They are most comfortable with their particular tools and methods for solving problems, and they are often reluctant to come out of their comfort zones.

Look, Understand, Make, Advance: A Human-Centered Design Framework

In 2010, the LUMA Institute was born out of the MAYA community to alleviate the design challenges resulting from the innate differences among people.[7] We observed that many people had the *characteristics* of innovators but lacked literacy and fluency around the basic *skill sets* of innovation. LUMA Institute's mission was to teach diverse individuals and groups a common lexicon and the skills needed for innovative, human-centered design. LUMA Institute began by conducting an in-depth study of the methods used by successful innovators, compiling a catalog of more than nine hundred techniques that spanned nearly a century. We then focused on the methods that people from various disciplines could learn and practice within a short time and that organizations could apply regardless of whether they were solving product, service, or policy-related challenges. These methods fit into categories that became the foundation for the organization's name and approach:

Looking for unvoiced and unmet needs

Understanding how to analyze and synthesize findings into systems and models

Making to envision future possibilities

Advancing to move invention from idea to solution in the real world

The methods within these categories are tools for establishing a common ground for problem-solving among innovators across disciplines. For example, team members learn to "look" through such methods as fly-on-the-wall observation or contextual inquiry. Stakeholder mapping, persona development, and concept mapping aid "understanding." Storyboarding, imagining alternative worlds, and rough-and-ready prototyping help a team focus on "making." Iterative loops of looking, understanding, and making "advance" the right ideas into real world solutions.

These tools do not belong to any one discipline, so they do not require special training or expertise; they help people operate at the intersections of disciplines while using the same language. Further, many common tools do more than just invite team members to contribute—the very nature of the tool *compels* them to contribute. It is important to note that these tools are not just theoretical. MAYA has successfully and repeatedly field-tested them over many years in our human-centered design and innovation practice.

Innovation Literacy in Action

Consider the case of industrial automation innovator Emerson Process Management. After working with MAYA and sending its project teams through literacy training and practice with the LUMA Institute methods, the company embarked on an effort to improve a product that had been in production for decades. Its "Delta V" automation solution overturned thirty-five years of accepted practice of asset management within the factory setting. An evolutionary change to one of its own products drove dramatic improvements in value to customers. At the end of this effort, a third-party analysis by the ARC Advisory Group documented an 82 percent reduction in factory worker time spent on routine tasks.[8]

Using the basic skill sets identified in the LUMA Institute framework allowed Emerson Process Management's teams to focus on the human element of the design challenge. They engaged real users of the product in an iterative design process, and those users were invested in the outcome and embraced the change. The LUMA Institute framework can take teams a long way toward making sure that unmet or unvoiced needs are solved in a way that is structurally sound and confers some future-proofing. As in the case of Emerson Process Management, many of the methods help to keep the focus on solving the right problems and making the *right things*.

Team Structures

If you are working on an incremental, evolutionary project, many classic forms of problem-solving may work just fine (figure 4.2). Moving from version 2.5 to version 2.6 is a "little change" that creative and clever individuals or classic team structures may achieve. Often the solution is to find a team of experts and run a proven project management process.

However, these classic structures fail when they hit the limit of individual expertise or even the limits of expert group innovation, particularly when the problems move across one form of expertise to another. One of the problems James Surowiecki notes in his book *The Wisdom of Crowds* is that teams of experts from any one domain tend to self-reinforce, exploiting the center of their bell curve rather than exploring its edges.[9] He also cites studies showing that experts cannot accurately predict how right or wrong they are (a process called "calibrating" their judgment).

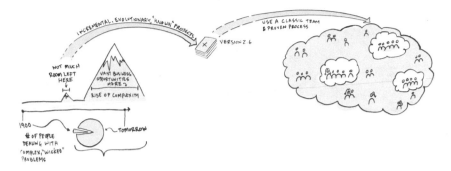

Figure 4.2
Incremental problems can likely be solved using classic team structures. Courtesy of MAYA.

Experts from a given discipline speak a different language than experts from other disciplines. They have a jargon that has evolved over a long period of time and helps reinforce a community of practice. Jargon can be useful for deep research into a given topic, but it becomes an obstacle when you have to solve a problem at the edges of two disciplines. Experts' perspectives are what makes their discipline so valuable but are *also* exactly what makes it so hard to understand them. When the problem spans technological, physiological, or social dimensions, finding a common ground can be complicated.

We believe most disagreements arise when people *think* they're talking about the same thing but actually are not. Sometimes the disconnect between disciplines means that each successive member of the team has less time to deal with challenges. Worse, the team continues to believe it is following best practices and wonders why a project failed to have an innovative outcome.

Building High-Performance Teams

Innovators on a journey to solve complex, interconnected, wicked problems need to take a different path than the traditional one described above. At MAYA, we have found that journey requires deeply interdisciplinary teams from the onset. We believe that interdisciplinary collaboration, when done right, can collapse seemingly intractable problems and shorten development cycles.

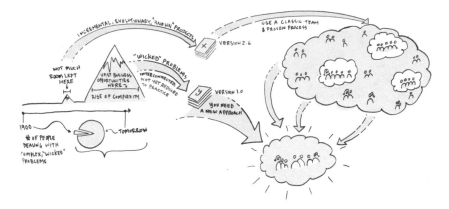

Figure 4.3
High-performance teams can tackle complex and "wicked" problems that foil teams
in traditional structures. Courtesy of MAYA.

Any system has a given minimum level of complexity that will have
to be managed either by the creators of the system in its design, or by the
end users during its use.[10] We can't always eliminate all the complexity in a
given problem space, but sometimes we can move it around from one discipline to another (figure 4.3). We can tame complexity. For instance, maybe
a problem has no obvious technical solution, but we can use the native way
people think (how cognition works) to solve it. Sometimes a different discipline has already found a best practice in coping with a difficult challenge,
but different language prevents others from seeing the patterns.

As Lucas points out in *Trillions*, talent within a discipline usually falls on
a bell curve:

> Those of average talent tend to huddle toward the center of their particular disci-
> plinary piece. That is where they will find safety in numbers among many others
> who share the assumptions and values that they have all been taught. But this
> is not how the superstars behave. Rather, they migrate toward the very edges of
> their puzzle piece. Why? Because they know that by doing so they will encounter
> other bold thinkers like themselves, exploring the unknown territory at the edges
> of other disciplines. So, the interstices between disciplines are always where the
> action is. It is where the best practitioners go to invent the future.[11]

The challenge in building innovative groups is not in creating teams of
people from different disciplines, but in finding those practitioners who
have migrated to the edges and can act as bridges back to the core experts
of a given domain.

The Life Cycle of a Team

We have observed in our innovation practice that well-formed, high-performance interdisciplinary teams are emergent entities in their own right—innovative teams have life cycles. Armed with the methods and literacy of innovation, a team needs both guidance and freedom to grow, explore, and mature. When teams are born, they may need more top-down leadership to point them in the right direction. This is an appropriate role for single-minded creative leadership. But when teams become toddlers, they may test boundaries. When they experience adolescence, they may become far more competitive and begin to think they know everything and doubt their elders. As they mature, collaboration and wisdom may come to the fore.

We have found this pattern to be true whether the life cycle spans weeks, months, or years. Supporting this life cycle in a repeatable way is critical to sustaining a culture of innovation. Innovative teams live rich lives, make the *right things*, then ultimately die. At that point, their constituent members recombine into new innovation entities.

Accelerated Trust

We have also learned that high-performance interdisciplinary teams depend on trust. Trust is the only way you can work with someone who sees the world differently. Suspending disbelief—not trusting your instincts for some decisions but rather trusting another person who sees the world differently—is often key to collaborative innovation. You may not be sure why your teammate feels so strongly about something, but you trust that she has her reasons, and you trust that she will, in turn, support you when you sound like you are tilting at windmills.

Accelerating trust can have a lasting impact on the collaborative and creative workings of a team. Creative leaders can help in this process. For example, they can create ways to increase the number of cycles of decision making between team members early in the team's life cycle. This helps members get a feel for how others will react in a given situation before more critical decision making is required.

Idea Flow

Building diverse teams and supporting their life cycle is not enough to ensure success. The way communication flows within a high-performance team—the "energy" being spent getting over the friction of teamwork—can

dramatically shape a team's performance. Sandy Pentland's research reinforces the value of collaborative innovation methods and draws focus to
the underlying science of idea flow.[12] He points out some key characteristics
that define communication in a successful team:

1. Everyone on the team talks and listens in roughly equal measure, keeping their contributions short and sweet.

2. Members face one another, and their conversations and gestures are
   energetic.

3. Members connect directly with one another, not just with the team
   leader.

4. Members carry on back-channel or side conversations within the team.

5. Members periodically break, go exploring outside the team, and bring
   information back.

In addition to Pentland's observations, we add the power of drawing to
promote idea flow. Sketching is the *lingua franca* among design disciplines.
Unlike renderings or fully functional prototypes, sketches are fast, cheap,
disposable, and easily allow members of a team to ask, "What if?" Anyone
can sketch; even a stick figure can be a visual form that others on the team
can interpret and extend.

The Double Helix

A number of MAYA's methods for developing high-performing teams foster
an increased amount of turn taking, high-energy interactions, cooperative
creation of work products, and empathy toward others' points of view. The
approach is combinatorial. Certain combinations of techniques together
are particularly well suited to increase and accelerate trust within the team.

One example combination is what we call the "double helix." The double
helix is a form of scenario planning in which two threads of innovation—an
advanced technology and a new business model—are developed indepen-
dently by two teams (figure 4.4). The ideas of each team inform the other as
they are tested in accelerated time at periodic "crossover" events. A series of
"rough-and-ready" prototypes of an entire system are built by the tech team
in successive levels of fidelity over a period of days, weeks, or months. At
the crossover events, the teams come together and simulate a year's worth
of activities in a week, and a week's worth in a day. These include tasks

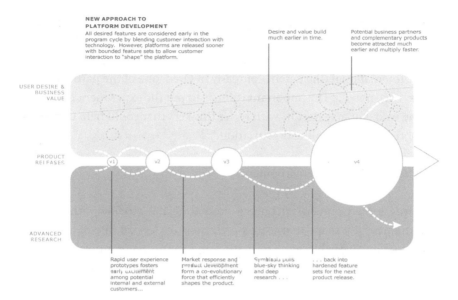

NEW APPROACH TO
PLATFORM DEVELOPMENT
All desired features are considered early in the
program cycle by blending customer interaction with
technology. However, platforms are released sooner
with bounded feature sets to allow customer
interaction to "shape" the platform.

Desire and value build
much earlier in time.

Potential business partners
and complementary products
become attracted much
earlier and multiply faster.

USER DESIRE &
BUSINESS
VALUE

PRODUCT
RELEASES

ADVANCED
RESEARCH

Rapid user experience
prototypes fosters
early excitement
among potential
internal and external
customers...

Market response and
product development
form a co-evolutionary
force that efficiently
shapes the product.

Symbiosis pulls
blue-sky thinking
and deep
research . . .

. . . back into
hardened feature
sets for the next
product release.

Figure 4.4

MAYA's double-helix framework uses simulation at crossover events to foster innovation. Courtesy of MAYA.

performed by representative users of the system as well as such factors as business externalities beyond the users' control. The teams sit "behind the glass" while all this is happening, observing the flow and impact of their ideas to advance the next round of technology and business innovation.

The method uses a technique called "Wizard of Oz prototyping" in which the nonfunctional aspects of the system are manipulated to seem as if they are already functioning, much like the man behind the curtain in the famous story, who pulled levers to create the illusion of a grand and powerful wizard. This form of gaming fosters a shared understanding of the emergent challenges within a complex problem, creating a framework in which team members see what each of them would do in a given circumstance and how users respond to high-pressure situations.

MAYA has used the double-helix technique across a wide range of challenges, from inventing entirely new ways of collaborating in times of war, with programs like the United States Army's Command Post of the Future initiative, to prototyping entirely new forms of connected factories for major consumer brands.[13] An article documenting one of our double-helix-style team events in the business journal *Fast Company* identified the value of this sort of petri dish approach for both the team and organization. Bonin Bough, vice president of global media and consumer engagement at Mondelez International (the parent company of Oreo, one of the participants in the event), noted, "The important thing is to put the experiment out, test it in the wild, but also [test it] with other thinkers that could help us explore and bring it back into the organization."[14]

The Value of Creative Environments

The way an environment encourages communication to flow within a team can dramatically shape that team's performance. A good rule of thumb we have found as we explore the formation and nurturing of innovative teams is this: If you can't change the individual, change the environment.

Herbert Simon, the Nobel Prize–winning economist, proposed that humans have bounded rationality.[15] He used the analogy of a pair of scissors to describe the human mind (figure 4.5). Scissors work by pushing each of two blades against each other. That's how they cut. You can't really consider a pair of scissors without understanding both blades and how they interact. In his analogy, one of the blades is the brain, with all of its processes and cognitive limitations. The brain is the blade running all the rules. But when the challenges that are encountered are too complex to "fit" within the bounding of the brain, the mind compensates by pushing against the other blade, which in this analogy is the environment. The mind can exploit the physical world, with its known and regular structures, to offload complexity.

Consider ants. A fairly limited set of rules is running on their neurons. But ants somehow manage to build cities, fight wars, and harvest crops. Where does that rich and emergent ability to solve complex problems come from? Ants, it turns out, push against that other blade—the world, the universe, physics. They build complex nests, forage, and navigate by leveraging environmental stimuli—sun position, light patterns, texture, and chemical clues—to enhance their cognitive abilities. Think of how much the environment impacts our lives. We live in the physical world. And so does a team. If we are building an innovative team and don't take the environment into consideration, we are setting ourselves up to fail.

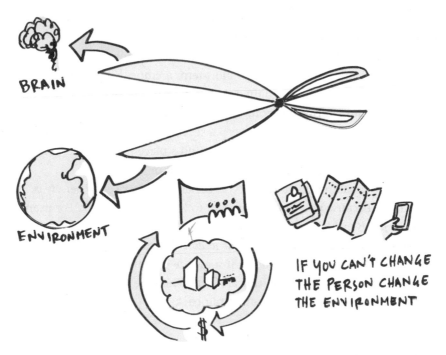

Figure 4.5
Herbert Simon's scissor analogy implies that if you can't change the person, change the environment. Courtesy of MAYA.

Where It Is, Is What It Is

When we first started MAYA, we embarked on a significant multiyear research study to understand how people got work done in the real world.[16] Our team shadowed office workers, sketched their spaces, and asked lots of questions. We used these photos and sketches as tools to elicit real-time feedback: "Why do you pile papers over there?" and "How do you find things that are important?" We followed them to meetings, we sat at their desks, and we watched how the environment took shape, and was shaped by, their work.

Over and over again, we found that workers used space as a sort of external memory. For example, we observed a person in his workspace who was looking at an airline site to make a reservation. He had a number of Post-it notes tacked to his computer screen to remind him of important tasks, while planning some of his activities using a calendar application. One glance told us a great deal about what he cared about at that moment, and we could tease out how he was organizing his efforts. This space was his external memory; the space remembered for him.

Many people have heard the classic idea that humans can retain about seven (plus or minus two) things in short-term memory.[17] While the reality may be a bit more complicated than that, it does appear that we have a limited capacity for short-term recall. Some say it is about a two-and-one-half second episodic "loop."[18] We can keep a seven-digit phone number in our heads perfectly fine, at least until we can find a place to write it down. But try to recall a hundred digits of π and we are out of luck.

We started calling office workers' external use of space "Where it is, is what it is," because people used location to "index" their work. For example: *This pile is for stuff I have to do right away; this pile is just fun reading. That pile is stuff I have to do but don't want to do.* We understand this principle intuitively; imagine how you'd feel if someone secretly entered your office at night and moved all your stuff around. When we ignore the environment, it is like tying our arms behind our back.

A Machine for Innovation

Imagine a typical conference room. There is a long table, a podium where someone will present from a slideshow, and a projection screen. More often than not, there is a small whiteboard behind that screen. Now imagine you are going to have a meeting in this room. You want big thinking. Someone lowers the projection screen and stands at the podium and holds forth. After fifty slides—with five or six bullet points per slide—it is now time to be innovative! But since, like everyone else, you have a limited short-term memory, you can't remember much of this presentation except maybe the loudest or most persuasive voice or the last few points. Meanwhile, that big table has given the team quite a bit of surface area to set up their laptops and start checking their Facebook accounts and emails.

Now consider a very different environment. In the southwestern part of North America, you can find Native American structures called "kivas." The kiva was used for community meetings and special ceremonies. The entire community would gather in the kiva, often entering from above by climbing through a hole and down a ladder into the space. The act of entering the kiva symbolized that something different and important was happening.

We found these community spaces inspiring. We began experimenting with kiva-style environments more than two decades ago. In contrast to

Figure 4.6
View of a MAYA Kiva showing whiteboard walls that move to create a 360-degree writing surface. Photo © Ed Massery.

that traditional conference room, the MAYA Kiva is a machine for innovation. You can still pull down a projection screen, but at the same time that someone explains an important point on a slide, someone else can draw what it means on the walls. Three-hundred-sixty degrees of whiteboard allow participants to collaboratively capture far more than the last seven things someone said. There is no central table. Everyone sits around the periphery with movable chairs and tables, within arms' reach of an area to draw (figure 4.6).

In the MAYA Kiva, ideas have a chance to live and breathe, and over the course of the day, team members are mixed together. Participants have to move out of the way as each part of the environment begins to fill up with ideas. People and positions naturally recombine as the team exercises the room and finds new forms of equilibrium. *Where* ideas end up on the wall turns out to be *what* they are.

In his book *Moonwalking with Einstein*, Joshua Foer notes that people have used vivid images in physical places for thousands of years. He reminds us that we can remember 1, 10, 100, 1000, or even 10,000 things when we use these sorts of "memory palaces."[19] Often, we find people walking back into a MAYA Kiva after having had a meeting there years before. They will walk over to a part of the wall and say something like, "Remember that iceberg we drew over here?" Everyone on the team will be brought back in time and recollect the idea. This is a powerful example of "Where it is, is what it is."

Supporting Project Evolution

Kiva-style spaces are valuable during the inception phase, but teams may need different environments during other phases of a given project. For example, during formation, it is important to be able to mock up ideas rapidly; providing prototyping areas facilitates this activity. During periods of contemplation and personal exploration, project spaces that allow for the layering of ideas over months support the team in synthesizing and generating ideas. During evaluation, usability labs help control variables and facilitate engagement with stakeholders. And well-designed social spaces facilitate interaction and idea flow across different project teams. The idea of Simon's scissors can be applied by providing spaces to support innovation activities throughout the team's life cycle.

Conclusion

The process of innovating doesn't just happen on its own. Despite that romantic notion of the independent rebel innovator, a smart problem-solver cannot go it alone in a world of rampant technological complexity and wicked problems. Arriving at tenable solutions today demands that we figure out how to overcome politics, turf wars, and human nature in order to morph diverse, innovative individuals into cohesive, high-performance teams. We believe that interdisciplinary collaboration done right can collapse seemingly intractable problems, help solve the right problems, and in the process shorten product-development cycles.

In this age of unbounded complexity and wicked problems, organizations can no longer count on the product-development processes that worked when all they needed to do was make *things right*. Today, companies need to figure out how to make the *right thing*, and that demands a new approach to innovation built upon collaboration, not competition or separation of disciplines. But real collaboration can be a struggle, combining as it does vastly different viewpoints, learned methodologies, goals, and even descriptive language. There are, however, optimal methods, team structures, and environments for making interdisciplinary collaboration much easier, more creatively satisfying, and more innovative. We believe that innovation today is about honing and applying individual skills in the service of high-performance, interdisciplinary teams working within a creative environment tuned specifically for its needs and maturity.

Notes

1. The opinions expressed in this article are the authors' own and do not reflect the view of the Boston Consulting Group.

2. Peter Lucas, Joe Ballay, and Mickey McManus, *Trillions: Thriving in the Emerging Information Ecology* (New York: Wiley, 2012).

3. C. West Churchman, "Wicked Problems," *Management Science* 14, no. 4 (December 1967): B141–B142.

4. To be fair, observers in every technological epoch have made similar claims about the pace and complexity of technological change, while speculating about the possible impacts. For a meditation on this phenomenon, see Joel Mokyr, Chris Vickers, and Nicholas L. Ziebarth, "The History of Technological Anxiety and the Future of Economic Growth: Is This Time Different?" *Journal of Economic Perspectives* 29, no. 3 (Summer 2015): 31–50.

5. Randal Goodall et al.,"Long-Term Productivity Mechanisms of the Semiconductor Industry," in *Semiconductor Silicon 2002: Proceedings of the Ninth International Symposium on Silicon Materials Science and Technology*, ed. H. R. Huff, L. Fábry, and S. Kishino (Pennington, NJ: Electrochemical Society, 2002), 1:125–144.

6. Michael Barr, "Real Men Program in C," *Embedded Systems Design* 22, no. 7 (2009): 3.

7. "MAYA Design Launches LUMA Institute to Teach Design Thinking to Everyone," press release, 23 September 2010, http://maya.com/news/maya-design-launches-luma -institute-to-teach-design-thinking-to-everyone.

8. "Emerson's Approach to Human-Centered Design Makes Automation Easier," ARC Advisory Group white paper, 2009.

9. James Surowiecki, *The Wisdom of Crowds* (New York: Anchor, 2004).

10. Dan Saffer, *Designing for Interaction: Creating Smart Applications and Clever Devices* (Berkeley, CA: New Riders, 2006).

11. Lucas, Ballay, and McManus, *Trillions*, chapter 6.

12. Alex "Sandy" Pentland, "The New Science of Building Great Teams," *Harvard Business Review* (April 2012): 60–70.

13. Harry Greene, Larry Stotts, Ryan Paterson, and Janet Greenberg, "Command Post of the Future: Successful Transition of a Science and Technology Initiative to a Program of Record," *Defense Acquisition Review Journal* 53 (2010): 3–26.

14. Rea Ann Fera, "An Oreo Experiment Reveals Mondelez's Approach to Innovation," *Fast Company*, 19 March 2014, http://www.fastcocreate.com/3027897/an-oreo -experiment-reveals-mondelezs-approach-to-innovation.

15. Herbert A. Simon, "A Behavioral Model of Rational Choice," in *Models of Man, Social and Rational: Mathematical Essays on Rational Human Behavior in a Social Setting*, ed. Herbert A. Simon (New York: Wiley, 1957), 99–118.

16. Joseph M. Ballay, "Designing Workscape: An Interdisciplinary Experience," *CHI '94: Proceedings of the SIGCHI Conference on Human Factors in Computing Systems*, ed. Beth Adelson, Susan Dumais, and Judith Olson (New York: ACM, 1994), 10–15.

17. George A. Miller, "The Magical Number Seven, Plus or Minus Two: Some Limits on Our Capacity for Processing Information," *Psychological Review* 63 (March 1956): 81–97.

18. Alan Baddeley, "The Episodic Buffer: A New Component of Working Memory?" *Trends in Cognitive Sciences* 4 (November 2000): 417–423.

19. Joshua Foer, *Moonwalking with Einstein: The Art and Science of Remembering Everything* (New York: Penguin, 2011).

5   Raising the NSF Innovation Corps

Errol Arkilic

In the spring of 2011, a small group of slightly irreverent and off-kilter program staff at the National Science Foundation (NSF) developed the Innovation Corps, better known as I-Corps. Combining a curriculum focused on assessing business potential with team building and mentorship, I-Corps helps budding entrepreneurs bring NSF-funded, technology-enabled projects to the market—in other words, to innovate. What today seems like an obvious extension of the NSF program portfolio was, at the time, a high-risk start-up led by a bunch of misfits with a small chance of success. I was the head misfit and led the program for the first two and a half years.

In what follows, I share some insights on what we hoped to accomplish, how we structured the program, and how I-Corps fits into the broader innovation ecosystem. I am, of course, only one person among the many who participated in the founding first years of the program; my comments are naturally colored by personal biases and foibles (of which there are many). Moreover, having since left NSF for the venture capital industry, my current perspective and interests are not necessarily the same as NSF's. My opinions, in other words, are my own.

A little about me: I arrived at NSF in 2003 with a background in systems engineering and entrepreneurship, both successful and not. After some time working on a project supported by the Department of Defense, I moved to Silicon Valley, where I learned about the dysfunctions of start-up companies firsthand. Around 2000, just at the cusp of the telecom bubble, I founded a start-up of my own. We had all the critical parts (I thought) of a high-flying start-up: really cool technology, a really smart technical team, a white-hot market, and a purchase order from one of the biggest players in the telecom space. Despite these seemingly positive markers, the company was a complete, epic failure. My second company (like the first, it was related to semiconductors) also failed. I had better luck with my third company, StrataGent LifeSciences, which licensed technology from Stanford and from the University of California. That company was acquired in 2007, and the acquiring company went public in 2014.[1] So that, at least, was a success. But like a lot of entrepreneurs, I've experienced more failure than success. Why? I've spent a lot of time pondering that question.

I was in the midst of this self-reflection when NSF recruited me for its Small Business Innovation Research (SBIR) program. SBIR is a sort of risk-seeking (strange, I know) publicly funded investment program that supports technology-based small businesses at the start-up phase. The program targets firms with fewer than five hundred employees and finances the commercialization of their research, an intentional divergence from NSF's traditional focus on noncommercial "pure" science grants to universities and other large institutions. At that time, the SBIR program annually invested about $100 million in high-tech start-ups with the goal of stimulating the economy through innovation. "America's seed fund" is how we thought of it at NSF.[2]

In the early 1980s and 1990s, NSF had very close relationships with the venture capital community, but the tech bubble in the late 1990s caused many of those ties to atrophy. During that time, investors were bringing so much capital to the table that they (the venture capitalists) saw NSF programs, with their support of $100,000 to $150,000, as a distraction. "Why would any of my companies deal with the federal government to obtain a measly $150K?" was the common refrain.

By 2003, however, the venture capital community had greatly sobered, and early-stage capital dried up. NSF wanted somebody—me—to help reestablish its ties to the venture capital community to increase the nation's effectiveness at commercialization. Both government and the private sector play a critical role in this process, and it is better for us all if we work together.

For the next eight years, I ran the software and services portfolio of NSF's SBIR program. During that time I supported about four hundred companies, usually in chunks of $150,000 to $500,000.[3] The NSF-funded software companies had various profiles, but the ones I liked the most were academic "spin-outs"—start-ups offering software products and services that had begun as someone's academic research project, usually in computer science or engineering.

This fascinating work gave me firsthand insight into almost every possible way that a company can fail. I learned that by far the most typical way for a company to fail is to build something that nobody cares about. I saw this over and over and over again. By seeing it in others, I also began to recognize it in my own previous failures. In every one of my own failed companies, I could have done a better job of understanding the target customer and the value I was planning to deliver. This concept, I now know, is called product-market fit.[4] Looking back, I realized that I could have investigated channels and revenue models. I could have worked on partnerships and cost structures. Most importantly, I could have developed a business model. Back in my failed start-up days, I focused on technology, features, IP, and financing instead of customers and markets. When I failed, I failed to deliver any real value, and wasted human and capital resources.

In almost every project I supported while at NSF, all of which involved high-risk technology, I came to understand that the biggest challenges start-ups face revolve around developing business models, not technology. My SBIR colleagues and I began to wonder: Can we do anything about it?

From our collective experience, some things stood out. We knew that innovation is about taking something novel, something creative, and generating value from it. Innovation is not invention, and it's not entrepreneurship. Innovation is about *creating value*—an economic term, and something we cared about a great deal in the SBIR program.[5] For that matter, it was in our name, and we spent a lot of time trying to figure out how to do a better job supporting our nation's ability to innovate. But what could we do? As a program mandated by Congress, SBIR faced many constraints.

Change was coming. In 2010 Subra Suresh, the former dean of MIT's School of Engineering, became NSF's new director. By now, in the wake of the 2008 recession, innovation was a hot topic in Washington, and especially at NSF. In his new position of director, Suresh wanted to increase the economic impact of NSF's basic research portfolio, and he wanted to use innovation to do it.[6] Because of his experience with MIT's Venture Mentoring Service (MIT VMS), a program designed to help academic entrepreneurs get their start-ups off the ground, he was convinced that a support program could be established at NSF to increase the chances of commercialization.[7] Academic research represented a significant source for innovation, and we could increase the probability of innovators' success through mentoring. This was the founding principle for I-Corps.

At this point, we thought we knew what we wanted to do; we just weren't sure exactly how we were going to pull it off. As it was initially conceptualized, the I-Corps program was intended to match mentors from across industry with academics through a series of "low-risk" engagements. It would serve as a sort of matchmaker between academic researchers and industry-aware mentors who had an interest in supporting the economic impact of NSF. We had in mind a public-private partnership, with NSF supplying the grantees capital, mentors, and a forum. It was going to combine the best aspects of MIT VMS and SBIR.

Outside NSF, the concept had support. Even before we had established our program, Desh Deshpande, a highly successful entrepreneur and investor, put up $1 million to pull it off as a pilot. The Kauffman Foundation followed shortly thereafter, promising to provide additional capital as soon as we had an outline for the program.[8]

There was only one problem. The person chosen to execute the program—me—didn't think it could be done. I had previous experience with an ill-fated matchmaker program within SBIR called Matchmaker.[9] The idea was to connect grantees and individuals in industry where we thought there would be a reasonable expectation of shared interests. The program was a disaster. In some rare cases the relationships blossomed, but the most common outcome was a resounding thud. Matchmaking is extraordinarily difficult, even when you think you know the people well. When a match goes badly, the person or organization that made the match looks like a fool to both parties. Failure breeds contempt. My scar tissue led me to believe that it would be difficult to pull off another mentor/matchmaking program at any scale. Federal government programs are nothing if they don't scale. My feeling at the time was that I-Corps was almost dead even before it started.

Then I had an idea.

I knew from experience that most companies successfully spun out of academic labs had people intimately familiar with the technology in a founding position of authority, usually graduate students turned founders. Only in very rare exceptions were the academic spin-outs led by Principal Investigators (PIs). It turns out that PIs don't necessarily make good entrepreneurs, but graduate students and postdocs sometimes do. Here was an idea: What if we brought together teams of experienced academic researchers (PIs) and their graduate students (who would be known as Entrepreneurial Leads, or

ELs) who would then recruit their own Mentors (no NSF matchmaking)? We would teach all three about the biggest risks facing the start-up, while providing them with $50,000 to explore the risks. The PIs would be there to provide technical support and to take the content back to the lab with them. The ELs would be there to take the project forward, should it prove worthwhile, and the Mentor would be there to help guide the journey. After a few conversations with my colleagues at the proverbial water cooler of NSF (in reality there were no water coolers, only water fountains), I knew we were on to something.

With the team structure identified, we knew the kinds of people we would be working with. The next question was what to teach them. What vital information would their journey impart to their roles as innovators?

During my time in SBIR, I had become a fan of Steve Blank's work. Blank, a serial entrepreneur, is also a gifted teacher. His popular book on start-ups, *The Four Steps to the Epiphany*, is based on his life experience on the front lines of innovation and a course he taught at Haas Business School at the University of California, Berkeley. In the book and elsewhere, Blank developed the concept of "customer discovery," which refers to the concept of discovering the product market by systematically developing and testing *hypotheses* and *exploring* customer needs and wants. It is, as Steve would say, "a big idea."[10]

I frequently recommended Blank's book to SBIR grantees, often asking them to focus on the concept of dividing a start-up into a "search phase" and an "execution phase." More recently (spring 2011), Blank began offering a new class at Stanford in which he combined customer discovery with Alex Osterwalder's template for developing new business models, the Business Model Canvas (figure 5.1).[11] I followed Blank's blog, where he wrote about the whole process in real time. From my standpoint, it could not have been more timely.

The attractive thing about Blank's curriculum was that it focused on exactly the thing that academic (and, indeed, all) entrepreneurs need help with: the concept of opportunity recognition. For the purposes of the I-Corps program, we asked ourselves, "How do we bring something to the table that allows us to clarify those unknowns so that founders don't build things that people don't care about and/or aren't accepted in the marketplace?" To address this problem, our program would, over the course of nine weeks, send our teams out to talk to one hundred potential customers.

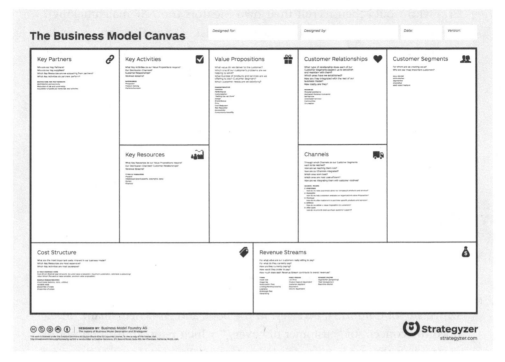

Figure 5.1
The Business Model Canvas is a management tool to help innovator-entrepreneurs
understand the range of social and technical factors that impact their potential business. Designed by Business Model Foundry AG.

They would have to identify customer segments, value propositions, channels, revenue models, and cost structure: a business model. At the end of
the nine weeks, they would have to make a decision: go or no go?

Forcing the teams to decide "go or no go" was a fundamental shift in how
NSF usually related to its grantees. When most academics wrap up a grant,
the conclusion is that some interesting and unforeseen phenomena have
presented themselves. The researchers find that there are opportunities that
might need more resources to explore. ("We've turned over this many rocks
and, surprise, there are more rocks to turn over, but we need more money!")
In the context of scientific exploration, this never-ending quest isn't bad.
In the context of a start-up, it is lethal. For I-Corps, we wanted something
different. At the end of this short, intense period, grantees would have to
make a decision about whether or not to scrap their business concept. Just

like a start-up, they had to decide: Given what they had discovered up to this point, was the endeavor still worth pursuing?

It is worth taking a moment to underscore just how much I-Corps deviated from NSF's business as usual. Granted, NSF had been funding start-ups and encouraging high-tech commercialization through its SBIR program since the 1970s; in that respect, the I-Corps initiative was nothing new. However, SBIR was funding start-up firms *that had already been started up*; it was providing start-up capital to firms and innovators that already had an entrepreneurial mindset. In contrast, I-Corps would be identifying traditional NSF grantees (i.e., academics working in universities), training them in principles of entrepreneurship, and encouraging the best grantees to start new commercial firms. With I-Corps, we were trying to launch start-ups and turn academics into innovators.

These were the founding principles. We literally wrote them on the whiteboard. At this point we reached out to Blank. I called him, in fact, with a rather unlikely pitch: "Steve, you don't know me, but I work for the government, and we want to copy your E-145 curriculum with some 'minor' tweaks. Instead of students, we want to apply it to PIs, Entrepreneurial Leads, and Mentors. Oh, and the teams will be from all over the country. Oh, and we want to do parts in person and parts online. Oh, and instead of eight teams, we want to teach twenty-seven teams at a time. Oh, and we want to launch this thing in three months." After a few minutes of him asking me about what NSF was, Blank's reply was short and sweet: "I'm in." We found our curriculum.

Blank asked for two things: coinstructors and space. For the coinstructors, he recruited Jon Feiber from Mohr Davidow, and I signed up John Burke from True Ventures; both were from leading venture capital firms specializing in early stage founders and teams. Having these two venture capitalists on board added instant credibility to our project, giving us an answer to an obvious (but in many ways misinformed) question: What does the government know about teaching researchers about entrepreneurship and innovation? Looking for a place to pilot I-Corps, we reached out to Stanford. When Stanford agreed to provide the classrooms, we were off and running.

With a curriculum, instructors, and space, all we needed were teams. Here, the key player was Babu DasGupta, at the time a Program Director in NSF's Engineering Directorate.[12] As the founding lead for the teams component of I-Corps, DasGupta was responsible for designing the solicitation.

We wanted an invitation-only program, with internal review and a relatively short (less than ninety-day) turnaround. To those unfamiliar with NSF's usual procedures this may not seem like much, but it presented a dramatic contrast with NSF's six-month, open, peer-reviewed solicitation process. One of the beauties of having support from the top is that things can be hurried along when necessary!

By April 2011, after about four weeks of intense effort, we issued our first solicitation: I-Corps would offer small grants ($50,000) to facilitate customer discovery. Invitation-only applications for the curriculum-centered program should come from three-person teams. The teams would include an NSF Principal Investigator, a student acting as the Entrepreneurial Lead, and an external business Mentor. The proposals would be reviewed internally. The process for vetting the teams took some experimentation, but after a few missteps, we established a protocol for getting teams into the pipeline. The key was to provide lightweight communication channels before a proposal landed on NSF's desk. We did this through a series of phone calls with the entire team and with the participation of the I-Corps instructors. The invitation-only applications were not new to NSF, but the team vetting process was, and at times it ruffled some feathers. Still, one of our goals with I-Corps was to establish an approach to simulate the pressure of a start-up culture and even put some stress on the team in a high-pressure interview environment, and so we persevered through the initial rough patches.[13]

In summer 2011, we awarded the first set of I-Corps grants to over twenty teams.[14] That's not bad for a program that had not even been envisioned five months earlier. Now the fun began. In September 2011, all the participants—teams, instructors, and I-Corps staff—gathered at Stanford (figure 5.2). Over the next three days, Blank, Feiber, and Burke provided what I-Corps now calls a "relentlessly direct" immersion course in customer discovery. Among other things, they kicked the teams out of the building for fifteen or so hours, forcing them to engage directly in customer discovery. Some thrived, most struggled. Over the next seven weeks, the process continued, albeit remotely, via WebEx.

All this culminated in week 9, when everyone returned to Stanford for a "lessons learned" day. We made a conscious decision to *not* describe this as a demo day. We wanted each team to focus on what they had learned, not what they had produced. We explicitly did not want a product/company pitch. We only wanted two things: Tell us what you learned, and tell us whether or not

Figure 5.2
Representatives from the first twenty-one I-Corps teams at Stanford in September
2011. Photo provided by Steve Blank.

you are going forward with taking the technology out of the lab. In other
words: go or no go? We emphasized that it was okay to kill the project.

The final session was awe-inspiring. Sure, most teams failed to uncover a
scalable business model. But all of the teams learned an enormous amount.[15]
They learned about the market, about the pressures of a start-up, about
their team members, and about themselves. Some of the team members
underwent what can only be called a transformation. And I'm not just talking about the grad students and postdocs. Both seasoned PIs and Mentors,
some of whom had started companies previously, told us that the process
fundamentally changed the way they thought about doing technology,
product, and market research. It was incredible and moving. As of fall 2018,
over thirteen hundred teams (and counting) have gone through the NSF
I-Corps program. One notable success, Arable Labs, has recently raised
$4.25 million to "bring predictive analytics to farming."[16]

By the time of the on-site lessons learned debriefing with the instructors,
we knew we were on to something—but now we had a new problem. NSF
wanted to offer I-Corps once a quarter. Because two of the three instructors

had day jobs as venture capitalists and the other already carried a respectable teaching load as an adjunct at Stanford and Berkeley, we needed to figure out how to copy the teaching team. Don Millard, then Program Director in the Education and Human resource Directorate at NSF, had a plan. We would put together a teaching faculty. We would train them using Blank's curriculum and house them in academic institutions. We would call these institutions "Nodes" (as in a network of nodes). Millard took the lead for the Nodes. Soon thereafter, Anita LaSalle, program director within the Computer Information Science and Engineering Directorate at NSF, would develop a program extension called "Sites" to foster a permanent ecosystem for innovation within select institutions.[17] That, however, is a story for another time.

By the winter of 2012, Georgia Tech and the University of Michigan were on board as the first official I-Corps Nodes.[18] Jerry Engel from Berkeley joined as the national faculty director. The National Collegiate Inventors and Innovators Alliance (NCIIA), now VentureWell, stepped in to facilitate and help deliver the program.[19] Our first mix of Node instructors, drawn from the Georgia Tech and Michigan faculty, injected an energy that revitalized our ragged band of misfits. As misfits themselves, they fit right in. Engel masterfully trained the new instructors on Blank's curriculum, while I nixed all their proposed improvements! With LaSalle and Millard by my side, I told the new instructors that if we didn't copy the program exactly, we would not be able to scale it. They came around, embraced the program, and we have not looked back since (figure 5.3).

That was in 2012. In the first year, we trained approximately seventy-five entrepreneurial teams.[20] Since then, I-Corps has become a living, breathing, evolving program at NSF that is spreading throughout the country and even around the globe. Other national and international institutions are adopting its practices, including (in the United States) the Department of Defense, the National Institutes of Health, and the Department of Energy, as well as institutions in Mexico, Singapore, and Ireland.[21] Many of the founding key players have moved on from NSF, but the effort moves on with a new cast of characters and challenges. NSF's next director, France Córdova, has embraced the program with enthusiasm and support from Congress.[22]

I-Corps' growth raises the obvious question: What have we learned? First, innovators can be cultivated. While not everyone will succeed, with the right nurturing and the right conditions, the potential becomes reality and innovators begin to form. Through practice, they hone their skills,

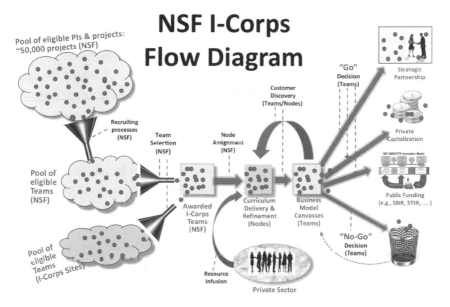

Figure 5.3

I-Corps creates an opportunity to commercialize ideas from the more than 50,000 projects NSF supports. The I-Corps award process and curriculum is designed to help would-be innovator-entrepreneurs to develop business models, to make "go/no-go" decisions, and to help those that have potential. Graphic provided by Anita LaSalle and Errol Arkilic, NSF.

and some of them go on to attain mastery. I-Corps' contribution has been to standardize that process of cultivation. The Nodes program makes this explicit: I-Corps instructors have to stick with the program. At NSF, we wanted to raise an army of innovators and raise them quickly. That requires a manufacturing mentality. If you have a process, it can be improved. Without a process, it is much more difficult to improve.

The second, more important point has to do with the definition of innovation. Innovation combines the elements of novelty and value. An innovator is someone who brings something novel to the world *and* in doing so, creates value. Novelty is easy, but new doesn't necessarily mean better. Value, too, is easy, and there is no shame in copycat businesses. Combining the two, however, is less easy. In fact, it is extraordinarily difficult, and most would-be innovations and innovators fail.

Combining the requirements for novelty and the creation of value creates a vast number of unknowns. When innovators launch into unknown spaces they are, in a real way, explorers. You can teach people to explore, but you cannot necessarily prepare them for what to do with what they find or guarantee that they will find anything of value. What you *can* teach is the scientific method, the best way known to explore the unknown. The scientific method as a process is pretty straightforward: Here's what I thought (hypothesis), here's what I did (experiment), here's what I learned (insight), and here's what I'm going to do next (new hypothesis). The insight at the center of I-Corps is that innovation skills can be taught as a form of the scientific method, which, as you might imagine, resonates with scientists and engineers. I-Corps has also proven that these skills can be taught in a replicable and scalable way. The skills can be practiced, and mastery comes from practice.

Will everyone who practices become a master? No. Nor will everyone who masters the process identify a key insight that is worth bringing to the market. Skills alone do not guarantee that an insight—whether around a product, a process, or a business model—will be forthcoming. Some well-respected scientists toil their entire lives without making a significant contribution to our body of knowledge. There are a lot of rocks to be turned over. Some would-be innovators (even those trained in a process) try their entire lives and never deliver anything of value to the market. Brutal, isn't it?

Should the vicissitudes of the outcomes dampen NSF's enthusiasm for supporting those who want to be technology-based innovators? That is, should we stop trying to make innovators? Absolutely, resoundingly no! NSF is a risk-seeking granting agency. It should take risks that the private sector simply cannot take.[23] These high-risk activities will lead mostly to failure. To be sure, most innovation projects do. But every once in a while there will be breakthroughs like NSFnet, the network that led directly to the internet.[24] All these investments (including the basic research funding around NSF-net) are investments in people, and investing in people is the ultimate long game, one whose payoff is measured in generations and decades. I don't know if innovators can be made, but I do know that humanity progresses. The young grow wise as the wise grow old. Ours is a future of opportunity, and the NSF I-Corps program is a modest investment in that future—one that continues to be imbued with hope and optimism.

Notes

1. Anubhav Arora, Itzhak Hakim, Joy Baxter, Ruben Rathnasingham, Ravi Srinivasan, Daniel A. Fletcher, and Samir Mitragotri, "Needle-Free Delivery of Macromolecules across the Skin by Nanoliter-Volume Pulsed Microjets," *Proceedings of the National*

Academy of Sciences of the United States of America, published online before print on 6 March 2007, http://www.pnas.org/content/104/11/4255.full?sid=2efe0bcc-8e13 -4bd2-a07a-c6a3f70e4de1. See also "Corium Sets IPO Terms, Valuing Biopharma at Around $184M," *Wall Street Journal*, 24 March 2014, accessed 31 May 2017, http:// www.wsj.com/articles/DJFVW00020140324ea3oto4lz.

2. With the passage of the Small Business Innovation Development Act of 1982, the SBIR program piloted at NSF was expanded to the eleven federal agencies with extramural research budgets greater than $100 million annually—for example, the Department of Defense, the National Institutes of Health, the Department of Energy, and the National Aeronautics and Space Administration. On the history of NSF's SBIR program, see Charles Wessner, ed., *An Assessment of the SBIR Program at the National Science Foundation* (Washington, DC: National Academies Press, 2008). On the SBIR programs across all agencies, see Josh Lerner, "The Government as Venture Capitalist: The Long-Run Impact of the SBIR Program," *Journal of Business* 72, no. 3 (July 1999): 285–318.

3. "Portfolio," *America's Seed Fund: SBIR.STTR*, accessed 22 September 2017, https:// seedfund.nsf.gov/portfolio/.

4. Steve Blank, *The Four Steps to the Epiphany: Successful Strategies for Products That Win* (S. G. Blank, 2013).

5. Advisory Committee on Measuring Innovation in the 21st Century Economy, "Innovation Measurement: Tracking the State of Innovation in the American Economy," January 2008, accessed 18 September 2017, http://users.nber.org/~sewp/SEWP digestFeb08/InnovationMeasurement2001_08.pdf.

6. "Carnegie Mellon Names NSF Director as Its Ninth President," press release, Carnegie Mellon University, last modified 5 February 2013, accessed 22 September 2017, https://www.cmu.edu/news/stories/archives/2013/february/feb5_ninthpresident.html.

7. "History," MIT Venture Mentoring Service, accessed 10 June 2017, http://vms.mit .edu/history-of-vms.

8. "I-Corps: To Strengthen the Impact of Scientific Discoveries," news release 11–153, National Science Foundation, 28 July 2011, accessed 10 June 2017, https://www.nsf .gov/news/news_summ.jsp?cntn_id=121225.

9. National Science Foundation: SBIR/STTR, "Matchmaker Program Technology Prospectus," FY 2000–06, accessed 18 September 2017, https://www.scribd.com /document/998956/National-Science-Foundation-matchmaker.

10. Blank, *Four Steps to the Epiphany*, chapter 3.

11. Alexander Osterwalder, "The Business Model Ontology: A Proposition in a Design Science Approach," PhD diss., University of Lausanne, 2004.

12. DasGupta's career spanned academia and industry. He came to NSF from the CONTECH division of the global manufacturing firm SPX Corporation.

13. With pride I can say that I-Corps runs all of its programs on time, and we established this discipline from the first interaction with the program: these vetting calls.

14. "NSF Innovation Corps Announces First Round of Awardees," news release 11–214, National Science Foundation, 6 October 2011, accessed 10 June 2017, https://www.nsf.gov/news/news_summ.jsp?cntn_id=121879&org=NSF&from=news.

15. Steve Blank, "Steve Blank Entrepreneurship and Innovation," accessed 10 June 2017, https://steveblank.com/.

16. Dean Takahashi, "Arable Labs Raises $4.25 Million to Bring Predictive Analytics to Farming," *VentureBeat*, 27 March 2017, accessed 20 September 2017, https://venturebeat.com/2017/03/27/arable-labs-raises-4-25-million-to-bring-predictive-analytics-to-farming/.

17. "NSF Promotes Inclusion in Tech Entrepreneurship through Eight I-Corps Sites," news release, National Science Foundation, 25 January 2017, accessed 10 June 2017, https://www.nsf.gov/news/news_summ.jsp?cntn_id=190921.

18. "I-Corps Sites," National Science Foundation, accessed 10 June 2017, https://www.nsf.gov/news/special_reports/i-corps/sites.jsp.

19. VentureWell is a nonprofit organization founded in 1995 with a similar goal of funding and training faculty and collegiate student innovators to create successful, socially beneficial businesses. On VentureWell's role in establishing the University Innovation Fellows, see chapter 3 in this volume.

20. Steve Blank, "Making a Dent in the Universe: Results from the NSF I-Corps," last modified 11 June 2012, accessed 22 September 2017, https://steveblank.com/2012/06/11/making-a-dent-in-the-universe-results-from-the-nsf-i-corps/.

21. "National Science Foundation Director Visits Mexico to Strengthen and Promote Scientific and Technological Collaboration," news release 15–022, National Science Foundation, 18 March 2015, accessed 18 September 2017, https://www.nsf.gov/news/news_summ.jsp?cntn_id=134505.

22. Jeffrey Mervis, "NSF's New Budget Reflects White House Priorities on Climate and Environment," *Science AAAS*, 5 February 2015, accessed 10 June 2017, http://www.sciencemag.org/news/2015/02/nsfs-new-budget-reflects-white-house-priorities-climate-and-environment.

23. Mariana Mazzucato, *The Entrepreneurial State: Debunking Public vs. Private Sector Myths* (London: Anthem Press, 2013).

24. "A Brief History of NSF and the Internet," National Science Foundation, accessed 13 November 2017, https://www.nsf.gov/news/special_reports/cyber/internet.jsp.

6 Making Innovators, Building Regions

Maryann Feldman

Innovation is the practice of blending and weaving different types of knowledge into something new, different, and unprecedented that also has social and economic value. In the course of blending and weaving, innovators deploy their own talents and skills, but they also draw upon local resources such as an educated workforce, local firms, and government partners. In short, innovators emerge from distinct places and do their work in a particular, local context.

Regions that succeed in fostering innovators are culturally vibrant and economically prosperous. Some of these places—such as California's Silicon Valley, Massachusetts's Route 128, and North Carolina's Research Triangle Park—are a source of endless fascination and imitation. These highly innovative locales are economically diverse and usually feature a mix of entrepreneurial start-ups and mature anchor firms that create well-paying jobs. These jobs, in turn, produce high incomes and a solid tax base that supports good local schools, public universities, and a robust infrastructure—each of which enable and feed the business sector in a mutually reinforcing cycle. Mayors, university presidents, and chambers of commerce are constantly striving to turn their towns into the next Silicon Alley, Silicon Desert, or Silicon Forest. But how do you do it? What are the ingredients? Are there recipes?

I have been studying the making of such innovative places for more than twenty-five years. I began my career with the 1994 book *The Geography of Innovation*.[1] In that work, I used an econometric model to identify the location-specific determinants of innovative regions, and I found that successful clusters require a research university, industrial R&D, skilled workers, and the presence of related industries. Because a disproportionate share of innovation occurs within cities, I followed this up with work at the city level, which helped me better understand microgeography.[2]

Over the span of my career, my work has contributed to a broad consensus that regions in which innovators flourish have several common ingredients. These include the participation of entrepreneurs, who invest in building infrastructure as they build their firms; local champions, who believe in a place and make investments; good universities, which educate and create graduates with new ideas; and benevolent, large "anchor" corporations that build and sustain the local resources and relationships that benefit their activities. Successful innovative economies also depend upon long-term and altruistic government investments in the interest of public welfare. A related ingredient is good governance, defined as the democratic process of building consensus to solve a collective problem, which simultaneously creates the social norms and institutions that convey place-specific advantages. Finally, innovative regions link into broader national and international networks, often through multinational firms with a local presence; these connections allow regions to draw on new knowledge and talent.

However, as my research has evolved, I have realized that my earlier work was too deterministic. My findings implied that policymakers could simply line up the appropriate inputs and then turn the crank—in the manner of an economic development sausage machine. In reality, municipal and regional leaders have found it difficult to replicate the success of iconic regions. Similarly, much of the abundant advice from well-heeled consultants on this topic is shortsighted and mimetic. For example, Silicon Valley provides many fruitful case studies, and policymakers often invoke it as an exemplar, but that model raises questions about social inequality, environmental injustice, and whether innovative regions can be replicated and sustained.[3] We must look elsewhere to diversify and broaden our outlook.

My subsequent research on the origins of the mid-Atlantic pharmaceutical industry forced me to reexamine my thinking. In the early 1990s, the corridor between Philadelphia and New York boasted the highest concentration of new product innovation in the United States.[4] The roots of this dense network of innovators were broad and deep. The early location was influenced by factors such as transportation, trade, migration, and settlement, as well as by cultural factors such as the Quaker love for science. The first domestic drug production began in colonial Philadelphia, which was then the country's largest city and also home to its first medical and pharmacy schools. The industry progressed in the nineteenth century with the establishment of firms such as Sharpe & Dohme and Smith, Kline & French. While the industry's success grew out of entrepreneurial efforts, it relied in large part on the building of institutions, notably professional and trade associations. Furthermore, government played a decisive role in developing the industry through protective tariffs, regulatory standards, and the post–World War I nationalization of German companies such as Merck.[5] In other words (as any good historian would tell you), I learned that the success of the mid-Atlantic pharmaceuticals industry was a situated, and therefore unique, event.

The lesson for those working to understand and build innovative regions is that models and conceptualizations must be attentive to specific social processes, history, and local context. Considering such contingencies does not mean denying the common characteristics shared by regions in which innovators flourish. It does, however, make clear that a cluster's key ingredients—entrepreneurs, local champions, anchor firms, universities, consensus around a technology, good government, and national and international networks—are necessary but not sufficient for success. The most critical factor in an innovative region is the temporal process of constructing shared meaning over time: the way local actors build institutions and create social capital *during* the sequential and dynamic process of creating an industrial cluster. This chapter explores key concepts in the economic geography of innovation (box 6.1) and describes how innovators and innovative regions grow together in a dynamic, self-sustaining, virtuous cycle.

The Tendency to Cluster

Scholarship exploring the characteristics of innovative places dates back at least 125 years. Writing in his magisterial *Principles of Economics* (1890), British economist Alfred Marshall noted the tendency of English manufacturing firms to geographically group themselves into "industrial districts."[6] For example, Marshall observed that Britain's pottery industry had clustered around Staffordshire, while Sheffield had become the center of the knife and cutlery trade. Marshall cited three reasons for this clustering: the infrastructure of related and supporting industries; the presence of deep, specialized, skilled labor pools; and the presence of nonmonetary externalities

Box 6.1

A Glossary of Economic Concepts

---

agglomeration. An economic snowball effect in which early technical or economic developments accumulate over time into a critical mass of wealth, talent, institutions, and know-how. Innovators can then draw upon these accumulated assets to build the next generation of technical and economic developments, creating a virtuous cycle.

externality. A cost or benefit that derives from economic activity that positively or negatively affects parties who are not direct participants in that activity. For example, air pollution is a negative externality resulting from the motorized transportation of goods between buyers and sellers; the resulting air pollution affects everyone, not just the buyers and sellers. Similarly, a beekeeper maintains a hive to sell the honey, yet the bees' cross-pollination is a positive externality for other nearby growers.

knowledge spillover. An exchange of information or ideas, especially among coworkers or competitors located in close geographic proximity. Spillovers can be positive externalities (e.g., when the exchange of ideas leads to new products) or negative externalities (e.g., when proprietary information accidentally leaks out).

tacit knowledge. The kind of know-how that is difficult to transfer to another person through writing or conversation. Tacit knowledge is typically uncodified and thus is acquired and transferred generally through direct experience. For example, an experienced carpenter simply knows the best way to frame a doorway; he or she learned this from other carpenters and past experience, and it is not written down.

transaction costs. The cost of doing business. Transaction costs (e.g., shipping, recruiting new employees) are usually reduced when the two parties involved in a transaction are located within close geographic proximity to each other.

---

that arise from accelerated knowledge exchange facilitated by geographic proximity.

Marshall maintained that related firms within a specialized industry clustered together because they drew from a deep local pool of skilled and specialized labor. These firms also shared knowledge and best practices via local market transactions (e.g., when metal suppliers provided advice to their knife-making customers) or through nonmarket knowledge spillovers (e.g., when a knife-maker left one firm and joined another, bringing along new knowledge and techniques). Because of the density and geographical proximity of workers with similar skill sets, economic actors—specifically, the firms, entrepreneurs, scientists, or workers—could more easily use formal and informal channels to solve problems. Experience with a technology or industry increased the stock of available knowledge locally, yielding better ideas. That is, economic agents benefited from easy communication, knowledge exchange, reduced transaction costs, and the serendipity of unexpected—but highly relevant—chance occurrences.

The phenomenon Marshall described has been observed at different times and across multiple geographic regions.[7] This phenomenon is heightened in a knowledge—versus an agricultural or industrial—economy, as innovation is a creative, cognitive activity that benefits from colocation. So-called knowledge spillovers, or the nonpecuniary transfers of knowledge, are a major reason why innovators cluster spatially. Knowledge spillovers are subtle; over time, individuals observe one another, copy ideas, and build up the stock of knowledge with new ideas, components, and design elements. These spillovers are what economists term an externality: they exist because knowledge, once created, is difficult to value and price. The most interesting aspect of this phenomenon is that knowledge is subject to increasing returns, meaning that its value increases as more people use it.

Information can easily transfer around the globe, but knowledge often remains place-specific for a number of reasons.[8] First, knowledge is difficult to codify and transfer without some loss of content.[9] Second, using knowledge relies on absorptive capacity, which requires significant and specific investments.[10] Third, the availability of specialized organizational structures, such as local communities of practice, can significantly lower the costs of transferring knowledge. In addition, knowledge spillovers often result from serendipity, which suggests unexpected outcomes. If an innovator knows what information is required, he or she can search for a source. Knowledge spillovers suggest new and unexpected ideas.

Of all different types of economic activity, technological innovation stands to benefit most from location and has the greatest potential to improve a region's economy. At the earliest stages of the product life-cycle model, when a nascent technology is undefined and shrouded by great uncertainty, there is considerable ambiguity about its potential to generate new business opportunities. Entrepreneurs, as social agents, help to create these opportunities by creating shared meaning about emerging technologies.[11]

As more individuals within a locality come to understand a technology, and draw on local resources to commercialize it, the potential for meaningful and valued breakthroughs increases.

Today's innovators know this intuitively. That's why—even as email, mobile phones, text messaging, and teleconferencing facilitate long-distance communication—software developers still flock to Silicon Valley, and why aspiring screenwriters still move to Hollywood. In short, location matters for the diverse set of people and institutions who contribute to innovation.

Entrepreneurs: Building a Cluster While Building a Firm

The attributes we associate with fully functioning clusters do not explain their existence, but rather *result* from their success. Indeed, many of the factors associated with successful clusters, such as the presence of readily available venture capital or active university involvement, lag rather than lead industrial viability.[12] Moreover, these factors are necessary yet not sufficient. While it is always difficult to attribute causality, there is evidence that cluster genesis is a social process.[13] In other words, innovation owes more to people and ideas than to institutional dynamics and political context. What matters most is the entrepreneurial spark that takes hold and transforms a region.

Entrepreneurs discover opportunities, take risks, mobilize resources, create new firms, and—in some cases—bring prosperity to a region. In addition, entrepreneurs simultaneously build the local institutions and shared resources that develop the cluster as they build their own firms.[14] What does this look like? Over time, entrepreneurs build a social consensus around the potential of a new idea or a new technology. New business models emerge, and the cluster collectively begins to represent something unique and not easily replicated by other places. Entrepreneurs compete for the talent and resources required to produce innovation, creating tension about how firms should interact with their local environment. For a given technology and place, the propensity of entrepreneurs and firms to share information may be a differentiating characteristic that drives cluster growth. Through their social networks, entrepreneurs trade knowledge, reduce uncertainty, and further reinforce a shared vision for an emerging technology and its business model.[15] The resulting local cohesion and culture also produce new firms at a faster pace, suggesting the salience of internal cluster dynamics on industry's ability to grow and realize critical mass in a given place.[16]

In short, entrepreneurs play a pivotal role in creating the institutions and building the regional capacities that enable regions to sustain economic growth.[17] Entrepreneurs are important actors in developing clusters as complex adaptive systems in which the external resources associated with clusters are developed over time. Entrepreneurs who adapt to both constructive crises and new opportunities create the factors and conditions that facilitate their business interests and, in turn, contribute to the development of further external resources.

Not only do entrepreneurs benefit from location, but they also influence how local communities are transformed. Further, entrepreneurs' efforts can affect the potential to become an innovative cluster and the region's prosperity.

Local Champions and Dealmakers

A cluster's most important entrepreneurs are often local champions—individuals with a strong dedication to the region in which they both live and work. This became apparent to me when I began to explore the town of Greenwood, located in the Mississippi Delta.[18] Greenwood, Mississippi, is a small city that is only known to me because my mother-in-law was born there. About one hundred years ago, Greenwood was a well-known commercial center for the cotton industry. After cotton production first mechanized, and then globalized, Greenwood fell into a downward spiral, crushed by falling tax revenues and limited subsequent investments. No federal or state government program offered a magical remedy. Greenwood became one of the poorest cities in the poorest state in the United States.

However, a local champion demonstrated that new ideas and innovation could bring hope to Greenwood. In 1987, Fred Carl, a local entrepreneur, founded the Viking Range Corporation, a cutting-edge professional kitchen appliance company (figure 6.1). While working as a building contractor, Carl realized that consumers wanted high-quality residential stoves that looked and cooked like commercial stoves. He identified this opportunity and created an entire new industry segment. Against the trend of offshoring production, Carl located manufacturing operations in his hometown of Greenwood, gathering financing from local investors. Carl also invested in

Figure 6.1
Local champion Fred Carl (center), former CEO of Viking Range Corporation in Greenwood, Mississippi. Photo: Jennings-Greenwood.

revitalizing a local hotel in Greenwood for hosting Viking's vendors, suppliers, and distributors. His efforts created 1,500 jobs with good benefits and educational opportunities; at its peak, Viking employed more workers than did the local hospital, often the largest employer in small and medium-size cities.

Carl's commitment to Greenwood created economic stability in a rural, struggling community, providing benefits both to his firm and his hometown. His story reinforces the idea that individual entrepreneurs—especially those with local ties—are crucial for building innovative places. These local champions create institutions and build the capacity of a local economy as they grow their firms.

Civic-minded local champions bring other champions along with them. Consider the case of Ewing Marion Kauffman.[19] Kauffman was born and raised in Missouri. After working as a salesman for a pharmaceutical firm, he established his own pharmaceutical company, Marion Laboratories, in Kansas City—an unfavorable place in the 1950s, as most of the industry was concentrated in the Philadelphia–New Jersey corridor. By the time the company merged with Merrell Dow in 1989, it had become a global pharmaceutical company with nearly $1 billion in annual sales and over 3,400 employees. Marion Laboratories was noted for its progressive employment practices, which included educational and training benefits, profit-sharing plans, and employee stock options. These policies were investments in people with substantial dividends. By developing new employee skills and spreading the firm's wealth, these policies built additional capacity while fostering a sense of employee attachment to the company. Moreover, since these activities were grounded in place, the local Kansas City community also benefited.

In 1966, Kauffman went further, creating the foundation that bears his name. Kauffman established the foundation using income from the company—rather than waiting until he sold the company and cashed out. The Kauffman Foundation has twin goals of promoting entrepreneurship while improving education, the arts, and social programs in Kansas City. Other prominent entrepreneurs in Kansas City have followed this pattern, notably brothers Henry and Richard Bloch, founders of the H&R Block tax services firm, and Joyce C. Hall, founder of Hallmark Cards. Kansas City's culture of local philanthropy has produced a vibrant community in what many consider the fly-over zone.

Local champions, like Greenwood's Fred Carl and Kansas City's philanthropists, are motivated by objectives that extend beyond profits. They take responsibility for the stewardship of a place and are dedicated to their local community. In other words, instead of seeking short-term profit maximization, they are dedicated to fostering long-term prosperity in their home communities and discovering new opportunities that may bring about new profit. Moreover, local champions can advocate for the types of government interventions that will not only help their individual firms but also promote their overall industry and their local place. Stories from these champions suggest that regions can become prosperous when entrepreneurs actively engage in extramarket activities.

Ted Zoller and I empirically extended the idea of local champions to examine dealmakers—individuals who are central to local networks.[20] By accumulating data on the composition of local boards of directors, we were able to build a statistical model of regional social capital that permitted a more rigorous examination of these dealmakers. We examined cases in which interlocking companies shared local board members. We found that strong local entrepreneurial networks are associated with successful entrepreneurial economies. Specifically, communities in which a larger number of individuals work together in various capacities have higher rates of new firm formation. Moreover, the regions with the densest, most cohesive, and most interconnected networks are the most successful at generating new start-ups. Dealmakers, such as the aforementioned Fred Carl and Ewing Marion Kauffman, are personally committed to a region; by brokering connections among other local businesspeople and lending their credibility to various projects, they help their local economies grow.

In determining whether a region will develop into an economic cluster, we have found that the local presence and actions of dealmakers are more important than the simple existence of a local network, or the total number of entrepreneurs within it. Well-connected individuals, like dealmakers, possess extensive experience building, advising, financing, and operating entrepreneurial firms. Because a dealmaker's span of influence exceeds a single entrepreneurial firm, his or her connections help diffuse the information, experience, and expertise required to develop high-growth entrepreneurial ventures. Furthermore, that influence may transcend the given region as a dealmaker uses his or her connections to import additional knowledge and social capital from outside the region into the regional ecosystem.

The idea that local social capital yields economic benefits is fundamental to theories of agglomeration and is central to claims about the virtues of cities. This claim has not, however, been evaluated using methods that permit confident statements about causality. In work with Tom Kemeny, we examined what happens to firms that become affiliated with one highly connected dealmaker.[21] We adopted a quasi-experimental approach, examining 325 firms in the life sciences and information technology sectors in twelve innovative regions in the United States. We selected firms that had added exactly one new individual to their board. Some of these individuals were highly connected and could be considered dealmakers, while other individuals were less well connected. After controlling for a variety of firm characteristics, we found that firms linked to one highly connected local dealmaker were rewarded with substantial gains in employment and sales. Our results suggest that dealmakers have an organizing effect on local social capital, yielding specific benefits for the firms to which they become affiliated. Connections to dealmakers are one way that firms can become better situated in a regional economy, permitting entrepreneurs to leverage regional social capital in ways that promote firm growth.

It may be tempting to look solely at big regional and institutional factors, but individual human agency, hometown loyalty, and social entrepreneurship have great influence in making innovative regions—regions that, in turn, make innovators. Local champions embody this commitment to place, leading to more regional support and a larger local network of innovators.

A Place to Educate

Many high-tech regions are associated with research universities. For example, Duke University, North Carolina State University, and the University of North Carolina at Chapel Hill comprise the three vertices of North Carolina's Research Triangle. Places with universities benefit from the presence of an important economic anchor unlikely to go bankrupt, merge, or move away. Moreover, universities are places of experimentation, exploration, unfettered inquiry, and open discourse. Universities take federal and private sector grant money and then turn it into new scientific findings and technologies. When entrepreneurs commercialize these findings, they often locate their start-ups near the university, and local areas keep the economic gains. Finally—as several of this volume's contributors demonstrate— universities and community colleges are key sites for training the workforce to be innovators.[22]

One of the strengths of the American system of higher education is the diversity of institution types, each of which plays a contributing role. Research universities contribute to innovation and technological change through research and education that increase the regional and national capacity for problem-solving. Liberal arts colleges are a creative force in the economy that foster tolerance and diversity, and they create better citizens and members of society. Community colleges do the yeoman's work of providing affordable education, access to opportunity, and a bridge to employment at a time when corporations have curtailed training programs. Of course, the ability of individuals to avail themselves of the offerings of higher education depends on their preparedness, which, in turn, depends on the efficacy of primary and secondary education. Nevertheless, despite certain limitations, colleges and universities of every stripe remain key institutions for training innovators. It is an intricate and complicated system, and the American economy has benefited throughout history from investments in university science and education.

Universities frequently transfer their technology to companies—both existing companies and start-ups—based on university licenses. A landmark piece of legislation was the 1980 Bayh-Dole Act, which granted universities the right to commercialize and realize profits from products that resulted from publicly funded research. American policymakers experienced a competitiveness crisis in the late 1970s, fearing that the American economy was losing its edge to foreign competitors from Japan and Europe. In light of declining federal support, universities sought new revenue sources and a means to demonstrate their economic relevance.[23] In response to the Bayh-Dole Act, virtually every university now has dedicated technology transfer offices and commercialization support organizations—such as incubators and accelerators. Many states have also initiated programs that attempt to leverage academic research to reap rewards within their own jurisdictions.[24]

Universities are necessary but not sufficient for technology-based innovation to occur within regions. While regional leaders frequently see universities as the engines of innovation, there are many counterexamples of prominent universities that have been unable to commercialize their research.[25] For instance, Johns Hopkins University has long had one of the top medical schools in the world, yet Baltimore did not become an early leader in the biotechnology industry.[26] Universities can certainly enact policies to promote entrepreneurship.[27] But often there are great differences between the norms and expectations of academic scientists, and the behaviors required to engage in commercial activity.[28] Academic culture is certainly changing, yet the net impact and social desirability of a more commercial orientation within the university remains unclear.[29]

Universities play an important role in processes of local economic development. While scholars and policymakers have generally focused on the direct impact of technology transfer, universities are important social spaces that train and equip innovators while promoting experimentation, creativity, and collaboration.[30]

The Role of the Government

As the agent of collective investment in capacity, the government has an important role in cultivating innovators and creating innovative economies. Federal, state, and local governments serve as agents of collective investment in capacity, but that role needs to be redefined.[31] Economic development occurs when individuals have the opportunity to actively contribute to the advancement of themselves and their community. The best economic development policy is predicated on a longer-term and more capacious perspective than suggested by the cookie-cutter approach to regional development, which often relies on attracting business through incentives. Continuously working toward measurable increases in regional capacity will harness the natural tendency of innovative activity to cluster spatially, leading to greater prosperity.

In market economies, the central government cannot dictate the actions of private companies. It may only offer incentives to encourage firm location decisions and investments in R&D. Of course, too great of a reliance on the private sector can lead to imbalances that favor corporate profits over citizens' rights. Moreover, businesses change ownership and/or management, fail, or relocate; in fact, large corporations have demonstrated little attachment to places.[32] In contrast, governments can reliably build local capacity by investing in higher education.

In the United States, the closest thing we have to a government-induced cluster is Research Triangle Park (RTP) in North Carolina, which resulted from state and local government actions (figure 6.2). From its beginnings in the 1920s, RTP has grown to be the largest research park in the world.[33] Its development has been a long undertaking, but the most critical processes took place as the industrial landscape developed. By articulating a vision and consistent policy, local government leaders built a successful cluster. Several governments have attempted to build clusters in market economies, but the results often look very different from what was originally intended.[34]

Despite this mixed record, the concept of local industrial clusters has gained great currency among government policymakers as an idea for generating economic growth and bringing vibrancy to places. Flying in the face of increased globalization, this idea argues that *places* serve a critical role in defining an innovative product, process, or business model and in so doing become a source of job creation and wealth. As a result, governments have strived to build self-sustaining clusters by coordinating and aligning the activities of local firms, universities, community colleges, government agencies, and trade associations. Local governments also provide information about new opportunities for entrepreneurs and workers, offer incentives to lower the costs of starting new businesses, and fill in missing elements

Figure 6.2
Planning Research Triangle Park, North Carolina. Photo: RTI International.

of the ecosystem that might have been neglected by the private sector. Government-sponsored economic development influences the micro-economic function of the economy by impacting the quality of inputs, bringing business opportunities to firms, and in turn, creating the conditions that enable long-run economic growth.

While the concept of economic development preoccupies our collective imagination, the term is often not well defined, or it is defined in a limited manner that does not accommodate the full range of places faced with restructuring and economic uncertainty. All too often, the emphasis is on innovation as an end to itself rather than as a mechanism to create prosperity and greater well-being.[35] In the absence of a definition, analysts often conflate economic development with economic growth, or they rely on private sector constructs (such as rate of return) that are inappropriate for government investments.

I define economic development as the development of capacities that expand economic actors' capabilities. These actors may include individuals, firms, or industries that are likely to exert their potential based on the development of capacities. Under this definition, economic development is best measured by the quality of jobs, the caliber of business practices, and the density of social capital. Development can also be regarded as fortifying autonomy and substantive freedom, which promotes individuals' participation in economic life.[36] In this sense, expanding capacities provides the basis for realizing individual, firm, and community potential which, in turn, advances society. This more expansive view of economic development articulates a new role for government as the primary agent of collective investment in capacity.

Beyond Borders: The Multinational Corporation

Large multinational firms are often a key ingredient in the successful functioning of high-tech regions. They serve as an economic anchor in local clusters, train skilled workers, import talent and knowledge from across the globe, and generate positive knowledge spillovers.

Yet too frequently, critics present economic development as an inherent policy trade-off: encourage either large corporations or smaller entrepreneurial establishments.[37] The widely held assumption that large corporations, especially multinationals, will out-compete smaller, entrepreneurial firms for scarce policy resources and attention reinforces this perspective. In this interpretation, the presence of prominent corporations in a given region will constrain opportunities for entrepreneurs. Larger corporations, after all, have the ability to hold regions hostage with incessant threats of relocation—a type of leverage that smaller firms cannot wield because of their limited economic footprint.[38] Recent work on multinational corporations has helped to shift the focus away from corporate relocation decisions to the institutional contexts and strategies that motivate both multinational companies and their host locations. For example, economic geographers have demonstrated that multinational corporations are often active change agents in their local clusters, providing access to international resources and leveraging their role as large generators and transferors of new technology.[39] Moreover, large multinationals require sophisticated local suppliers, which helps to build robust value chains.

It is important to recognize that large firms are not static entities; rather, they constantly shift their strategies in ways that impact the places in which they locate.[40] Over time, a large firm might shift from an insular, proprietary orientation toward an open posture that is more supportive of nascent entrepreneurs.[41] But equally likely is a reverse sequence, in which a corporation moves from an engaged to a more withdrawn position. Commonplace events, such as large corporate mergers and acquisitions, can greatly affect how and when multinational firms might act in ways that support or hinder local entrepreneurial development. Nichola Lowe and I have demonstrated, for example, that corporate mergers can entail generous severance packages and market-promoting opportunities designed to motivate entrepreneurship through outsourcing and technology licensing. In an era of diminished government investment, policy can incentivize corporations to consider the impacts of their strategies on local economic vitality.[42]

Moreover, large multinationals cultivate innovators in a variety of ways. They do so directly through R&D labor and employee training, and indirectly through spin-offs when internally generated innovations do not match core businesses or when entrepreneurial employees leave to start their own firms. As their technology needs evolve, multinationals also drive innovation among their suppliers and subcontractors. Multinationals also influence curricula and create industrial fellowships at nearby universities to train their workforces. Finally, multinationals create corporate cultures that come to be emulated across the region.

These capacities, in turn, create opportunities and draw more innovative people to these clusters. Often the strategy of providing incentives to attract multinational firms is myopically seen at odds with the development of an innovative entrepreneurial economy, yet in reality, when employed judiciously, these strategies may be used together to further economic development.[43] The local capacity created by multinational corporations benefits the region by tapping into global networks. These networks have to become self-sustaining to become successful—creating a hotbed for innovative minds to gather in community.

Conclusion

In a service economy increasingly dominated by digital communication, cloud computing, and other "virtual" products, it may seem quaint to suggest the crucial importance of geographic location in the pursuit of innovation. Yet digital product companies such as Apple, Facebook, and Google are each investing billions in new Silicon Valley campuses.[44] Clearly, place still matters.

Innovators emerge from—and help build—innovative places. In turn, the particular local circumstances influence the work of those same innovators. During my career as an economic geographer, I have argued that innovative places *can* be constructed. There is a natural tendency for innovative activity to cluster geographically, and that tendency can be encouraged by economic agents in both the private and public sectors, especially when they work together. In my view, there are several necessary ingredients for a high-tech region—entrepreneurs, local champions, universities, multinational anchor firms, and smart government policies—but there is no one-size-fits-all recipe.

That is why policies to build innovative clusters often fail; policymakers typically lack an adequate understanding of the particular context and history of a given location. Most importantly, the literature indicates that an innovation ecosystem—at whatever level considered—should form a coherent logic. This suggests that emulating any one part of a system may not produce the desired result; each locality is more than the sum of its respective parts.[45]

Successful industrial clusters are social collaborations that require substantial time to mature. Given this reality, policymakers should first examine their local context, including industry structure and infrastructure in their region. Policymakers should also consider their region's strengths and weaknesses as they strategize about how best to overcome disadvantages and reinforce advantages. They should also intentionally create connections among existing firms to generate positive externalities and knowledge spillovers across industries. Finally, policymakers should invest in building regional capacity in order to promote a diverse array of economic activities.

All that said, there is not one master plan or blueprint for building regions that nurture innovators. Innovation, economic development, and the realization of human potential are admirable but often elusive goals. The appropriate "recipe," then, is to focus on local ingredients and be patient.

Notes

1. Maryann Feldman, *The Geography of Innovation* (Boston: Kluwer Academic Publishers, 1994).

2. Maryann Feldman and David B. Audretsch, "Innovation in Cities: Science-Based Diversity, Specialization, and Localized Competition," *European Economic Review* 43 (1999): 409–429; Barak S. Aharonson, Joel C. Baum, and Maryann Feldman, "Desperately Seeking Spillovers? Increasing Returns, Industrial Organization, and the Location of New Entrants in Geographic and Technological Space," *Industrial and Corporate Change* 16 (2007): 89–130.

3. For a more skeptical view of Silicon Valley through the prism of race, class, gender, and environmental injustice, see David N. Pellow and Lisa Sun-Hee Park, *The Silicon Valley of Dreams: Environmental Justice, Immigrant Workers, and the High-Tech Global Economy* (New York: New York University Press, 2002); and Glenna Matthews, *Silicon Valley, Women, and the California Dream: Gender, Class, and Opportunity in the Twentieth Century* (Redwood City, CA: Stanford University Press, 2003). On various attempts to replicate Silicon Valley, see Stuart W. Leslie, "Regional Disadvantage: Replicating Silicon Valley in New York's Capital Region," *Technology and Culture* 42 (April 2001): 236–264; Stuart W. Leslie and Robert H. Kargon, "Selling Silicon Valley: Frederick Terman's Model for Regional Advantage," *Business History Review* 70 (winter 1996): 435–472. On attempts to replicate the "MIT model," see Pfotenhauer (chapter 11) in this volume.

4. Maryann Feldman and Y. Schreuder, "Initial Advantage: The Origins of the Geographic Concentration of the Pharmaceutical Industry in the Mid-Atlantic Region," *Industrial and Corporate Change* 5 (1996): 839–862.

5. There is a current perception that regulation harms industrial activity and prohibits innovation, but pharmaceutical entrepreneurs advocated for the 1906 Pure Food and Drug Law to promote professional standards and drive charlatans out of the industry.

6. Alfred Marshall, *Principles of Economics* (London: Macmillan and Co., 1890).

7. Historians, economists, and sociologists have called these places "ecosystems," "regional clusters," "places of invention," and the like. See Michael E. Porter, "Clusters and the New Economic Competition," *Harvard Business Review* (November–December 1998): 77–90; and Arthur P. Molella and Anna Karvellas, eds., *Places of Invention* (Washington, DC: Smithsonian Institution Scholarly Press, 2015).

8. Eric von Hippel, "'Sticky Information' and the Locus of Problem Solving: Implications for Innovation," *Management Science* 40 (1994): 429–439.

9. Richard R. Nelson and Sidney G. Winter, *An Evolutionary Theory of Economic Change* (Cambridge, MA: Belknap Press, 1982).

10. Wesley M. Cohen and Daniel A. Levinthal, "Absorptive Capacity: A New Perspective on Learning and Innovation," *Administrative Science Quarterly* 35 (1990): 128–152.

11. Maryann Feldman, "The Entrepreneurial Event Revisited: Firm Formation in a Regional Context," *Industrial and Corporate Change* 10 (2001): 861–891.

12. Feldman, "Entrepreneurial Event Revisited."

13. Pontus Braunerhjelm and Maryann Feldman, *Cluster Genesis: Technology-Based Industrial Development* (Oxford: Oxford University Press, 2006).

14. Maryann Feldman, Johanna Francis, and Janet Bercovitz, "Creating a Cluster While Building a Firm: Entrepreneurs and the Formation of Industrial Clusters," *Regional Studies* 39 (2005): 129–141.

15. AnnaLee Saxenian, *Regional Advantage: Culture and Competition in Silicon Valley and Route 128* (Cambridge, MA: Harvard University Press, 1994).

16. Stuart S. Rosenthal and William C. Strange, "Geography, Industrial Organization, and Agglomeration," *Review of Economics and Statistics* 85 (2003): 377–393.

17. Feldman, Francis, and Bercovitz, "Creating a Cluster While Building a Firm."

18. Maryann Feldman, "The Character of Innovative Places: Entrepreneurial Strategy, Economic Development, and Prosperity," *Small Business Economics* 43 (2014): 9–20.

19. Maryann Feldman and Alexandra Graddy-Reed, "Local Champions: Entrepreneurs' Transition to Philanthropy and the Vibrancy of Place," *Handbook of Research on Entrepreneurs' Engagement in Philanthropy*, ed. Marilyn L. Taylor and Arvin Gottlieb (Northampton, MA: Edward Elgar, 2014), 43–72.

20. Maryann Feldman and Ted D. Zoller, "Dealmakers in Place: Social Capital Connections in Regional Entrepreneurial Economies," *Regional Studies* 46 (2012): 23–37.

21. Thomas Kemeny, Maryann Feldman, Frank Ethridge, and Ted Zoller, "The Economic Value of Local Social Networks," *Journal of Economic Geography* 16, no. 5 (2016): 1101–1122.

22. For example, see Fasihuddin and Britos Cavagnaro (chapter 3), Arkilic (chapter 5), Pfotenhauer (chapter 11), Russell and Vinsel (chapter 13), Carlson (chapter 16), and Fisher, Guston, and Trinidad (chapter 18) in this volume.

23. Maryann Feldman, Alessandra Colaianni, and Kang Liu, "Commercializing Cohen-Boyer, 1980–1997," DRUID working paper no. 05–21 (2005), http://www .academia.edu/1382748/Commercializing_Cohen-Boyer_1980-1997.

24. Maryann Feldman, Allan M. Freyer, and Lauren Lanahan, "On the Measurement of University Research Contributions to Economic Growth and Innovation," in *Universities and Colleges as Economic Drivers: Measuring Higher Education's Role in Economic Development*, ed. Jason E. Lane et al. (Albany, NY: SUNY Press, 2012), 97–128.

25. See Margaret Pugh-O'Mara, *Cities of Knowledge: Cold War Science and the Search for the New Silicon Valley* (Princeton, NJ: Princeton University Press, 2005).

26. Maryann Feldman and Pierre Desrochers, "Truth for Its Own Sake: Academic Culture and Technology Transfer at Johns Hopkins University," *Minerva* 42 (2004): 105–126; Maryann Feldman and Pierre Desrochers, "Research Universities and Local Economic Development: Lessons from the History of the Johns Hopkins University," *Industry and Innovation* 10 (2003): 5–24.

27. Janet Bercovitz and Maryann Feldman, "Entrepreneurial Universities and Technology Transfer: A Conceptual Framework for Understanding Knowledge-Based Economic Development," *Journal of Technology Transfer* 31 (2006): 175–88.

28. Janet Bercovitz and Maryann Feldman, "Academic Entrepreneurs: Organizational Change at the Individual Level," *Organization Science* 19 (2008): 69–89.

29. Elizabeth Popp Berman, "Why Did Universities Start Patenting? Institution-Building and the Road to the Bayh-Dole Act," *Social Studies of Science* 38 (2008): 835–871.

30. For example, the University Innovation Fellows program has developed a number of student-led initiatives to help foster innovation and creativity on college campuses; see Fasihuddin and Britos Cavagnaro (chapter 3) in this volume.

31. Maryann Feldman, Theodora Hadjimichael, Lauren Lanahan, and Tom Kemeny, "The Logic of Economic Development: A Definition and Model for Investment," *Environment and Planning C: Government and Policy* 34 (2016): 5–21.

32. Barry Bluestone and Bennett Harrison, *The Deindustrialization of America: Plant Closings, Community Abandonment, and the Dismantling of Basic Industry* (New York: Basic Books, 1984); Jefferson Cowie, *Capital Moves: RCA's Seventy-Year Quest for Cheap Labor* (Ithaca, NY: Cornell University Press, 1999).

33. Albert N. Link, *A Generosity of Spirit: The Early History of the Research Triangle Park* (Durham, NC: Research Triangle Foundation, 1995).

34. Leslie and Kargon, "Selling Silicon Valley."

35. For a similar critique, see Godin (chapter 9) in this volume.

36. Amartya Kumar Sen, "Democracy as a Universal Value," *Journal of Democracy* 10 (1999): 3–17.

37. Maryann Feldman and Nichola Lowe, "Firm Strategy and the Wealth of Regions," working paper, 2017.

38. Susan Christopherson and Jennifer Clark, *Remaking Regional Economies: Power, Labor, and Firm Strategies in the Knowledge Economy* (New York: Routledge, 2007).

39. Simona Iammarino and Philip McCann, "The Structure and Evolution of Industrial Clusters: Transactions, Technology, and Knowledge Spillovers," *Research Policy* 35 (2006): 1018–1036; John Cantwell, ed., *Transnational Corporations and Innovatory Activities* (New York: Routledge, 1994); John Cantwell and Simona Iammarino, "Multinational Corporations and the Location of Technological Innovation in the UK Regions," *Regional Studies* 34 (2000): 317–332.

40. Wanda Orlikowski, "Improvising Organizational Transformation over Time: A Situated Change Perspective," *Information Systems Research* 7 (1996): 63–92.

41. Juan Alcácer and Wilbur Chung, "Location Strategies and Knowledge Spillovers," *Management Science* 53 (2007): 760–776.

42. Feldman and Lowe, "Firm Strategy and the Wealth of Regions."

43. Nichola Lowe and Maryann Feldman, "Breaking the Waves: Innovating at the Intersections of Economic Development Policy," working paper, 2017.

44. See Bill Rigby and Alistair Barr, "Will Apple, Google, Facebook, and Amazon Fall Victim to the 'Campus Curse'?" *San Jose Mercury News*, 18 May 2013, accessed 25 April 2017, http://www.mercurynews.com/business/ci_23335912/will-apple-google-facebook-and-amazon-fall-victim.

45. On the complexities of replication, see Pfotenhauer (chapter 11) in this volume.

7   Innovation for Every American

Jenn Gustetic

On 23 March 2015, President Barack Obama hosted his fifth annual White House Science Fair. In the State Dining Room, Tiye Garrett Mills, a seventeen-year-old high school senior from Denver, waited nervously to see if she would be one of the lucky exhibitors to share her project with the president and astronaut Leland Melvin (figure 7.1). Mills had engineered a low-cost desktop scanner that allowed nearly anyone to produce fast, accurate, and cheap images of plant leaves that helped professional botanists identify new species. For Mills, "citizen science"—public participation in scientific research by nonprofessional volunteers—was more than a path to college; it had helped her overcome a personal struggle with depression and anxiety. When President Obama approached, Mills mustered her practiced confidence. She told the president that with her simple process, ordinary citizens could upload "images of leaf venation systems into their computers, and we could pick the best ones, and use them online for the registry."[1]

Mills is representative of a uniquely American archetype—the amateur scientist, the do-it-yourself garage tinkerer, the self-made entrepreneur—that has permeated our culture from Benjamin Franklin's kite experiments to Quirky's armchair inventors.[2] Until recently, these individuals have tended to operate on the fringes of the scientific and technological establishment, with little or no government support. Over the last decade, however, citizen science projects are just one way in which the federal government has enticed Americans from diverse backgrounds to become innovators. Incentive prizes, challenges, crowdsourcing, open data, public deliberations, and a number of other policy tools are democratizing and expanding access to innovation.

I have spent the past ten years helping to scale these kinds of programs and policies, both from the private sector and as a civil servant at NASA and

Figure 7.1
Tiye Garrett Mills discusses her project with astronaut Leland Melvin at the 2015
White House Science Fair. Photo: Jenn Gustetic.

the White House Office of Science and Technology Policy (OSTP). To me,
being an innovator can mean many things. It can be a part-time hobby or
a full-time job. It can mean competing against other inventors and entrepreneurs to develop the best-performing solution to a defined problem. It
can mean building open platforms to enable collaborative innovation. It can
mean building a business and creating more jobs in your community. It can
also mean having your voice heard in the design of science and technology
programs that may impact you and your family.

This chapter explores policy initiatives developed during the Obama
administration that drive an inclusive innovation philosophy.[3] I explore
strategies for overcoming very real hurdles to scaling inclusive innovation across the federal government. I also share stories of problems and
people impacted by new federal innovation initiatives. Finally, I make some
tentative observations about how these approaches continue to be supported during the early days of the Trump administration. I argue that the

bipartisan policy tools I describe can be employed by any nation's government to give a wider and more diverse group of citizens broader access to innovation.

Expanding the Federal Government's Innovation Mandate

The federal government has played a crucial role in fostering innovation since the founding of the United States. In an effort to stimulate the practical arts among its citizens, the framers ensured that Article I, section 8, of the Constitution provided the protection of intellectual property through the establishment of a patent system. In the early 1800s, federal armories led the way in developing the precision manufacturing techniques for mass-producing muskets and breech-loading rifles. Those techniques spread to private industries and made the United States a leading manufacturer of consumer products, such as sewing machines, typewriters, bicycles, and automobiles. From the Lewis and Clark expedition to the Mars Curiosity rover, the federal government has also led the way in underwriting basic scientific research and exploration. Our government makes these investments in science and technology because they have traditionally helped the United States maintain its place of scientific, military, and economic leadership among nations.[4]

The federal government, moreover, has an important role in increasing access to innovation beyond the usual suspects. The government has traditionally awarded scientific research grants and technology procurement contracts to university-based scientists and large corporations, institutions where women, African Americans, and other minorities are underrepresented.[5] In recent decades, the federal government has developed a collaborative innovation strategy with specific diversity goals, such as policies that facilitate contracting with small businesses, especially minority- and women-owned firms.

However, fostering inclusive innovation through policymaking is complex, and implementing those policies across federal agencies is even more difficult. Beyond a clear strategy, it requires a long-term effort to change culture *within* the government. In my nearly ten years in government, I have learned that many federal employees operate within what I call a "happy place" in directing their programs. This does not mean they are striving for the easiest way to do their jobs. It means that given the approaches known to them, and the results they are expected to achieve, they naturally have limited experience engaging with the full domain of people and ideas that might positively impact their programs. They are rightfully uncertain that novel policy prescriptions will deliver results. It is difficult to become confident trying a new policy approach when it is unclear whether the benefits of trying something new will outweigh the costs. If the barriers to experimentation are seen as too high, most federal employees will not attempt these new approaches.

This workplace inertia limits the overall ability of the federal government to bring new ideas and people into problem-solving and program delivery. This is where targeted strategy and policy actions at the highest levels of government can help to reduce perceived and actual barriers, which encourages more federal employees to use approaches that promote increased access and diversity.

During the Obama administration, the OSTP conscientiously chose to experiment with new policy approaches to advance the government's science and technology goals, which included a commitment to diversify and expand access to innovation. We attempted to build innovation ecosystems that connected all levels of government—the White House, federal agencies, and the Congress—with the scientific and technical community. We sought sustainable culture change and hoped that OSTP's enthusiasm for policy experimentation would trickle down and reverberate across the federal government.[6] The following sections describe foundational strategies for advancing the Obama administration's goal of creating a more diverse cross section of innovators.[7]

The Open Government Initiative

The *first* memorandum issued by President Obama was the Open Government Memorandum, which stressed the importance of transparency as a guiding principle of his administration. Two additional principles highlighted in the memo—participation and collaboration—were foundational to OSTP's mission of expanding access to innovation opportunities to all Americans:

Government should be participatory. Public engagement enhances the government's effectiveness and improves the quality of its decisions. *Knowledge is widely dispersed in society, and public officials benefit from having access to that dispersed knowledge.*

Government should be collaborative. Collaboration *actively engages Americans in the work of their government.* Executive departments and agencies should use innovative tools, methods, and systems to cooperate among themselves, across all levels of Government, and with nonprofit organizations, businesses, and individuals in the private sector.[8]

I witnessed a new passion ignited in federal employees as a result of this memorandum. The Open Government Initiative provided an organizing umbrella for many innovative policy approaches and gave the interagency working groups charged with its implementation explicit permission to try them.

In addition, a 2009 directive from the Office of Management and Budget required each federal agency to publish a biennial open government plan on the Open Government website.[9] Over a seven-year period, these continuously evolving public reports drove accountability for implementing new open government initiatives, updating old policies, and improving current practices.

One particularly noteworthy project was the creation of data.gov as a one-stop shop for coordinating the government's open data work. As an expression of the administration's transparency goals, all federal agencies were mandated to publish their nonsensitive data to the website. The free availability of government-generated data created business opportunities for savvy innovators. Simple Energy of Boulder, Colorado, for instance, has developed a business that allows consumers to compare their energy consumption with their neighbors, leveraging open data from the US Census. The company has raised $8.9 million in start-up capital and employs twenty-seven people.[10]

A Strategy for American Innovation

Parallel with the Open Government Directive, the White House's National Economic Council and OSTP issued a preliminary Strategy for American Innovation (SAI). First published in 2009, this strategy sought to "harness the inherent ingenuity of the American people and a dynamic private sector" to ensure a "broad-based" and sustained recovery from the 2008 recession (figure 7.2).[11] While presidential administrations dating at least to the Nixon era have championed innovation, the Obama administration's Strategy for American Innovation was unique for championing investments in

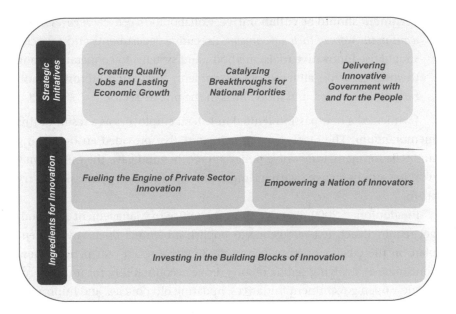

Figure 7.2
The framework for the Obama White House's Strategy for American Innovation.
Source: Obama White House, *Strategy for American Innovation*, 2015, accessed 17 July
2017, https://obamawhitehouse.archives.gov/sites/default/files/strategy_for_american
_innovation_october_2015.pdf.

the "'building blocks' of long-term economic recovery and growth," including research and development, education, and infrastructure. The SAI also
sought to create "a national environment supportive of entrepreneurship"
while "harnessing innovation" to address national priorities such as health
care and energy.[12] This strategy provided OSTP with a clear set of priorities and empowered the staff to more actively steer innovation policy than
many previous administrations.

The SAI also encouraged the government's increased use of incentive
prizes and grand challenges as important tools to engage more citizen-
innovators in addressing national priorities. Incentive prizes, crowdsourcing, and other forms of "open innovation" had enjoyed a renaissance in
corporate and philanthropic circles during the early 2000s. In a nutshell,
a patron would offer a well-advertised financial prize to the individual or
team that could solve a carefully defined technical problem. For example,
in 2004 Burt Rutan and financier Paul Allen claimed the $10 million Ansari

X Prize when they became the first commercial, nongovernment team to launch a reusable manned spacecraft 100 km into space twice within two weeks. Similarly, Procter & Gamble's "Connect + Develop" initiative crowdsourced ideas from outside inventors to launch popular new products, such as the Swiffer Duster.[13]

Given the proven success of crowdsourcing in the private sector, the SAI advocated for increased federal use of incentive prizes and other forms of open innovation. With traditional government grants and competitively bid R&D contracts, researchers were paid up front for their work, whether it panned out or not. With inducement prizes, the sponsor paid only for ideas that actually solved the problem. As an added bonus, these contests tended to multiply and diversify the amount of talent and capital directed at a given problem. In other words, the SAI advocated for incentive prizes as a financially prudent procurement strategy that expanded the pool of citizen-innovators working on the government's scientific and technological challenges.[14]

The 2011 and 2015 revisions of the SAI described progressively more ambitious policies, but the latter report noted the government's overall low adoption rate of novel innovation methods designed to include more Americans. Consequently, the administration committed to developing an "Innovation Toolkit," which would capture the best practices, case studies, policy guidance, and training resources for a wide variety of innovation strategies.[15]

The Federal Innovation Toolkit

Federal employees are tasked with driving scientific and technical advancements and delivering on their programs day after day. Given the breadth and depth of possible policy approaches, agencies can struggle to understand the variety of tools they might want to utilize and how those tools can be scaled. To encourage the adoption of novel innovation policies, the White House defined its priority set of scalable approaches in an Innovation Toolkit included with the 2015 Strategy for American Innovation as a means to deliver innovative government with and for the American people. Many of the included tools—such as open data, incentive prizes, high-impact multisector collaborations, and regional initiatives—inherently seek to expand access to innovation. Similarly, a 2016 report from the Government Accountability Office encouraged agencies to expand public involvement in innovation via additional policy approaches such as citizen science, crowdsourcing, idea generation, and open dialogues.[16] Granted, most of these ideas did not originate within the Obama administration. However, the administration was unique in its support for scaling these approaches as part of a comprehensive innovation agenda.

Congress as Innovator

While the White House was articulating policy through strategies, directives, and tool kits, Congress also passed legislation enabling agencies to engage a broader section of society in innovation. New congressional *authorizations* gave agencies expanded permissions to conduct business in new ways, while *appropriations* allocated budget resources for innovation projects at specific agencies.

For example, in December 2010, Congress passed the America COMPETES Reauthorization Act, which provided *all* federal agencies broad authority to conduct prize competitions to spur innovation. Prior to this legislation, only a few federal agencies, including NASA and the Department of Energy (DOE), had the authority to conduct competitions. This new authorization ignited a surge in the use of prizes among federal agencies. At the Challenge.gov prize clearinghouse, tens of thousands of citizen "solvers" offered potential solutions to government challenges—from ideas for healthier school lunches (Department of Agriculture) to devices for capturing energy from ocean waves (DOE).[17] By June 2017, Challenge.gov had featured more than 750 prize competitions and challenges valued at more than $250 million—conducted under COMPETES and other authorities—from over one hundred federal agencies, departments, and bureaus.[18] A subsequent reauthorization of the COMPETES Act in December 2016 explicitly encouraged the use of citizen science and crowdsourcing across the federal government. Congress believed these approaches would spur a cost-effective acceleration of scientific research, promote hands-on STEM learning, and connect "members of the public directly to federal science agency missions and to each other."[19]

In 2016, Congress also reauthorized the Small Business Innovation Research (SBIR) and Small Business Technology Transfer (STTR) programs through September 2022. Since 1982, these programs have required that all federal agencies spending more than $100 million annually on external R&D set aside a percentage of their grant and contract funds for small businesses, especially minority and women-owned firms. Essentially, these programs help finance the commercialization of cutting-edge technologies, while ensuring that the government looks beyond the usual corporate and university suspects in procuring government-sponsored research. The previous 2011 reauthorization mandated annual 0.1 percent increases in SBIR's reserve percentage, from 2.5 percent in 2011 up to 3.2 percent in 2017. SBIR and STTR Programs at eleven research-intensive agencies—including the Department of Defense, NASA, and the National Science Foundation—now set aside approximately $2.2 billion annually to finance the development of cutting-edge technologies by a much broader spectrum of innovators.[20] Moreover, the Obama administration advocated for increased commercialization support for SBIR/STTR grantees through boot camps such as NSF's I-Corps (see chapter 5).

Scaling Access: Innovation Policy in Action

The Open Government Policies and Strategy for American Innovation provided high-level frameworks to enhance access to innovation; however, tool-specific policy guidance and communities of practice were critical to their implementation and scaling across government. The following sections highlight a series of successful impact stories, in which incentive prizes, citizen science, crowdsourcing, and other federal policy tools encouraged a broad cross section of Americans to become innovators.

Incentive Prizes

Incentive prizes help "make" new innovators by expanding and diversifying the pool of citizens that engage with the government's challenges. For example, in 2013, Elena Lucas was a low-level analyst at a Fortune 200 company.[21] Today, she is a cofounder and CEO of UtilityAPI, a clean-energy start-up that provides a data service to evaluate energy usage and savings for homes and businesses. How did Lucas go from analyst to entrepreneur in just under five months? Her drive, intelligence, and savvy made her a perfect CEO, but a government incentive prize helped kick-start her company.

In 2013, former NASA rocket scientist Michael Contreras was working in the DOE's Solar Energy Technologies Office on a two-year fellowship sponsored by the American Association for the Advancement of Science.

Contreras observed a growing market demand for energy data products, but few new start-ups. Empowered by the SAI and the COMPETES Act, Contreras developed the SunShot Catalyst incentive prize to attract entrepreneurs to the emerging solar data sector. The challenge: create a secure and standardized data infrastructure for the evolving solar energy economy—a twenty-first-century digital update on the information once manually collected from nineteenth-century gas and electrical meters. By employing an incentive prize, DOE hoped to make solar energy faster to deploy, more accessible, and more affordable for Americans everywhere by actively engaging IT innovators from across the country.

In fall 2014, Elena Lucas and Daniel Roesler attended a "jamathon" hosted by Contreras's SunShot Catalyst Prize Competition. During this multiday collaborative brainstorming session, contestants developed potential solutions and business plans; attendees were not required to have any background in the solar domain or even coding skills.[22] Though Lucas had never worked on a software start-up, she was intrigued by the opportunity to participate in the prize competition, which had removed all barriers to entry. "Creating a company, with the jamathon as a catalyst, [was] a great alternate to the 9–5," Lucas shared with me over email. "I was attracted to entrepreneurship, as I'm sure many other millennials are, because we don't have job security." "Prize competitions," she continued, "are a strong way to encourage entrepreneurship and small business creation, which are a huge generator of jobs." Lucas also appreciated how prize competitions like SunShot Catalyst "set rules and criteria that make it possible for participants to see a way to winning. It's a clear path that doesn't hold as much ambiguity as starting a VC-funded start-up."[23]

In September 2014, Lucas and Roesler submitted their concept alongside 137 ideas from other innovators in the ideation phase of Catalyst's multistage competition (figure 7.3). Their idea used a customer's utility data and bill history to estimate their savings if they converted to solar. UtilityAPI was selected as one of 17 semifinalists; with their prize—crowdsourced software development credits—they built a prototype. On 21 May 2015, Lucas and Roesler unveiled their prototype and pitched their idea to a panel of judges at a demonstration day in San Francisco. UtilityAPI and four other teams were awarded $30,000 to advance their early-stage solutions toward commercialization.[24] According to Lucas, "having the validation of SunShot Catalyst" on their website as well as the government's "non-dilutive

Figure 7.3
Winners of the second cycle of the SunShot Catalyst Energy Innovation Prize were chosen out of 19 finalist start-ups that demonstrated their energy software solutions in Philadelphia on 10 December 2015. Credit: US Department of Energy (https://energy.gov/eere/sunshot/sunshot-catalyst-energy-innovation-prize).

funding" attracted private sector investors and helped UtilityAPI raise additional capital.[25] Between August 2014 and October 2015, UtilityAPI grew from two to eleven employees in its Oakland, California, office.[26]

Lucas did not expect to become an entrepreneur. She grew up in St. Clair Shores, Michigan, outside Detroit; her father had been sporadically employed since she was in third grade. She paid for her undergraduate and graduate degrees through student loans. Entrepreneurship became a possibility when she moved to the Bay Area and was exposed to its start-up community. The DOE's government-sponsored jamathon and Catalyst prize helped her get started on her entrepreneurial journey. She remarked to me, "My parents are on food stamps. I'm a job creator. It's been a wild ride."[27]

According to a DOE report to OSTP, three of the five teams who won $30,000 in the "seed" round successfully achieved their product milestones and were awarded an additional $70,000, for a total of $100,000 in DOE prize money per team. Furthermore, the five finalist companies collectively raised over $1 million from private investors in 2015.[28] Lucas and UtilityAPI also leveraged the Catalyst Prize win into an additional $762,000 SunShot grant from the DOE.[29]

Lucas's experience with the SunShot Catalyst Prize Competition is illustrative of the many benefits of offering incentive prizes: reaching beyond the "usual suspects" for solutions; increasing the number of solvers tackling a problem; generating multiple novel approaches; and minimizing financial risk for the government. Incentive prizes also provide more entrepreneurial opportunities for women. In a study of more than 166 science challenges involving over 12,000 scientists, open innovation scholars Karim Lakhnai and Lars Bo Jeppesen found that "female solvers—known to be in the 'outer circle' of the scientific establishment—performed significantly better than men in developing successful solutions." By "removing barriers to entry to non-obvious individuals," prizes increase access to innovation.[30]

Before 2009, NASA, DOE, and the Department of Defense had limited authority to offer incentive prizes, and they conducted only a handful of competitions.[31] However, between 2009 and 2015, the federal government conducted more than 440 prize competitions and challenges.[32] How did the use of this approach scale so significantly?

To complement the White House's Strategy for American Innovation (2009) and Congress's reauthorized COMPETES Act (2010), the White House issued a formal policy framework in 2010 to guide agency leaders in using prize competitions and challenges to advance their core missions. This memorandum further directed the General Services Administration (GSA) to "make available a web-based platform for prizes and challenges within 120 days."[33] So in September 2010, the GSA launched Challenge.gov, a one-stop shop where entrepreneurs and citizen solvers can find public-sector prize competitions and challenges. In 2011, the administration also launched the Center for Excellence for Collaborative Innovation (CoECI), a NASA-led, governmentwide team that provided agencies with guidance and support in implementing their prize competitions and challenges. Between 2011 and 2015, CoECI launched 156 crowdsourced challenges, including sixteen challenges for other agencies.[34]

These policy, programmatic, and legislative actions were critical to accelerating the use of prizes across the government, but they would be useless without federal employees who were willing to try them. Led by the GSA, more than six hundred federal employees have established a community of practice for prizes and challenges. This community enables employees across various federal agencies to help one another to implement progressively more ambitious and creative applications of these tools. On 16 December 2016, the federal prize community published a tool kit on Challenge.gov that assembled six years of accumulated expertise in conducting government-sponsored prize competitions.[35] This tool kit seeks to enable a greater number of employees to conduct even more prizes in the future.

Through a series of high-level meetings, the White House also spotlighted successes in order to increase awareness of incentive prizes, advance the state of practice in the government, and facilitate the initiation of high-impact, cross-sector competitions. For example, on 7 October 2015, the White House hosted "All Hands on Deck: Solving Complex Problems through Prizes and Challenges" to catalyze the next generation of ambitious prizes.[36] This event included more than 150 federal agency managers, along with state and local government leaders, representatives from foundations and other nongovernmental organizations (NGOs), and private-sector supporters.

Elena Lucas was a featured prizewinner at this event; she shared her experience of participating in the DOE's SunShot Catalyst, which helped her become an innovator. A few months later she was named to *Forbes* magazine's "30 under 30" watch list for young entrepreneurs in the energy sector.[37]

Citizen Scientists and Crowdsourcing

Tiye Garrett Mills, the high-schooler behind the leaf scanner, is just one of thousands of volunteers participating in citizen science and crowdsourcing projects supported by the federal government across the country. Mills is an example of a scientific "maker"; she modified hardware and collected biological samples in the field. However, many citizen scientists contribute to discoveries from the screens of their home computers.

In 2011, NASA astrophysicist Marc Kuchner sat in his office pouring over huge amounts of data from the agency's Wide-field Infrared Survey Explorer (WISE), a space-based telescope. Computer searches had already identified many objects seen by the WISE survey as potential dusty "debris" disks; these disks are exciting because they are the signposts of planetary systems. But software cannot reliably distinguish debris disks from stars blended with other infrared-bright sources, such as galaxies, interstellar dust clouds, and asteroids. While writing a book about science communication, Kuchner had become interested in citizen science approaches. What if NASA could leverage the human eye's ability to distinguish patterns by crowdsourcing the classification of these images? How much more science could be accomplished? Kuchner set up a partnership with Zooniverse, a community of scientists, software developers, and educators who collectively develop and manage citizen science projects via the internet. After a long struggle to find funding, the Disk Detective project opened for participation in January 2014.

In March 2014, just two months after Disk Detective launched, even Kuchner was surprised by the degree to which volunteers were getting involved. Kuchner started receiving emails from volunteers who complained about seeing the same object over and over again. "We thought at first it was a bug in the system," he explained, "but it turned out [the volunteers] were seeing repeats because they had already classified every single object that was online at the time." In less than one year from the project's start date, over 28,000 citizen scientist volunteers using DiskDetective.org had logged one million classifications of potential debris disks and disks surrounding young stellar objects. By November 2016, the community had completed two million classifications. These citizen scientists come from all over the globe—Argentina, Japan, Germany, Hungary, Australia—and they include students from Michael Blake's sixth-grade classroom in Aylesbury, Canada.[38]

The volunteers' involvement did not stop at online classifications. "Many of the project's most active volunteers are now joining in science team discussions," Kuchner explained, "and the researchers encourage all users who have performed more than 300 classifications to contact them and take part."[39] When volunteers identify objects of interest, scientists try to reserve time on telescopes in order to conduct follow-up observations and gather more data. At Disk Detective, the citizen scientists propose telescope times and help interpret new data. In one case, when the need for follow-up observations was particularly urgent, Hugo Durantini Luca drove twelve hours from his home to help out during an observation period in the mountains of Argentina (figure 7.4).

It would arguably not be "science" if this project were not generating new publications to share its discoveries. However, as of November 2016, the Disk Detective team has published two papers describing discoveries made by citizen scientists in collaboration with their professional colleagues. The first paper, which includes eight citizen scientists as coauthors,

Figure 7.4
Disk Detective Hugo Luca drove twelve hours to help out with an observing run in
Argentina. Photo: Marc Kuchner, NASA.

was accepted by the *Astrophysical Journal* in July 2016. The second paper,
which includes eight additional citizen scientists as coauthors, was accepted
by *Astrophysical Journal Letters* in September 2016.[40]

The team expects to wrap up the current project sometime in 2018, with
an expected total of about three million classifications and perhaps four
thousand disk candidates. This project is an incredible example of how
crowdsourced, public participation expands access to science and innovation. "More scientific breakthroughs are out there," writes Disk Detective
volunteer Jonathan Holden, "and citizen science is a chance for a person
like me who isn't a 'rocket surgeon' (yes, I'm being silly) to be a part of a
scientific community, discovering strange new worlds."[41]

So how did the use of citizen-science and crowdsourcing approaches spread across the federal government? Whereas prize competitions were scaled via top-down policy, programmatic, and legislative action early on, the citizen-science and crowdsourcing approaches grew through a bottom-up, grassroots movement. Universities, nonprofits, states, and local communities had been using these approaches for decades in a number of different domains and often without any federal role. For example, the Citizen Science Association, an NGO, has grown to over four thousand members from over thirty countries.[42] The federal government had to complement existing efforts while discerning its proper role in scaling its crowdsourcing and citizen science endeavors.

During the Obama administration, these two innovation tools achieved success through the following ingredients: catalogs and platforms to centrally list projects, policy and strategy documents to provide permission for their use and address common issues, tool kits and communities to share knowledge and advance best practices, and programmatic coordinators at the GSA to manage all of these components. From 2012 to 2016, citizen science and crowdsourcing initiatives were deliberately cultivated in the White House and across federal agencies. A "Champions of Change" event brought citizen science advocates to the White House; they encouraged federal employees to use these approaches to enhance scientific research, address societal needs, and provide hands-on STEM learning and literacy.[43] Elsewhere a group of five like-minded federal innovators formed the Federal Community of Practice for Crowdsourcing and Citizen Science (CCS), which has since grown to more than three hundred federal members from more than forty agencies.[44] To prove that citizen science can be done anywhere, the White House installed a citizen science rain gauge in the First Lady's vegetable garden. And in September 2015, OSTP announced key policy, programmatic, and legislative actions to help scale the use of citizen science and crowdsourcing. This included publication of the Federal Crowdsourcing and Citizen Science Toolkit, which was collaboratively developed by more than 120 people from twenty-four agencies.[45]

Finally, the Obama administration consolidated leadership for its citizen science and crowdsourcing efforts within the GSA. In April 2016, in conjunction with the sixth White House Science Fair, the White House announced that GSA had partnered with the Woodrow Wilson International Center for Scholars, a trust instrumentality of the US government, to launch CitizenScience.gov as the new hub for citizen science and crowd-sourcing initiatives in the public sector.[46] When it launched, the website's searchable database listed more than three hundred citizen science and crowdsourcing projects supported by twenty-five federal agencies.

December 2016 brought a surprise holiday gift to this community when Section 404 of the reauthorized COMPETES Act (S.3084) granted new authority for federal agencies to conduct citizen-science and crowdsourcing projects.[47] Much as the 2010 COMPETES Act stimulated the use of incentive prizes and challenges, this latest reauthorization could be transformative for scaling citizen science and crowdsourcing. This new authority requires biannual reporting, which will build a much more robust case for the effectiveness of these tools, establish a long-term role for the GSA in administering CitizenScience.gov, and formalize the citizen science coordinator role at all federal agencies. These grassroots efforts in citizen science have earned recognition. In 2017, the CCS was recognized as a finalist for the Harvard Kennedy School's prestigious Roy and Lila Ash Innovation in American Government Award.[48]

The citizen science and crowdsourcing community is currently more active in the academic and private sector, but its members regularly collaborate with their public sector counterparts to assist the scientific community. Many more researchers are beginning to understand the value that volunteers can bring to their work. One study estimated that the in-kind contributions of 1.3 million to 2.3 million citizen science volunteers to biodiversity research have an economic value of up to $2.5 billion per year.[49] Participation in citizen science efforts also promotes science literacy, STEM education, and inclusion. By their very nature, citizen science and crowdsourcing broaden access to innovation across the country—and around the world. Moreover, advocacy groups, such as the Citizen Science Association, strive to foster diversity and inclusion, not just among volunteers but within the scientific field itself.[50]

Hybridized Innovation Tools

We should expect to see more projects that combine multiple methods from the Innovation Toolkit as federal employees and partners learn to create hybrid applications for maximum impact. For example, in 2014, as the Ebola public health crisis spread across Sub-Saharan Africa, the United States Agency for International Development (USAID) employed a combination of approaches—including hackathons, an ideation challenge, and a traditional innovative procurement contract—to "rapidly source and develop potential solutions" to combat the challenges faced by health care workers. These challenges included inadequate personal protective equipment, difficulty in tracking person-to-person transmission, the absence of rapid point-of-care diagnostics, and a need to accommodate traditional burial ceremonies involving direct contact with a deceased body. As USAID reported to Congress, the Fighting Ebola Grand Challenge "sourced over 1,500 ideas and potential solutions."[51]

Similarly, a hybrid innovation strategy—leveraging both open data and citizen science—helped advance the social justice mission of Fair Policing, an organization that studies what it means to grow up policed in New York City. This project sought a way for local youth to produce their own knowledge, rather than just accepting the perspectives of adults. Professor Brett Stoudt of the John Jay College of Criminal Justice, CUNY, described how this impacted one youth co-researcher from the project, Keshan Harley: "[Harley] stood before a New York City council to share the data he helped generate. That data gave him a sense of authority and power to define his own narrative to policy makers that day." Stoudt commented that Harley later told him, "This research transformed the I into We. It took a thousand individual experiences and collectivized it. Too often, young people like myself feel powerless against the injustices we face in life. Above all, the participatory research process empowers young people."[52]

While the Fighting Ebola Grand Challenge was federally led with external partners, the Researchers for Fair Policing project had little government involvement other than the provision of open data. As policymakers consider which combination of tools could best be applied for different problems, they should also discern the appropriate role for the federal government. As innovation experts Raymond Tong and Karim Lakhani suggest in their 2012 paper, the federal government can serve as host, coordinator, or contributor for these innovation tools, depending on the involvement of external partners.[53]

Conclusion: A Future That Engages Even More Americans

During the Obama administration, the scaling of tools such as open data, incentive prizes, citizen science, and crowdsourcing expanded access to innovation by enticing a broader cross section of Americans to become innovators. On Challenge.gov alone, more than 250,000 solvers have participated in over 750 challenges for more than $250 million in prize money since 2010. CitizenScience.gov lists four hundred projects from twenty-six federal agencies. Every day, millions of consumers (and thousands of employees) depend on companies such as Trulia, Carfax, and Accuweather that are built upon open government data. Overall, in recent years, literally hundreds of thousands of people have become engaged in innovation opportunities that may not have been available were it not for the federal government.

These tools help make a new and more diverse crop of innovators who are motivated to participate for different reasons. Many citizen scientists engage because of the inherent value they see in contributing to scientific discovery. Others engage because of the new connections to their community and the like-minded friends they make by participating in citizen science projects. Prize competitions, in contrast, attract garage inventors driven by interesting technological challenges, as well as the allure of monetary rewards and the public spotlight. Hackers who engage with government open data and maker communities that advance the open hardware movement are alternatively motivated by community values of collaboration and openness.

Historically, the federal government has not actively engaged with citizen scientists, garage inventors, hackers, and makers. Just as individual federal employees operate in their "happy places," the federal government has traditionally engaged mostly with other *organizations*, such as companies and research institutions. Citizen scientists, garage inventors, hackers, and makers are *groups of individuals*, which makes it structurally and philosophically more difficult for the government to understand when and how best to engage them.

As more and more impact stories emerge from government-sponsored initiatives, the case for democratizing, diversifying, and expanding access to innovators of all stripes is stronger today than ever before. Thanks to policy, programmatic, and legislative support in recent years, the biggest question for future government policy makers will not be *if* these approaches have value, but *how* best to use them, while increasing the diversity and inclusion of participants.

At the end of the day, project managers across government must develop an understanding of when and how to use these tools to achieve the nation's science and technology goals. Scale will be limited by awareness of these approaches, the perceived and actual barriers to implementing them, and the existence of a learning culture to expand and develop even more effective applications. Each administration has an important role in defining the value they see for expanding access to innovation. Similarly, each administration must decide which policy tools they find most appealing, then formulate and execute strategies for scaling them.

It remains to be seen how the innovation tool kit will continue to scale during the Trump administration. There are a number of factors that contribute to scale, including legal frameworks, communities of practice, and budget availability for projects. Early signs throughout 2017 and 2018 point to most of these building blocks continuing to be supported at most federal agencies. Importantly, Congress's permanent authorizations for both incentive prizes and citizen science certainly provide a sustainable legal basis for those approaches. Active communities of practice and the continued clear coordination role for GSA are critical for longevity. GSA's work to publish a more expansive web-based innovation tool kit will also be important. Since most of the hundreds of projects that have been executed to date are largely agency- and program-based efforts, top-down support from the White House is incredibly enabling, as demonstrated during the Obama administration, but it is not the only factor for scaling the use of these approaches by federal agencies. As we approach the end of 2018, I am encouraged that these approaches continue to be leveraged and supported in the Trump administration as bipartisan tools that help to execute the priorities of administration initiatives such as the Office of American Innovation (OAI), whose stated objective is to "bring together the best ideas from Government, the private sector, and other thought leaders to ensure that America is ready to solve today's most intractable problems, and is positioned to meet tomorrow's challenges and opportunities."[54] In March 2018 explicit support was demonstrated by the OAI in a White House blog post that echoed much of the sentiment about prize competitions that was espoused in the Obama administration stating both, "The Trump administration looks forward to future engagement with federal agencies and the private sector to leverage prizes and challenges to improve the quality of life for Americans now and in the future" and "prizes and challenges can enable federal agencies to reach beyond the 'usual stakeholders' to increase the number of perspectives working to develop solutions for a specific problem."[55] Thus, the first two years of the Trump administration have continued to enable and encourage the use of some types of open innovation in the federal government, both as means to involve a diverse cross section of Americans in solving problems and as a way to improve the quality of life for Americans.

The Obama administration's coordinated support of innovative science and technology policies set the stage to help a broader and more diverse cross section of Americans to become innovators. In the process, the administration established a firm foundation of knowledge, infrastructure, best practices, and high-impact case studies to fuel the fire for many years to come, including today where open innovation approaches continue to be supported by the Trump administration.

Notes

1. Quoted in Jenn Gustetic, Lea Shanley, and John McLaughlin, "Citizen Science Is Everywhere, Including the White House," *Obama White House Office of Science and Technology Policy Blog*, 25 March 2015, https://obamawhitehouse.archives.gov/blog /2015/03/25/citizen-science-everywhere-including-white-house.

2. Jack Hitt, *Bunch of Amateurs: Inside America's Hidden World of Inventors, Tinkerers, and Job Creators* (New York: Broadway Books, 2013). For more on Quirky, see Hintz (chapter 10) in this volume.

3. It is important to note that many of the approaches and principles advanced by the Obama administration built upon the efforts of the previous administrations. The Clinton administration's Reinventing Government efforts and the George W. Bush administration's e-government efforts sought to increase the effectiveness of government through progressively more advanced technology environments. Also, many of the innovation tools described in this chapter were in use by the federal government before the Obama administration. For example, NASA's largest incentive prize program, the Centennial Challenges Program, was initiated in 2005 after being authorized by Congress. Government-sponsored citizen science projects have been around for decades. Finally, other governments have been using prizes to spur innovation for centuries. For example, beginning in 1714, the British Parliament offered awards to inventors who could solve the problem of measuring longitude at sea.

4. A. Hunter Dupree, *Science in the Federal Government: A History of Policies and Activities to 1940* (Cambridge, MA: Belknap Press, 1957).

5. See Cook (chapter 12) and Carlson (chapter 16) in this volume.

6. In an interview with Harvard Business School, OSTP's deputy director (2009–2017) Tom Kalil said, "My view is that in order to get something really big done, you usually need to create an ecosystem. Because we interact not only with the senior policy

leadership of the White House, but also with scientists, engineers and entrepreneurs inside and outside the federal government, we're able to accelerate the rate at which information flows between different communities. We're able to identify some ideas that we think are promising and then bring them to the attention of people who might not have heard them before." Linda A. Hill and Allison J. Wigen, *Tom Kalil, Deputy Director for Technology and Innovation*, Harvard Business School case study N9-417-021, 2016, 9.

7. Previous administrations also had policies to encourage such behavior, but this chapter focuses on actions I witnessed from 2008 to 2016.

8. Obama White House, "Memorandum—Transparency and Open Government," last modified 21 January 2009, accessed 17 July 2017, https://obamawhitehouse .archives.gov/the-press-office/2015/11/16/memorandum-transparency-and-open -government (emphasis added).

9. Obama White House, "Memorandum—Open Government Directive," last modified 8 December 2009, accessed 29 August 2018, https://obamawhitehouse.archives .gov/open/documents/open-government-directive.

10. "Impact," General Services Administration's (GSA) Data.gov program, accessed 17 July 2017, https://www.data.gov/impact/#energy.

11. Obama White House, "President Obama Lays Out Strategy for American Innovation," last modified 21 September 2009, accessed 17 July 2017, https://obamawhite house.archives.gov/the-press-office/president-obama-lays-out-strategy-american -innovation.

12. Hill and Wigen, *Tom Kalil*, 4.

13. See Henry Chesbrough, *Open Innovation: The New Imperatives for Creating and Profiting from Technology* (Boston: Harvard Business School Press, 2003); on incentive prizes, including the Ansari X Prize, see *"And the Winner Is…": Capturing the Promise of Philanthropic Prizes*, McKinsey & Company, 2009, accessed 15 June 2017, http://www.mckinseyonsociety.com/downloads/reports/Social-Innovation/And_the _winner_is.pdf; on Procter & Gamble, see Larry Huston and Nabil Sakkab, "Connect and Develop: Inside Procter & Gamble's New Model for Innovation," *Harvard Business Review* 84, no. 3 (March 2006): 58–66.

14. See Obama White House, "Strategy for American Innovation," 2015, accessed 17 July 2017, https://obamawhitehouse.archives.gov/the-press-office/2015/10/21/fact -sheet-white-house-releases-new-strategy-american-innovation; on the rationale for prizes, see Eric S. Hintz, "Creative Financing: The Rise of Cash Prizes for Innovation Is a Response to Changing Business Conditions—and a Return to a Winning Strategy," *Wall Street Journal*, 27 September 2010, R8.

15. Obama White House, "Strategy for American Innovation," 2011, accessed 29 August 2018, https://obamawhitehouse.archives.gov/sites/default/files/uploads

/InnovationStrategy.pdf. On the Innovation Toolkit, see Obama White House, "Strategy for American Innovation," 2015, accessed 17 July 2017, https://obamawhitehouse.archives.gov/sites/default/files/strategy_for_american_innovation_october_2015.pdf, 109–110.

16. "Open Innovation: Practices to Engage Citizens and Effectively Implement Federal Initiatives," Government Accountability Office, 2016, accessed 17 July 2017, http://www.gao.gov/assets/690/680425.pdf, 9.

17. "Recipes for Healthy Kids," Department of Agriculture, accessed 15 June 2017, https://www.challenge.gov/challenge/recipes-for-healthy-kids/; "Wave Energy Prize," Department of Energy, accessed 15 June 2017, https://www.challenge.gov/challenge/wave-energy-prize-2/.

18. "About," General Services Administration's (GSA) Challenge.gov program, accessed 17 July 2017, https://www.challenge.gov/about/.

19. "15 US Code § 3724—Crowdsourcing and Citizen Science," Cornell Law Legal Information Institute, accessed 17 July 2017, https://www.law.cornell.edu/uscode/text/15/3724.

20. On SBIR's history, see "Birth and History of the SBIR Program," accessed 17 July 2017, https://www.sbir.gov/birth-and-history-of-the-sbir-program. On the 2011 reauthorization, see "15 US Code § 638—Research and Development," Cornell Law Legal Information Institute, accessed 17 July 2017, https://www.law.cornell.edu/uscode/text/15/638.

21. Elena Lucas, "How I Went from Junior Analyst to Tech CEO in One Year," *Women 2.0 Blog*, 19 February 2015, http://women2.com/stories/2015/02/19/went-junior-analyst-tech-ceo-one-year.

22. For Elena Lucas describing her jamathon experience, see "Winners and Prizes 1," 2015, accessed 17 July 2017, https://www.youtube.com/watch?list=PLd9bGuOJ3nHmi8ezudkvqyEtZ3r7WB5Q&v=NCBDWLJjblY, especially at 1:26–2:30.

23. Elena Lucas, email to the author, 27 November 2016.

24. On the ideation phase of the competition, see "SunShot Announces Solar Start-Up Winners of Catalyst Demo Day," Department of Energy, 2015, accessed 17 July 2017, http://energy.gov/eere/articles/sunshot-announces-solar-start-winners-catalyst-demo-day.

25. "Winners and Prizes 1," from 3:06 to 3:31.

26. Elena Lucas, "Tips for Finding Funding and Power Networking at Pitch Events," *Elena Lucas Blog*, 24 October 2015, http://www.elenalucas.co/blog/tips-for-finding-funding-and-power-networking-at-pitch-events.

27. Elena Lucas, email to the author, 27 November 2016.

28. Obama White House, "Implementation of Federal Prize Authority: Fiscal Year 2015 Progress Report," 2016, accessed 17 July 2017, https://obamawhitehouse.arch ives.gov/sites/default/files/fy2015_competes_prizes_report.pdf, 10.

29. "Utility API Secures $760,000 Award from US Department of Energy SunShot Initiative to Fund Solar Energy Growth and Expansion," UtilityAPI, 16 November 2015, https://site.utilityapi.com/blog/utilityapi-secures-762000-award-from-us-department -of-energy-sunshot-initiative-to-fund-solar-energy-growth-and-expansion.

30. Lars Bo Jeppesen and Karim Lakhnai, "Marginality and Problem Solving Effectiveness in Broadcast Search," *Organization Science* 21 (2010): 1016.

31. Some of the government's pilot prize competitions were quite influential. For example, beginning in 2004, the Defense Advanced Research Projects Agency (DARPA) hosted a series of Grand Challenges that accelerated the development of autonomous (i.e., driverless) vehicles. See "Grand Challenge Overview," DARPA, accessed 16 June 2015, http://archive.darpa.mil/grandchallenge04/overview.htm.

32. Tom Kalil, Dave Wilkinson, and Jenn Gustetic, "Celebrating the Five-Year Anniversary of Challenge.gov with More Than 20 New Prizes," *Obama White House Office of Science and Technology Policy Blog*, 6 October 2015, https://obamawhitehouse .archives.gov/blog/2015/10/06/celebrating-five-year-anniversary-challengegov-more -20-new-prizes.

33. Obama White House, "Guidance on the Use of Challenges and Prizes to Promote Open Government," last modified 8 March 2010, accessed 17 July 2017, https:// obamawhitehouse.archives.gov/sites/default/files/omb/assets/memoranda_2010 /m10-11.pdf, 2.

34. Obama White House, "Implementation of Federal Prize Authority," 17.

35. Christofer Nelson, Jenn Gustetic, and Kelly Olson, "Incentivizing Innovation: A New Toolkit for Federal Agencies," *Obama White House Office of Science and Technology Policy Blog*, 15 December 2015, https://obamawhitehouse.archives.gov /blog/2016/12/15/incentivizing-innovation-new-toolkit-federal-agencies. For details on the tool kit, see "Challenges and Prizes Toolkit," accessed 17 July 2017, http:// www.challenge.gov/toolkit.

36. For event materials and recordings, see "Prizes in the Public Sector," accessed 17 July 2017, https://spi.georgetown.edu/prizes.

37. Christopher Helman, "Forbes 30 under 30 in the Energy Sector," *Forbes*, 6 January 2016, http://www.forbes.com/pictures/mef45ehhll/elena-lucas-27/-341a09d618e0.

38. For classification status updates, see "@Diskdetective," 28 October 2016, https:// twitter.com/diskdetective/status/792031042924412929, and "@Diskdetective," 28 October 2016, https://twitter.com/diskdetective/status/792404481031929856.

39. "Volunteer 'Disk Detectives' Top 1 Million Classifications of Possible Planetary Habitats," NASA, 6 January 2015, accessed 17 July 2017, https://www.nasa.gov /content/goddard/volunteer-disk-detectives-top-1-million-classifications-of-possible -planetary-habitats.

40. "Our First Paper and the First Debris Disk with a White Dwarf Companion," *Disk Detective*, 4 August 2016, https://blog.diskdetective.org/2016/08/04/our-first- paper-and-the-first-debris-disk-with-a-white-dwarf-companion/; "Our Second Paper and a New Kind of M Dwarf Disk," *Disk Detective*, 21 October 2016, https://blog .diskdetective.org/2016/10/21/our-second-paper-and-a-new-kind-of-m-dwarf-disk/.

41. Johnathan Holden, "NASA Press Release Announces Citizen Scientists Help Discover Oldest Stellar Nursery," *Flash Boiler Blog*, 29 October 2016, http://flashboiler .blogspot.com/2016/10/nasa-press-release-announces-citizen.html.

42. "The Power of Citizen Science," Citizen Science Association, accessed 17 July 2017, http://citizenscience.org/. Likewise, Scistarter lists 1,300 projects in its project database, only three hundred of which have a federal role (http://scistarter.com /finder).

43. Obama White House, "Champions of Change: Citizen Scientists," accessed 17 July 2017, https://obamawhitehouse.archives.gov/champions/citizen-scientists; see "Champions of Change: Citizen Science," 23 June 2013, https://www.youtube.com /watch?v=PLau1ZFA8z8, for video of the event.

44. "Federal Crowdsourcing and Citizen Science Community of Practice," GSA's CitizenScience.gov program, accessed 29 August 2018, https://www.citizenscience .gov/about/community-of-practice/#.

45. Tom Kalil and Dave Wilkinson, "Accelerating Citizen Science and Crowdsourcing to Address Societal and Scientific Challenges," 30 September 2015, https:// obamawhitehouse.archives.gov/blog/2015/09/30/accelerating-use-citizen-science -and-crowdsourcing-address-societal-and-scientific.

46. Obama White House, "Addressing Societal and Scientific Challenges through Citizen Science and Crowdsourcing," last modified 30 September 2015, accessed 29 August 2018, https://obamawhitehouse.archives.gov/sites/default/files/microsites /ostp/holdren_citizen_science_memo_092915_0.pdf.

47. Jeffrey Mervis, "Update: Surprise! Innovation Bill Clears House, Heads to President," *Science*, 16 December 2016, http://www.sciencemag.org/news/2016/12/update -surprise-innovation-bill-clears-house-heads-president.

48. "Ash Center Announces Finalists and Top 25 Programs for Innovations in American Government Award," Harvard Kennedy School Ash Center, 2 May 2017, https://ash.harvard.edu/news/ash-center-announces-finalists-and-top-25-programs -innovations-american-government-award.

49. E. J. Theobald et al., "Global Change and Local Solutions: Tapping the Unrealized Potential of Citizen Science for Biodiversity Research," *Biological Conservation* 181 (2014): 236–244, doi:10.1016/j.biocon.2014.10.021.

50. "Mission, Vision and Goals," Citizen Science Association, accessed 17 July 2017, http://citizenscience.org/association/about/vision-mission-goals/.

51. Obama White House, "Implementation of Federal Prize Authority," 12. On winning solutions, see Donald G. MacNeil, "Contest Seeks Novel Tools for the Fight against Ebola," *New York Times*, 12 December 2014, http://www.nytimes.com/2014/12/13/health/ebola-contest-brings-ideas-for-cooling-suits-and-virus-repellents.html?_r=0.

52. "Open Science and Innovation: Of the People, by the People, for the People," 30 September 2015, https://www.youtube.com/watch?t=4&v=J17uBahTdDE (quotes start at 1:06:20). For more on the "Being Policed" project, see http://publicscience project.org/researchers-for-fair-policing/.

53. Raymond Tong and Karim R. Lakhani, "Public-Private Partnerships for Organizing and Executing Prize-Based Competitions," Berkman Center for Internet and Society Research Publication Series, research publication no. 2012–13, https://papers.ssrn.com/sol3/papers.cfm?abstract_id=2083755, 19.

54. "Presidential Memorandum on the White House Office of American Innovation," 27 March 2017, https://www.whitehouse.gov/the-press-office/2017/03/27/presidential-memorandum-white-house-office-american-innovation.

55. "The Trump Administration Supports Fostering Innovation by Leveraging Prizes and Challenges," 20 March 2018, https://www.whitehouse.gov/articles/trump-administration-supports-fostering-innovation-leveraging-prizes-challenges/.

II Critics

Innovation is not an inherent social good. Automation and driverless vehicles have profound trade-offs, including the elimination of blue-collar jobs.[1] Facebook and Twitter erase privacy by design, while foreign governments utilize the platforms to manipulate elections.[2] Innovators, moreover, are not naturally virtuous. Engineers and entrepreneurs devote their talents to building clever smartphone apps even as vital infrastructures in underserved communities—such as Flint, Michigan's lead-tainted water supply—fall apart.[3] Meanwhile Silicon Valley firms are rife with sexual harassment, misogyny, and discrimination.[4]

Yet as recently as five years ago, few publicly questioned the imperative to innovate. It is not hard to see why. The initiatives described in the previous section provide convincing evidence that the nation requires more innovators to meet real societal challenges. By emphasizing individual and collective improvement, champions of innovation receive widespread social and political support. But attitudes are changing as the daily news cycle highlights the negative consequences of innovation and the questionable practices of innovators.

This section brings together experts who critically analyze the assumptions, methods, equity, and efficacy of the innovator imperative. Its contributors work primarily in academic disciplines that include history, science and technology studies (STS), and economics. Some of these critics believe that innovation, properly deployed, can be a force for good. They identify the imperative's flaws to improve the enterprise. A more strident subset of contributors argues that innovation is a destructive and irredeemable ideology. They reject "innovation" as a buzzword; they condemn the economic neoliberalism, racism, and sexism of pro-innovation institutions; and they espouse alternative values such as maintenance, care, and continuity. Overall, these critics invite readers to question their assumptions about innovation. They ask *why* the demand for innovators is so ubiquitous, whether innovation's champions are meeting their promises, and at what cost?

An increasingly common refrain is that innovation is a meaningless concept.[5] Today the term describes everything from "disruptive toothbrushes" to microenergy programs in rural India.[6] Still, "innovation" remains a convenient shorthand because of its conceptual breadth. As we saw in part I, experts ambiguously use the term to signify the diffusion of new ideas, institutional change, and technology commercialization. But what does "innovation" really mean? And *why* is it assumed to be a good thing?

In chapter 9, "How Innovation Evolved from a Heretical Act to a Heroic Imperative," historian Benoît Godin excavates the surprising history of the idea of innovation. From its semantic origins in ancient Greece through the Enlightenment, an "innovation" was an unwelcome novelty that upset the established social order. In fact, the word "innovator" was an epithet that branded someone as a religious heretic or political revolutionary. By the late nineteenth century, innovation began to signal technological and economic progress, but it was only after World War II that economists, politicians, and technologists embraced the concept as an imperative for remaking people, institutions, and nations.

Given the emerging backlash against innovation, Godin's chapter raises the possibility that innovation could return to its original, negative connotation. He implicitly chides those who valorize innovators as the panacea for society's problems; after all, yesterday's innovators were routinely castigated as dangerous subversives. Godin offers no solutions for the current innovation obsession. Moreover, his study privileges innovation's linguistic changes over the social and technological changes the term now describes; thus, practitioners might dismiss his account as an etymological curiosity. Nonetheless, his *longue durée* analysis destabilizes the twenty-first-century conviction that innovation is synonymous with progress.

Critics also question the efficacy and sustainability of innovator initiatives. These programs deploy impressive statistics on the number of students they train and the start-ups they launch. However, a fundamental instability resides beneath these metrics of success. For example, UIF depends on grants and corporate sponsorships that necessitate perpetual fund-raising (Fasihuddin and Britos Cavagnaro, chapter 3). Similarly, federal programs such as I-Corps (Arkilic, chapter 5), which is currently ascendant, and Obama-era citizen science efforts (Gustetic, chapter 7), which are moribund, are vulnerable to changing political agendas.[7] Skeptics ask, *why* should we support innovator initiatives if their track record is so precarious?

In chapter 10, "Failed Inventor Initiatives, from the Franklin Institute to Quirky," historian Eric S. Hintz describes a 150-year pattern of fragility among organizations that support would-be innovators. Hintz uncovers the history of precursors to programs such as Quirky, a recent start-up that combined crowdsourcing with in-house design expertise to "make invention accessible." Until its bankruptcy in 2015, Quirky was hailed as a completely new and participatory approach to innovation. However, Hintz shows that the Franklin Institute, a Philadelphia technical society founded in 1824, offered—but failed to sustain—a remarkably similar set of services. In fact, Hintz finds that nearly every inventors' association since the Franklin Institute has collapsed within a decade of its founding.

Hintz argues that professional communities like Quirky and the Franklin Institute are vital to the success of individual innovators; however, a founder's confidence and claims of novelty are not enough to guarantee an initiative's survival. He asserts that attention to history can help innovators better understand innovation as a social and institutional process, a theme taken up by W. Bernard Carlson in part III (chapter 16). Of course, champions of innovation might retort that the churn of these short-lived initiatives is actually a sign of success, the natural consequence of innovation's "creative destruction."[8] Conversely, the record of failure that Hintz documents may indicate that the innovator imperative is built on false premises.

Critics of innovation also dispute the essentialism of the best practices proffered by innovation experts. Efforts to emulate successful models of local and regional innovation (Feldman, chapter 6) have expanded on an international scale. At the same time, scholars have shown repeatedly that attempts to replicate technological and cultural practices in new settings are always context dependent. Initiatives for implementing American innovation models are complicated by the motivations of local imitators, by the experts hired to implement those models in foreign environments, and by the sheer messiness of change on the ground.[9] *Why*, then, are universal models for cultivating innovation and innovators in high demand?

In chapter 11, "Building Global Innovation Hubs: The MIT Model in Three Start-Up Universities," science studies scholar Sebastian Pfotenhauer describes a global imperative for innovation that looks to the United States for experts and best practices. He explores how technical universities in Singapore, Russia, and the United Arab Emirates have partnered with the Massachusetts Institute of Technology (MIT) to import the university's blueprint for training innovators, founding start-ups, and commercializing new discoveries. However, Pfotenhauer shows that these groups interpreted the same "model" in dramatically different ways: Singapore adopted the MIT model as a radical break from current engineering practices, Russia sought a "counter model" that bridged to its traditional institutions, and the United Arab Emirates sought regional capacity-building for economic development.

Pfotenhauer claims that there are no fixed models of innovation. In fact, he argues that the "MIT model" is not really a model at all but rather an ambiguous ideal that can accommodate the needs of multiple stakeholders while maintaining the legitimacy conferred by the MIT brand. Regardless, the mutability of the MIT model challenges the authority of American innovation expertise and suggests that other "best practices" of innovation may be just as tenuous.

A growing chorus of critics also has questioned the lack of diversity among innovators. In response, most programs that champion innovation have explicit diversity goals to cultivate a more representative cross section of innovators. Yet, despite major demographic shifts in the US population and the ongoing efforts of these groups, technological innovation remains a largely white, male enterprise.[10] *Why* do gender and racial disparities persist in the innovation workforce?

In chapter 12, "The Innovation Gap in Pink and Black," economist Lisa Cook explains that the underrepresentation of women and African Americans in the innovation economy is rooted in entrenched discrimination. For centuries, women and African Americans were denied equal access to education and the technical professions. Drawing on a wealth of empirical data, Cook demonstrates that women and African Americans are less likely to earn an advanced STEM degree, less likely to receive a patent, and less likely to commercialize those patents than their white, male counterparts. Consequently, women and African Americans are less likely to enjoy the higher employment rates, wages, and capital gains of the innovation economy, further exacerbating inequality.

Cook asserts that careers in innovation are pathways to personal wealth, national economic growth, and societal advancement, but that innovation's benefits are not equitably distributed. Her data indicates that the structural origins and impacts of these disparities cannot be solved with rhetorical Band-Aids or surface-level initiatives. She concludes with a call for further research to assist organizations that are working to confront innovation's racial and gender gaps (Sanders and Ashcraft, chapter 17). She is optimistic that innovation can be an engine of progress for women and African Americans; however, her own analysis also lends credence to the critique that the innovation economy is inherently unequal.

The harshest critics of innovation claim that its champions perpetuate a rigged system that privileges disruption and profit over stewardship and the common good.[11] They warn that an obsession with gadgetry and schemes for disruptive change obscure how we actually live and work with technology. We are, in fact, surrounded by and dependent upon infrastructures and legacy technologies such as electrical grids, roads, and sewer systems that require democratic governance and public investment. Innovator initiatives, however, can instill misleading notions about careers in science and technology and direct attention and resources away from the knowledge, practices, and values required for a healthy society. *Why*, then, do we privilege novelty and disruption over alternate values such as maintenance and care?

In chapter 13, "Make Maintainers: Engineering Education and an Ethics of Care," Andrew L. Russell and Lee Vinsel focus on the innovator imperative's corrosive effect on universities and the way it distorts how engineers are trained. Today's engineering students are now required to take courses in entrepreneurship and to complete capstone senior design projects that direct them to invent new things, such as robots and electric cars. Russell and Vinsel warn that these educational trends are misleading students, since most professional engineering work is comprised of the mundane—but critically important—labor of inspections, repairs, adjustments, and incremental improvements to existing systems.

Russell and Vinsel denounce the proliferation of "innovation-speak" and revile the corporate impulses that have co-opted modern universities. They condemn programs like UIF and I-Corps, and argue for alternative values of maintenance and care. In this respect, they anticipate contributors in part III such as Natalie Rusk, who seeks to blend creativity with caretaking (chapter 15), and the "responsible innovation" programs at Arizona State University described by Erik Fisher, David Guston, and Brenda Trinidad (chapter 18). Champions of innovation would counter that maintenance depends upon new innovations. Moreover, they worry that maintenance can devolve from responsible stewardship into a defense of the status quo, or worse, a retrograde obstacle to progress.

By asking *why*—and to what effect—innovation has become a societal imperative, the contributors assembled in this part work against boosterism for innovation. As they question the efficacy of innovator training initiatives and the presumption that innovation is a social, cultural, political, and economic good, they offer differing answers to whether America needs more innovators.

Contributors agree, however, on the reasons why innovation increasingly is on trial. First, innovation experts seem to overpromise and underdeliver. Likewise, innovation's champions are no closer to resolving long-term challenges such as the underrepresentation in the technology economy of women, African Americans, and other minority groups. Finally, as innovation has become a dominant societal goal, innovators appear less as insurgent advocates for progress than as contributors to an economic system that serves only a fraction of the world's population.

Seen together, however, the critics' flaws also become visible. The relatively homogenous backgrounds of innovation's critics—each of this section's contributors holds a PhD and works for a university or research center—invites the charge that they are "ivory tower" academics. Moreover, most are willing to teach in institutions whose missions support the training of innovators. By rejecting innovation as an empty ideology, the harshest critics dismiss its capacity for progressive social change, the very quality that attracts the champions. Finally, these contributors offer several critiques but few practical solutions.

In short, the following chapters emphasize diagnosis over treatment. They reveal how the values, practices, and inequities of innovation are perpetuated. By examining the limitations and fallacies of innovator initiatives, these critics set the stage for reform.

Notes

1. Martin Ford, *Rise of the Robots: Technology and the Threat of a Jobless Future* (New York: Basic Books, 2015).

2. Tim Wu, *The Attention Merchants: The Epic Scramble to Get Inside Our Heads* (New York: Knopf, 2016); Craig Timberg and Elizabeth Dwoskin, "Russian Content on Facebook, Google, and Twitter Reached Far More Users Than Companies Previously Disclosed, Congressional Testimony Says," *Washington Post*, 30 October 2017.

3. Donna Riley, *Engineering and Social Justice* (Williston, VT: Morgan & Claypool, 2008); Steve Kolowich, "The Water Next Time: Professor Who Helped Expose Crisis in Flint Says Public Science Is Broken," *Chronicle of Higher Education*, 2 February 2016, https://www.chronicle.com/article/The-Water-Next-Time-Professor/235136.

4. Anna Wiener, "Why Can't Silicon Valley Solve Its Diversity Problem?" *New Yorker*, 26 November 2016, accessed 17 July 2017, http://www.newyorker.com/business/currency/why-cant-silicon-valley-solve-its-diversity-problem.

5. For example, see Michael O'Bryan, "Innovation: The Most Important and Overused Word in America," *Wired*, accessed 17 July 2017, https://www.wired.com/insights/2013/11/innovation-the-most-important-and-overused-word-in-america/.

6. Alexander George, "Dental Disruption: Is the Toothbrush Ready to Be Reinvented?" *Popular Mechanics*, 27 May 2015, https://www.popularmechanics.com/technology/gadgets/a15706/toothbrush-test/; Bigsna Gill, "Lighting a Billion Lives: A Local Approach to a Global Problem," *Sustainability* 8, no. 5 (2015): 245–253.

7. In 2016, I-Corps' budget increased by over 200 percent. "NSF Announces $8M in New Funding for I-Corps Nodes," SSTi, 12 January 2017, accessed 30 April 2018, https://ssti.org/blog/nsf-announces-8m-new-funding-i-corps-nodes. The OSTP staff, in contrast, is down from approximately 130 under President Obama to approximately thirty under President Trump, and it still lacks a director. Jeffrey Mervis, "Trump's White House Science Office Still Small and Waiting for Leadership," *Science*, 11 July 2017, accessed 30 April 2017, http://www.sciencemag.org/news/2017/07/trump-s-white-house-science-office-still-small-and-waiting-leadership/.

8. According to economist Joseph Schumpeter, "the perennial gale of creative destruction" is the "process of industrial mutation that incessantly revolutionizes the economic structure from within, incessantly destroying the old one, incessantly creating a new one." Schumpeter, *Capitalism, Socialism, and Democracy* (New York: Harper & Brothers, 1942), 81–86.

9. Harry Collins, *Changing Order: Replication and Induction in Scientific Practice* (Chicago: University of Chicago Press, 1985); Sheila Jasanoff, ed., *States of Knowledge: The Co-Production of Science and Social Order* (New York: Routledge, 2004).

10. Adams Nager, David M. Hart, Stephen Ezell, and Robert D. Atkinson, "The Demographics of Innovation in the United States," Information Technology and Innovation Foundation, 24 February 2016, https://itif.org/publications/2016/02/24/demographics-innovation-united-states.

11. Langdon Winner, "The Cult of Innovation: Its Colorful Myths and Rituals," *langdonwinner.com*, 12 June 2017, accessed 10 September 2018, https://www.langdonwinner.com/other-writings/2017/6/12/the-cult-of-innovation-its-colorful-myths-and-rituals.

9 How Innovation Evolved from a Heretical Act to a Heroic Imperative

Benoît Godin

> It were good, therefore, that men in their innovations would follow the example of time itself; which indeed innovateth greatly ... but quietly, by degrees scarce to be perceived.
> —Francis Bacon, *Of Innovation*, 1625

Innovation is a concept that everyone understands spontaneously—or thinks he understands. Every theorist talks about innovation in glowing terms, everyone likes to be called an innovator, every firm claims it innovates, and every government espouses programs to make whole nations innovative. Yet it has not always been so. For most of history, "innovation" has been a dirty word.

There are many words and concepts that we use with no knowledge of their past. Such concepts are taken for granted and their meaning is rarely questioned. Innovation is such an *anonymous concept*.

Today, the concept of innovation is wedded to an economic ideology, so much so that we forget it has mainly been a political—and contested—concept for most of history. Before the twentieth century, innovation did not have anything to do with creativity and progress. And there was no economic or social theory of innovation. The concept instead had a "negative history," to use French historian Pierre Rosanvallon's phrase: a history of contestations, refutations, denigrations, and denials.[1] "Innovation" was a term that conservative opponents of change used to describe deviance—a vice, something explicitly forbidden by law and used as a linguistic weapon by the opponents of change.

The history of "innovation" is an untold story of myths and conceptual confusions that both innovation experts and historians have misunderstood.

Most attribute innovation's conceptual origin to the twentieth-century economist Joseph Schumpeter.[2] Others, following the historian George Pocock, attribute a typology of innovators to the Renaissance statesman Niccolò Machiavelli.[3] A few historians trace innovation's lineage to ancient Greece but fail to distinguish the term's meaning from mere "novelty."[4] Still other scholars suggest that what some called innovation in the past was not real innovation.[5] Throughout these accounts, anachronism is omnipresent. "Social innovation," for example, is often claimed to be a recent alternative to industrial or technological innovation; in fact, the notion of social innovation appeared one hundred years before the phrase "technological innovation."[6]

The history of the concept of innovation raises critical questions about how and why innovation has become a valorized force of social progress. When exactly did the concept of innovation first emerge? How could people of the previous centuries constantly innovate but at the same time deny they innovate?[7] Finally, through what *route* did innovation change in meaning, and why?

Over the last ten years, I have traced the meanings of innovation across centuries.[8] I have searched for antecedents of the modern concept of innovation in Greek and Roman sources from ancient times. I have collected over five hundred documents with titles containing "innovation," from the Reformation to the late nineteenth century, including pamphlets, public speeches, sermons, and legal proclamations. I have also studied hundreds of titles from the twentieth century, when the idea of innovation crystallized in modern theories. In a second phase, I have supplemented these titles with searches through hundreds of other texts online.[9]

In this chapter, I study how thoughts about innovation in early modern society gave rise to innovation theory in the twentieth century. I describe how, when, and why a pejorative word with negative moral connotations shifted to a much-valued concept. I offer a history of the concept of innovation going back to antiquity, a history that takes the *use* of the concept seriously, from polemical to instrumental to theoretical (figure 9.1). I argue that innovation acquired a positive meaning because of its instrumental function to the political, social, and material change that could "create, even sanctify," a progressive future.[10] I further contend that innovation has become a basic value of twenty-first-century society because the concept

Figure 9.1
Frequency of the term "innovation" over time (Google Ngram).

of innovation itself contributes to defining society, both as an idea and in practice.

Subversives and Heretics: 2500 BCE to the Sixteenth Century

The word "innovation" was coined in the late 1200s, but the concept underlying the term originated in both Greek and Roman antiquity with distinct meanings that evolved and intermingled in the preceding millennia (table 9.1). The ancient Greek *kainotomia*, meaning "introducing change into the established order," had a negative political connotation from its very emergence. The word is a combination of *kainos* ("new") and the radical *tom* ("cut; cutting"). It described changes that were subversive, or revolutionary as we say today. Such were Plato's and Aristotle's meanings, the former focusing on cultural innovation (games, music) and its effect on society, and the latter on changes to political constitutions. Aristotle, for example, wrote dismissively of innovation that

> [if] people abandon some small feature of their constitution, next time they will with an easier mind tamper with some other and slightly more important feature, until in the end they tamper with the whole structure.... The whole set up of the constitution [is] altered and it passed into the hands of the power-group that had started the process of innovation.[11]

Certainly there were a few positive uses of the concept in classical Greece. The polymathic scholar Xenophon, for example, interpreted *kainotomia*

Table 9.1
Origins of the Word "Innovation"

	France	England	Italy
Innovation	1297	1297	1364
Innovate	1315	1322	14th century
Innovator	1500	1529	1527

Sources: *The Oxford English Dictionary* (Oxford: Clarendon Press, 1989); O. Bloch and W. Wartung, *Dictionaire étymologique de la langue française*, 5th ed. (Paris: Presses Universitaires de France, 1968); C. Battisti and G. Alessio, *Dizionario Etimologico Italiano* (Florence: Barbèra, 1952); M. Cortelazzo, *Dizionario etimilogico della lingua italiana* (Bologna: Zanichelli, 1979).

literally in his writings on political economy. Xenophon's use of "innovation" is interpreted as "making new cuttings," namely, opening new mine galleries, with the objective of increasing the revenues of the city of Athens.[12] But in general, the concept of innovation appeared infrequently during Greek antiquity and usually in a negative connotation.

"Heresy" ascended to central importance in religious and political life in the Middle Ages and the Renaissance and the concept is central to understanding "innovation." As St. Isidore of Seville (ca. 570–636) put it in *The Twenty Books of Etymologies*, "*Haeresis* is called in Greek from choice…because each one chooses that which seems to him to be the best…since each [heretic] decides by his own will whatever he wants to teach or believe." Isidore continues: "Whoever understands scripture in any sense other than that which the Holy Spirit, by whom it was written, requires…may…be called an heretic."[13] By the early thirteenth century, Robert Grosseteste, bishop of Lincoln and first chancellor of Oxford, gave what became the standard definition of heresy: "[1] an opinion chosen by human faculties, [2] contrary to sacred scriptures, [3] openly held, [4] and pertinaciously defended [preached]."[14]

For a long period in Western history, the innovator was a heretic and called as such.[15] Medieval and Renaissance writers spoke of both heresy and innovation in terms of evil, sickness, and disease; they spoke of innovators as flatterers and seducers eager for novelty. Opponents of both heresy and innovation accused the enemy of similar acts: rebellion, civil wars, instability, and disorder. The vocabulary of royal proclamations against heresy and heretics was similar to that against innovation and innovators.[16] Both heresy and innovation shared the idea of liberty or "private opinion" or "private design," especially in religious conflicts and debates.[17]

Soon this idea of innovation as deviant liberty traveled from the religious to other spheres of society. For example, accusations of "private design" abound in politics, such as the royalist Robert Poyntz on the abuse of parliaments—one of the first political pamphlets to carry a form of "innovation" in the title.[18] In sum, "innovation" was the secularized term for heresy and included the religious, political, and social "heretic" or deviant. The concept served as a linguistic weapon or label in the arsenal of those opposed to change: clerics, monarchists, and conservatives alike.

While the Greek meaning of "innovation" was negative, the concept made its entry into Latin vocabulary with a more positive inflection. In contrast to the Greeks, the Romans had no word for "innovation," although they had many words for "novelty" (*novitas, res nova*). In addition, the verb *novare* carried a pejorative meaning similar to *kainotomia/mein*, depending on the context. Yet from the fourth century, Christian writers and poets coined *in-novo*, which means "renewing," in line with other Christian terms of the time: regeneration, reformation, renovation.[19] *Innovo* had no future connotation as such, although it signaled movement toward a "new order." *Innovo* referred to a return to the past: going back to purity or the original soul.[20]

Revolution and *renewing* are the two poles of a spectrum of meanings that defined innovation in the following centuries, both in dictionaries and lay discourses. *Renewing* pointed to the past (return to the old, changing or renewing the old), and *revolution* pointed to the future (introducing something new, entirely new). For example, Catholic popes in the fifteenth century used "innovation" in a legal context as renewing previous statutes, and Machiavelli did so in the sense of imitation. In spite of his revolutionary political morality, Machiavelli's understanding of "innovation" was introducing new laws similar to those of great rulers of the past. On the other hand, reformers and counterreformers from the sixteenth century used the concept of innovation as an accusation for changing things with "revolutionary" consequences.

Innovation as Polemic: The Sixteenth to the Eighteenth Century

The Reformation was a key moment in the history of the concept of innovation. Catholics accused Protestant reformers of innovating. The Puritans served the same argument to the Anglican Church and accused it of bringing the church back to Catholicism. The word served both sides of the debate—reformers and counterreformers.

Innovation lost its positive valence of renewal when it moved to the politico-religious sphere of the Reformation. As Protestant reformers such as Martin Luther and John Calvin separated themselves from the Catholic Church, royal and ecclesiastical authorities started using innovation in discourse. In 1548, Edward VI, king of England and successor to Henry VIII, issued a *Proclamation against Those That Doeth Innouate*. The proclamation first places innovation in context, then admonishes subjects not to innovate, and finally imposes punishments on offenders:

> Certain private Curates, Preachers, and other laye men, contrary to their bounden duties of obedience, both rashely attempte of their owne and singulet witte and mynde, in some Parishe Churches not onely to persuade the people, from the olde and customed Rites and Ceremonies, but also bryngeth in newe and strange orders...according to their fantasies...is an evident token of pride and arrogance, so it tendeth bothe to confusion and disorder....
>
> Wherefore his Majestie straightly chargeth and commandeth, that no maner persone, of what estate, order, or degree soever he be, of his private mynde, will or phantasie, do omitte, leave doune, change, alter or innovate any order, Rite or Ceremonie, commonly used and frequented in the Church of Englande....
>
> Whosoever shall offende, contrary to this Proclamation, shall incure his highness indignation, and suffer imprisonment, and other grievous punishementes.[21]

The proclamation was followed by the *Book of Common Prayer*, whose preface enjoined people not to meddle with the "folly" and "innovations and new-fangledness" of some men.[22]

It was precisely in the context of the Reformation that the concept entered everyday discourse. The English Puritan Henry Burton was an emblematic writer. Every later argument on innovation would be found in his pamphlet *For God and the King* (1636), the sum of two sermons preached on obedience to God and the king in times of "innovations tending to reduce us to that Religion of Rome."[23] Innovators were those who transgressed the disciplinary order and intended to change it for evil purposes, namely, bringing the Protestant Church back to Catholic doctrine and discipline. Innovating was considered a private liberty—like heresy—that crept imperceptibly and, with time, led to dangerous consequences.[24] Archbishop William Laud and his supporters produced replies that opposed Burton's argument entirely: *we* are not innovating but bringing the church back to purity. Burton was brought to the court, was put into prison, and had his ears cut off.

In the mid-seventeenth century, King Charles prohibited innovation again.[25] The proclamations required bishops to visit parishes to enforce the ban; instructed bishops, university scholars, and schoolmasters to take an oath against innovations; and ordered trials to prosecute the "innovators."[26] Advice books and treatises for princes and courtiers supported this understanding and included instructions not to innovate. Books of manners urged people not to meddle with innovation.

This was only the beginning. First, the pejorative meaning of innovation expanded in the political realm, with monarchists of the seventeenth and eighteenth centuries accusing the republicans of being "innovators." Such was the accusation made against Henry Neville in England and his pamphlet *Plato Redivivus: or, a Dialogue Concerning Government* (1681).[27] Innovation was revolutionary and violent. No republican—not even the most famous Protestant reformers or the French revolutionaries—thought of applying the concept to his own project. In contrast, and precisely because the word was morally connoted, the monarchists used and abused the word and labeled the republican as an innovator.[28] This linguistic practice continued until the French Revolution, when a general disgust of novelty emerged on the idea of innovation.[29]

Second, innovation widened its meaning to the social. The social reformer or socialist of the nineteenth century was called a social innovator, as William Sargant put it in *Social Innovators and Their Schemes*.[30] His aim was to overthrow the social order, namely, private property. Innovation was a *scheme* or *design* in a pejorative sense—much as it was a conspiracy in political literature (described with words such as "project," "plan," "plot," or "machination").[31]

Everyone shared this essentially social and political representation of innovation. Natural philosophers from Francis Bacon onward never referred to innovation as what was certainly the most innovative project in science: the experimental method. Equally, very few artisans and inventors talked of their inventions in terms of innovation.

To the ruling classes, the concept of innovation served to discipline people and regulate society. To writers and pamphleteers, "innovation" was a word used to exploit emotions. In *Studies in Words*, literary scholar C. S. Lewis speaks of a "tendency to select our pejorative epithets with a view not to their accuracy but to their power of hurting…not to inform…but to annoy."[32] He also writes that "we call the enemy not what we think he is but what we think he would least like to be called."[33] From antiquity through most of the nineteenth century, "innovator" was one of the most odious epithets one could hurl at a rival.

Rehabilitating Innovation as Progress: The Eighteenth to the Nineteenth Century

The concepts of innovation and revolution each changed meaning and started to be used in a positive sense at about the same time. The "spirit of innovation," a pejorative phrase of the previous centuries, became one of praise. This occurred gradually over the nineteenth century and received full hearing in the twentieth century.

Two rehabilitations of the concept served the purpose. One was a semantic redescription: people began producing reflexive thoughts on the meaning of innovation and concluded that the concept admitted different interpretations. Innovation was neutral. There were good and bad innovations. Yet innovation was still a word of accusation—the "war cry of the fools," as Jean d'Alembert put it; a "damned word," as the Fourierist Victor Considérant claimed.[34] Yet innovation could also be a good and useful thing. As philosopher Jeremy Bentham wrote in *The Book of Fallacies*:

> *Innovation* means a *bad* change, presenting to the mind, besides the idea of a *change*, the proposition, either that change in general is a bad thing, or at least that the sort of change in question is a bad change.... [But] to say all new things are bad, is as much as to say all things are bad, or, at any event, at their commencement; for of all the old things ever seen or heard of, there is not one that was not once new. Whatever is now *establishment* was once *innovation*.[35]

Here was a second rehabilitation, an instrumental one. During and after the French Revolution, "innovation" repeatedly came to describe a means to political, social, and material *progress*. "We must never fear to innovate, when the public good is the result of innovation," wrote one anonymous commentator, "every century having other morals and new usages, every century must have new laws."[36] The aristocratic revolutionary Charles Pigott similarly argued, "If it had not been for this happy spirit of innovation, what would be the state of mechanics, mathematics, geography, astronomy, and all the useful arts and sciences?"[37]

Nineteenth-century scholars began to rewrite the story of the Reformation and the French Revolution in terms of innovation and even began to speak of innovators in superlative terms—declaring, for example, that the "Government of the Church by bishops is an innovation," the British constitution "owes its beauty to innovation," and "the great charter and the bill of rights are innovations."[38] To Alexis de Toqueville, innovation even became a source of national pride:

> The American must be fervent in his desires, enterprising, adventurous, and above all, innovative. This spirit can be found in everything he does: he introduces it into his political laws, his religious doctrines, his theories of social economy, and his private industry; it remains with him wherever he goes; be it in the middle of the woods or in the heart of cities.[39]

This sentiment was codified by Auguste Comte, who asserted that innovation was a distinguishing quality between animalism and humanity.[40] Finally, writers also discussed the feelings of the people toward innovation. For example, anthropologists looked at how the "primitives" reacted to innovation, as opposed to the moderns.[41] The dichotomy of tradition-innovation/*conservateur-novateur* became a common framework for understanding the past, the present, and the future.

Yet the transition of "innovation" from negative to positive was not sudden. First, the neutral use of the concept coexisted with the pejorative before the nineteenth century.[42] Second, the pejorative use of "innovation" continued to share a place with the positive over the nineteenth century.[43] One had to wait until the twentieth century for a complete reversal in the representation of innovation.

Theorizing and Enhancing Technological Growth: The Twentieth Century to Today

Early in the twentieth century, "innovation" increasingly appeared as a common and positively connoted word in law, education, literature, arts, sciences, medicine, and the social sciences. Innovation began to be cast in terms of a vocabulary of *initiative*, together with *entrepreneurship* and *creativity*.[44]

Two discourses encapsulated this change in a story that is essential to innovation as a phenomenon: a theoretical discourse and a public (government) discourse. Theorists began to study innovation and, in doing so, embraced a eulogistic view of innovation, or "pro-innovation bias," as the sociologist Everett Rogers put it. The aim was to understand innovation in order to serve the practical: the development of strategies and policies to accelerate and get more out of innovation.

Beginning in the 1940s, theoretical thoughts on innovation appeared, and theories of innovation began to multiply. Psychological, sociological, and economically oriented theories followed one after the other from economic historians, anthropologists, sociologists, educators, political scientists, management theorists, engineers, mainstream economists, and evolutionary economists.[45] What had been called "change" (e.g., social change) and "modernization" before now became "innovation." Everyone and everything was studied through the lens of innovation, from the individual to organizations to nations. Innovation was "any thought, behavior, or thing that is new because it is qualitatively different from existing forms," suggested the anthropologist Homer Barnett in one of the very first theories of innovation in the twentieth century.[46] To Everett Rogers, innovation was "an idea perceived as new by an individual" or another "unit of adoption."[47] The definitive transition from the negative to the positive connotation occurred after World War II. Ironically, those governments that contested innovation in the past started promoting innovation and producing reflexive thoughts on innovation as a policy tool. One after another, international organizations and governments embraced innovation as a solution to economic problems and international competitiveness.[48]

At that precise moment, the dominant representation of innovation shifted to that of the economy: *technological* innovation—a phrase that emerged after World War II—as commercialized invention.[49] Technological innovation served economic *growth*. It became a tool to reduce *lags* or *gaps* in productivity between countries and was conducive to industrial *leadership*. Theorists developed a linear model in which basic scientific *research and development* (R&D) led to innovation, and innovation to prosperity.[50] Statisticians developed sophisticated metrics to support the idea—for example, officials administered innovation surveys to firms and collected the results into "innovation scoreboards" that served as so-called evidence-based information to policymakers. Innovation became a basic concept of economic policy. In a matter of decades, science policy shifted from technology policy to *innovation policy*, and indicators on science and technology were relabeled *indicators of innovation*. In all these efforts, academic consultants supported the governments by imagining models of innovation by the dozens as a way to frame and guide policies. *Model* itself became an integral concept in the literature on innovation.

Two theoretical perspectives particularly—economics (technology) and policy—served a new ideology, and the theorists rapidly got the government's hearing. To paraphrase Kevin Sharpe on revolutions, the study of innovation—particularly "innovation studies" (i.e., the management, policy, and economics of innovation) established a cultural dominance that contributed to political discourse.[51] "Innovation studies" became part of the political culture that was essential to its ascendency and was instrumental in its creation and survival.

To be sure, many of the twentieth-century terms of innovation's semantic field—such as "change"—were in place in the previous centuries. But innovation now suggested intentional or *planned change*. It necessitated strategy and investment. Innovation also retained the idea of revolution. There were *major* innovations, so it was said, and they were the most studied innovations because of their revolutionary impacts on society.

In spite of these continuities, a new vocabulary emerged during the twentieth century. Innovation now suggested *originality* in three senses. First, innovation connoted difference, departure.[52] Second, innovation was creativity in the sense of *combination*. Innovation recombined ideas or things in a new way.[53] Third, innovation referred to origin, namely, being *first* to originate (initiate) or use a new practice. To economists, innovators were the first to commercialize a new invention. This connotation owed its existence to the market ideology. For example, the organizational theorist David Teece discussed the "strategies the firm must follow to maximize its share of industry profits relative to imitators and other competitors."[54]

For sociologists, innovation was the first adoption of a new practice within a group or a community; it included the economists' focus on technology, plus a far larger range of practices. This meaning was heavily influenced by governmental institutions' objective of modernizing agriculture and diffusing new techniques among farmers.[55] It gave rise to a whole vocabulary on innovators *versus* laggards.[56]

Both the sociologists' and economists' vocabulary encapsulated the fundamental representation of innovation in the twentieth century. Innovation was a source of revolutionary change (terms used were *major, structural, systemic, paradigmatic*), hence the need to support innovators (*change agents, entrepreneurs*) and turn everyone, even laggards, into innovators. This was an ironic reversal. To Machiavelli, "all human affairs are ever in a state of flux and cannot stand still," and therefore there was a need for (political) innovations to stabilize the world.[57] In contrast, to twentieth-century moderns, the world was too stable and needed revolutionary innovations.

Originality was only one basic concept of the semantic field of innovation. There were also counterconcepts that emerged during the twentieth century. For example, innovation was contrasted with imitation. Imitation was not considered original or creative. When discussing the strategies of firms, technology theorist Chris Freeman limited and contrasted "the traditional strategy" [use of invention] as "essentially non-innovative, or insofar as it is innovative it is *restricted* to the adoption of process innovations, generated elsewhere but available equally to all firms in the industry."[58] To Freeman and his colleagues, innovation "excludes simple imitation or 'adoption' by imitators."[59] Such a view was contested. To a few others, like Charles Carter and Bruce Williams, a firm "may be highly progressive [innovative] without showing much trace of originality [research]. It may simply copy what is done elsewhere.... It is nonsense to identify progressiveness with inventiveness."[60] As Barnett put it, the imitator does something new (if not original) "instead of doing what he is accustomed to do."[61]

Another counterconcept to innovation was *invention*. Invention was the (often mental, sometimes manual) act of creating technology, literature, or art. Innovation was putting invention to work. As Schumpeter, among others, put it, "Innovation is possible without anything we should identify as invention and invention does not necessarily induce innovation."[62] Yet invention played the role of a basic concept to innovation at the same time. While science and innovation were two separate things to natural philosophers of past centuries, they were now part of the same process. Invention (or science or research; these terms are not always distinguished in the literature) was the first step in the *process* of innovation. Innovation started with basic research, followed by applied research, and then development. This view gave rise to what is known as the "linear model of innovation," a much-criticized view but one that remains in the background of policies and theories.[63]

However, the most basic concept of the semantic field was "action" or action-related concepts (box 9.1). Action went hand in hand with another concept, *usefulness/utility*, which is typically described in terms of *progress*

Box 9.1

Innovation as Action

---

introduction. The presenting of something new to the world. This concept first appeared among anthropologists and sociologists but is most popular among economists and management scholars.

application. Assimilation, transformation, exploitation, translation, implementation; applying (new) knowledge in a practical context. Innovation is the application of ideas, inventions, and science.

adoption. Acceptance, utilization, and diffusion; adopting a new behavior or practice. These concepts are mainly used by sociologists.

commercialization. The bringing of a new good to the market. Used concurrently with "introduction" or "application," this concept applies to industrial innovation.

(modernization, advancement, development), *economic growth* (productivity, competitiveness, profits), *organizational efficiency*, and *social needs*.

By the end of the twentieth century, innovation was no longer an individual liberty but a collective process. To be sure, the twentieth century had its individual heroes—namely, the entrepreneurs. Yet entrepreneurs were only one part of the process of innovation: a *total* process as some called it, or a socioeconomic process. Jack Morton, former research director at Bell Laboratories, who brought the transistor from invention to market, and an author of numerous articles and a book on innovation, suggested:

> Innovation is not a single action but a *total* process of interrelated parts. It is not just the discovery of new knowledge, not just the development of a new product, manufacturing technique, or service, nor the creation of a new market. Rather, it is *all* these things: a process in which all of these creative acts, from research to service, are present, acting together in an integrated way toward a common goal.[64]

From the mid-twentieth century, innovation has been studied as a "process," a sequential process in time.[65] Herein lies a semantic "innovation," an innovation that has had a major impact on the modern representation of innovation. Innovation was no longer a thing or a single act but a series of events or activities (called *stages*) with a purpose. The theorists have made themselves "innovative ideologists" and brought in a new definition of innovation, in reaction to earlier ones.[66]

The nuance between "innovation" as a verb and innovation as a process is not as clear-cut as it might appear at first sight. This is not unlike the distinction between "innovation" as substantive or verb. In fact, "innovation" is an abstract word that admits of two meanings: action (introduction of something new) and result/outcome (the new). For example, sociologists use "innovation" as a substantive but focus on the verb (diffusion). Similarly, economists stress the verb form (commercialization). Be that as it may, innovation as a process has contributed to giving the concept of innovation a very large function: innovation encompasses *every* dimension of an invention, from generation (initiation) to diffusion. To the sociologists, the process is one from (individual) adoption to (social) diffusion; to the economists, from invention to commercialization; to management schools, from (product) development to manufacturing. Everywhere, this process is framed in terms of a sequence (with stages) called *models*.

By defining innovation as *process*, it became a counterconcept to science—and more particularly to basic research—as a dominant cultural value of the twentieth century. Technological innovation sprang from a tension between science (for its own sake) and society, or an aspiration to action. It emerged as a category in the twentieth century because in discourse, action, and policy, it was useful to include a larger number of people (other than scientists) and activities (other than science or basic research). Innovation is a *process* that includes several people and activities, so it was claimed. Science or research was only one step or factor in the process of innovation, and often not even a necessary step.

Conclusion

Today, the word "innovation" seeps into almost every sentence.[67] The spontaneous and dominant representation of "innovation" is technological innovation. There also is a profound absence of reflexivity in the imperative to innovate; innovation is always good. Innovators are the panacea to every socioeconomic problem. One need not inquire into the society's problems; innovation is the a priori solution.

The present myths of innovation result from a lack of historical understanding of the *concept* of innovation. Concepts are context-bound.[68] As political philosopher Gordon J. Schochet has suggested, "Civil Societies require common or shared vocabularies that contain their identities and act as centralizing and nearly sovereign forces."[69] These shared vocabularies accrete and change over time.

As the nineteenth century ended, the word "innovation" had accumulated four characteristics that made it a powerful (and pejorative) term. From the Greeks, the representation of innovation had retained its subversive (revolutionary) character. The Reformation added a heretic dimension (individual liberty), and the Renaissance a violent overtone. Together, these characteristics led to a fourth one: innovation as conspiracy (designs, schemes, plots). Yet in spite of these connotations, "innovation" seems to have escaped the attention of intellectual or conceptual historians. Many concepts of change (crisis, revolution, progress, modernity) have been studied in literature, but innovation has not. Is "innovation" only a word—a mere word—in the vocabulary of adherents to the *status quo*—churches, kings, and their supporters—and devoid of sociological meaning?

In a certain sense, it is. Before the twentieth century, no theory of innovation existed. Innovation was a concept of limited theoretical content, a linguistic weapon used against one's enemy. In another sense, innovation is not devoid of sociological meaning. The opponents of innovation in the seventeenth and eighteenth centuries provided the first image of innovation and innovators, one that lasted for centuries. What constitutes innovation and who is an innovator were defined by the enemies of innovation and innovators. It is against this pejorative image or representation that innovators had to struggle in the nineteenth century when they started making use of the concept in a positive sense.[70]

The history of the concept of innovation is not different from that of many other concepts, such as curiosity, imagination, originality, and, in the world of action, revolution.[71] In his study on the idea of happiness, Robert Mauzi suggests that some ideas belong "at the same time to thought, to experience and to dreams."[72] Before the twentieth century, the idea of innovation belonged to experience, but very rarely to thoughts and dreams. The innovator himself makes no use of the word. For centuries it was not innovation itself that shocked humanity but the word describing it.[73] The novelty (the "innovation") of the twentieth century was to enrich the idea of innovation with thought, dreams, and imagination. Innovation took on a positive meaning that had been missing until then, and became an obsession.[74]

Innovation is a synthetizing concept, like civilization, that is defined by way of associations and analogies to existing concepts.[75] Of these, four are fundamental. The first is the concept of change. Intentional change (scheme, design, and the like) gave rise to planned change, which has been a common definition and synonym of "innovation" over the twentieth century. Another concept is heresy, which gave way to thinking about innovation as intention or liberty, and then *initiative* or *initiation*. A third concept is revolution, which led to the idea of revolutionary or major innovation and the metaphor of "creative destruction."[76] A fourth is combination. Before innovation was equated to creativity in the twentieth century, there was *combination*, a concept from philosophy and the doctrine on the association of ideas in the eighteenth century.[77] Combination brings ideas, things, and exciting inventions into a new whole, which is precisely how innovation is defined in many theories today, although usually more as a slogan than a substantial concept.[78]

The changing fortune of innovation over the centuries sheds light on the values of a time. In the seventeenth and eighteenth centuries, the uses of the concept were essentially polemical. It served as a linguistic weapon, attaching a pejorative label to the innovators. In contrast, from the nineteenth century onward, innovation started to refer to central values of modern times: progress and utility. As a consequence, many people started appropriating the concept for their own ends. Innovation became quite a valuable buzzword, a magic word. Yet there is danger here that innovation as a "rallying-cry" may become "semantically null."[79] Like the term "revolution" before it, "innovation" "may soon cease to be current, emptied of all meaning by constant overuse."[80]

Notes

1. Pierre Rosenvallon, *Pour une histoire conceptuelle du politique* (Paris: Seuil, 2003), 43–45.

2. John M. Staudenmaier, *Technology's Storytellers: Reweaving the Human Fabric* (Cambridge, MA: MIT Press, 1985), 56; Norbert Alter, *L'innovation ordinaire* (Paris: PUF, 2000), 8.

3. John G. A. Pocock, *The Machiavellian Moment: Florentine Political Thought and the Atlantic Republican Tradition* (Princeton, NJ: Princeton University Press, 1975).

4. Armand d' Angour, *The Greeks and the New: Novelty in Ancient Greek Imagination and Experience* (Cambridge: Cambridge University Press, 2011).

5. To Anthony Milton, the innovation the English Puritans accused the bishops of in the seventeenth century is not real innovation because it was symbolic or minor, as we say today—a myopia shared centuries ago by, at least, Jacques Bossuet. Anthony Milton, *Catholic and Reformed: The Roman and Protestant Churches in English Protestant Thought, 1600–1640* (Cambridge: Cambridge University Press, 1995); Jacques Bénigne Bossuet, *Opuscules de M. Bossuet, évêque de Meaux*, tome 5 (Paris: Le Mercier; Dessaint & Saillant; Jean-Th. Herissant; Durand; Le Prieur, 1751), 225.

6. Julie Cloutier, *Qu'est-ce que l'innovation sociale?* (Montreal: Crises, 2003).

7. I suggest that the paradox, as David Zaret calls it, is best explained linguistically. Zaret, *Origins of Democratic Culture: Printing, Petitions, and the Public Sphere in Early-Modern England* (Princeton, NJ: Princeton University Press, 2000), 37–43, 254–257.

8. Benoît Godin, *Innovation Contested: The Idea of Innovation over the Centuries* (London: Routledge, 2015).

9. I have used archival databases such as Perseus Digital Library, British History Online, Early English Books Online (EEBO), Eighteenth Century Collection Online (ECCO), Gallica (Bibliothèque Nationale de France), the ARTFL Project, and Google Books (Ngram). This chapter is based on the analysis of these documents, concentrating on documents of English and French origin.

10. Gordon J. Schochet, "Why Should History Matter? Political Theory and the History of Discourse," in *The Varieties of British Political Thought, 1500–1800*, ed. J. G. A. Pocock (Cambridge: Cambridge University Press, 1993), 322. Innovation is a linguistic construct that is maintained, as Reinhart Koselleck says of modern concepts, by continuous expectations toward the future, about how the future should be. Koselleck, *Futures Past: On the Semantics of Historical Time* (New York: Columbia University Press, 2004).

11. This quote is from Aristotle, *Politics*, X, xii, 1316b, trans. R. Rackham, Loeb Classical Library (Cambridge, MA: Harvard University Press, 2005). For Plato's understanding of cultural innovation, see Plato, *Laws*, VII 797b, trans. Trevor J. Saunders (London: Penguin Classics, 2004); Plato, *Laws*, Plato in Twelve Volumes, vols. 10 and 11, trans. R. G. Bury (Cambridge, MA: Harvard University Press; London: William Heinemann, 1967–1968). For Aristotle's interpretation, see Aristotle, *Politics*, trans. T. A. Sinclair, rev. Trevor J. Sauders (London: Penguin Classics, 2004).

12. Xenophon, *Ways and Means*, in *Xenophon in Seven Volumes*, ed. E.C. Marchant, G.W. Bowersock, trans. Constitution of the Athenians (Cambridge, MA: Harvard University Press, 1925). Plutarch's biography of Greeks and Romans is another example of positive uses. Plutarch, *Lives*, trans. Bernadotte Perrin (Cambridge, MA: Harvard University Press, 1919). Additionally, in his *Histories*, Polybius coins *kainopoein*, the meaning of which is "making new," a term that he applies to himself as inventor of

a new kind of history. Polybius, *Histories*, trans. Evelyn S. Shuckburgh (London, New York: Macmillan, 1889).

13. Isidore of Seville, *Etymologies*, in *Heresy and Authority in Medieval Europe*, ed. Edward Peters (Philadelphia: University of Pennsylvania Press, 1980), 49–50.

14. Peters, *Heresy and Authority*, 167.

15. According to James S. Preus, during the Middle Ages "innovation and heresy are practically synonymous.... We frequently find [the innovator and heretic] accusing each other of innovation." Preus, "Theological Legitimation for Innovation in the Middle Ages," *Viator* 3, no. 3 (1972): 2.

16. Paul L. Hughes and James P. Larkin, *Tudor Royal Proclamations* (New Haven, CT: Yale University Press, 1964), 1:57–60, 181–186.

17. See, for example, England and Wales, Sovereign (Elizabeth I), *By the Queene a proclamation for proceeding against Iesuites and secular priests, their receiuers, relieuers, and maintainers* (London: Robert Barker, 1602); England and Wales, *England* (London: Bonham Norton and Iohn Bill, 1626); England and Wales, Sovereign (Charles I), *His Maiesties declaration to all his louing subiects, of the causes which moued him to dissolue the last Parliament published by His Maiesties speciall command* (London: Bonham Norton and Iohn Bill, 1628).

18. Robert Poyntz, *A VINDICATION OF MONARCHY and the Government long established in the Church and Kingdom of England, Against The Pernitious Assertions and tumultuous Practices of the Innovators during the last Parliament in the REIGN of CHARLES the I* (London: Roger Norton, 1661).

19. Gerhart B. Ladner, *The Idea of Reform: Its Impact on Christian Thought and Action in the Age of the Fathers* (Cambridge, MA: Harvard University Press, 1959).

20. The Vulgate—the Latin translation of the Bible—was influential here. In 382, Pope Damasus I commissioned St. Jerome to produce a "standard" version of the *Vetus Latina*, which made use of *innovo* in a spiritual context in the books of Job, Lamentations, Psalms, Wisdom.

21. England and Wales, Sovereign (Edward VI), *A proclamation against those that doeth innouate* (London: In aedibus Richardi Graftoni regij impressoris, 1548).

22. Church of England, *The booke of common prayer and administracion of the Sacramentes, and other rites of the Churche: after the use of the Churche of England* (London: in officinal Edouardi Whitchurche [and Nicolas Hill], 1549).

23. Henry Burton, *For God and the King* (Amsterdam: Theatrum Orbis Terrarum, 1636; Norwood, NJ: W. J. Johnson, 1976).

24. This chain reaction or slippery slope argument goes back to Plato, Aristotle, and Polybius. It was served regularly against innovators, from the Reformation

onward. "All Innovations in Government are Dangerous," wrote an anonymous writer against the English republican Henry Neville. It is "like a Watch, of which any one piece lost will disorder the whole." W.W., *Antidotum Britannicum: or, a counterpest against the Destructive Principles of Plato Redivivus…, against ALL INNOVATORS* (London: Richard Sare, 1681), 172.

25. England and Wales, Sovereign (Charles I), *The King's Proclamation on Religion*, in *The Constitutional Documents of the Puritan Revolution, 1625–1660*, ed. S. R. Gardiner (Oxford: Clarendon Press, 1906).

26. Church of Scotland, *Act of the Commission of the General Assembly, Against Innovations in the Worship of God* (Edinburgh, 1707).

27. Henry Neville, *Plato redivivus, or, A dialogue concerning government wherein, by observations drawn from other kingdoms and states both ancient and modern, an endeavour is used to discover the present politick distemper of our own, with the causes and remedies…*, 2nd ed. (London: Printed for S. I. and sold by R. Dew, 1681).

28. Anonymous, *Antidotum Britannicum*; Thomas Goddard, *Plato's Demon: or, the State-Physician Unmaskt, Being a Discourse in Answer to a Book call'd Plato Redivivus* (London, H. Hill, 1684).

29. See, e.g., François Dominique de Reynaud de Montlosier, *De la monarchie française, depuis son établissement jusqu'à nos jours; ou recherches sur les anciennes institutions françaises, leur progrès, leur décadence, et sur les causes qui ont amené la révolution et ses diverses phases jusqu'à la déclaration d'empire; avec un supplément sur le gouvernement de Buonaparte, depuis ses comencemens jusqu'à sa chute; et sur le retour de la maison de Bourbon*, 3 vols. (Paris: H. Nicolle/A. Édron/Gide fils, 1814).

30. William L. Sargant, *Social Innovators and Their Schemes* (London: Smith, Elder and Co., 1858).

31. This connotation remained in vocabulary until late in the nineteenth century—although some writers discuss social innovation using the positive idea of (social) *reform*. For example, in 1888, a popular edition of the *Encyclopedia Britannica* included a long article on communism which begins as follows: "Communism is the name given to the schemes of social innovation which have for their starting point the attempted overthrow of the institution of private property." *Encyclopedia Britannica*, 9th ed., s.v. "Communism" (New York: Henry G. Allen, 1888), 211–219.

32. C. S. Lewis, *Studies in Words* (Cambridge: Cambridge University Press, 1960, 1967), 323.

33. Ibid., 122.

34. Jean le Rond d'Alembert, *Histoire des membres de l'Académie française, morts depuis 1700 jusqu'en 1771, pour servir de suite aux éloges imprimés & lus dans les Séances*

publiques de cette Compagnie, tome 3 (Amsterdam: Moutard, 1786); Victor Consi-dérant, *Destinée sociale* (Paris: Librairies du Palais-Royal, 1834), 1:312.

35. Jeremy Bentham, *The Book of Fallacies: From Unfinished Papers of Jeremy Bentham* (London: John and H. L. Hunt, 1824), 143–144, 218.

36. Anonymous (Comte de M***), *L'innovation utile, ou la nécessité de détruire les Par-lements: Plan proposé au Roi* (Paris: La Gazette infernale, 1789), translated from the original French.

37. Charles Pigott, *The Jockey Club or a Sketch of the Manners of the Age* (London: M. D. Symons, Paternoster-Row, 1792), 171.

38. François Dominique de Reynaud de Montlosier, *De la monarchie française, depuis son établissement jusqu'à nos jours; ou recherches sur les anciennes institutions fran-çaises, leur progrès, leur décadence, et sur les causes qui ont amené la révolution et ses diverses phases jusqu'à la déclaration d'empire; avec un supplément sur le gouvernement de Buonaparte, depuis ses comencemens jusqu'à sa chute; et sur le retour de la maison de Bourbon*, 3 vols. (Paris: H. Nicolle/A. Édron/Gide fils, 1814); Louis Blanc, *Histoire de la Révolution française* (Paris: Librairie internationale, Lacroix et Cie 1847; 1878); Edgar Quinet, *La Révolution* (Paris: Félix Alcan, 1891); Jean-Marie Dubeuf, *Revue rétrospec-tive des principaux faits, innovations et événements acquis à l'histoire depuis le règne de Napoléon III* (Caen: Emile Alliot et Co., 1866);

39. Alexis de Tocqueville, *De la démocratie en Amérique I* (Paris: Gallimard, 1835; 1992), 201, translated from the original French.

40. Auguste Comte, *Cours de philosophie positive*, tome quatrième (Paris: Bachelier, 1839), 558–559, 636, 642.

41. Arthur Comte de Gobineau, *Essai sur l'inégalité des races* (Paris: Pierre Belfond, 1853).

42. Louis de Rouvroy Saint-Simon, *Mémoires*, tome 11 (Paris: Chéruel, 1697–1700).

43. Hubbard Winslow, *The Dangerous Tendency to Innovations and Extremes in Edu-cation*, delivered before the American Institute of Instruction (Boston: Tuttle and Weeks, 1835); Richard Frederick Littledale, *Innovations: A Lecture Delivered in the Assembly Rooms, Liverpool, April 23rd, 1868* (Oxford: A. R. Mowbray; London: Simp-kin, Marschall & Co., 1868).

44. One of the firsts, if not the first to talk of innovation in terms of "initiative" is Gabriel Tarde in *Les lois de l'imitation* (Paris: Seuil, 1890; 2001).

45. Benoît Godin, *Models of Innovation* (Cambridge, MA: MIT Press, 2017).

46. Homer G. Barnett, *Innovation: The Basis of Cultural Change* (New York: McGraw Hill, 1953), 7.

47. Everett M. Rogers, *The Diffusion of Innovation* (New York: Free Press, 1962), 13; Everett M. Rogers, *Diffusion of Innovation*, 3rd ed. (New York: Free Press, 1983), 11.

48. OECD, *Government and Technical Innovation* (Paris: OECD, 1966); OECD, *The Management of Innovation in Education*, Center for Educational Research and Innovation (CERI) (Paris: OECD, 1969); OECD, *Gaps in Technology: Comparisons between Countries in Education, R&D, Technological Innovation, International Economic Exchanges* (Paris: OECD, 1970); Keith Pavitt and S. Wald, *The Conditions for Success in Technological Innovation* (Paris: OECD, 1971); US Department of Commerce, *Technological Innovation: Its Environment and Management* (Washington, DC: USGPO, 1967); UK Advisory Council for Science and Technology, *Technological Innovation in Britain* (London: HMSO, 1968); Keith Pavitt and W. Walker, "Government Policies towards Industrial Innovation: A Review," *Research Policy* 5 (1976): 11–97.

49. A few exceptions before that date are Thorstein Veblen, *Imperial Germany and the Industrial Revolution* (London: Macmillan, 1915), 118, 128–129; Bernhard J. Stern, "Resistance to the Adoption of Technological Innovations," in *Technological Trends and National Policy*, Report of the Subcommittee on Technology to the National Resources Committee (Washington, DC: USGPO, 1937), 39–66; Joseph A. Schumpeter, *Business Cycles: A Theoretical, Historical, and Statistical Analysis of the Capitalist Process*, vol. 1 (New York: McGraw Hill, 1939).

50. Simon Kuznets, *Six Lectures on Economic Growth* (Glencoe, IL: Free Press, 1959); Keith Pavitt, "Research, Innovation and Economic Growth," *Nature* 200, no. 4903 (19 October 1963): 206–210. This model was first explained in detail in the report on science to the US president from Vannevar Bush, director of the Office of Scientific Research and Development; Vannevar Bush, *Science: The Endless Frontier* (Washington, DC: United States Government Printing Office, 1945)—but without the word "innovation." Economic historian Rupert Maclaurin, secretary to one of the four committees, assisted Bush (William Rupert Maclaurin, *Invention and Innovation in the Radio Industry* [New York: Macmillan, 1949]).

51. Kevin M. Sharpe, *Reading Revolutions: The Politics of Reading in Early Modern England* (New Haven, CT, and London: Yale University Press, 2000), 6–7; Benoît Godin, "'Innovation Studies': The Invention of a Specialty," *Minerva* 50, no. 4 (2000): 397–421; Benoît Godin, "'Innovation Studies': Staking the Claim for a New Disciplinary 'Tribe,'" *Minerva* 52, no. 4 (2014): 489–495.

52. Alphonse de Candolle, *Histoire des sciences et des savants depuis deux siècles, d'après l'opinion des principales académies ou sociétés scientifiques* (Genève: II Georg, 1873), 56.

53. Barnett, *Innovation*.

54. David J. Teece, "Profiting from Technological Innovation: Implications for Integration, Collaboration, Licensing, and Public Policy," *Research Policy* 15 (1986): 285–305.

55. Subcommittee on the Diffusion and Adoption of Farm Practices, *Sociological Research on the Diffusion and Adoption of New Farm Practices: A Review of Previous Research and a Statement of Hypotheses and Needed Research* (Lexington: Kentucky Agricultural Experimental Station and Department of Rural Sociology, University of Kentucky, 1952).

56. Rogers, *Diffusion of Innovation*.

57. Niccolò Machiavelli, *The Prince and the Discourses* (New York: McGraw-Hill Humanities, Social Sciences and World Languages, 1950); see *The Prince* I, 6; see also *The Discourses*, II, preface.

58. Chris Freeman, *The Economics of Industrial Innovation* (Harmondsworth, UK: Penguin Books, 1974), 257.

59. SPRU, *Success and Failure in Industrial Innovation: A Summary of Project SAPPHO* (London: Centre for the Study of Industrial Innovation, 1972), 7.

60. Charles F. Carter and Bruce R. Williams, *Investment in Innovation* (London: Oxford University Press, 1958), 108.

61. Homer G. Barnett, "The Innovative Process," in *Alfred L. Kroeber: A Memorial*, Kroeber Anthropological Society Papers (1961), 25:25–42. Seventy years earlier, Tarde discussed imitation in similar terms: "Le plus imitateur des hommes est novateur par quelque côté." (The most imitative man is to a certain extent an innovator [Tarde, *Les lois de l'imitation*, 46]).

62. Schumpeter, *Business Cycles*, 84–85.

63. Benoît Godin, "The Linear Model of Innovation: The Historical Construction of an Analytical Framework," *Science, Technology, and Human Values* 31, no. 6 (2006): 639–667; Benoît Godin, "In the Shadow of Schumpeter: W. Rupert Maclaurin and the Study of Technological Innovation," *Minerva* 46, no. 3 (2008): 343–360.

64. Jack A. Morton, "The Innovation of Innovation," *IEEE Transactions on Engineering Management* (1968) EM-15 (2): 57–65 (emphasis added).

65. Maclaurin, *Invention and Innovation*, 208; Subcommittee on the Diffusion and Adoption of Farm Practices, *Sociological Research on the Diffusion and Adoption of New Farm Practices*.

66. Quentin Skinner, "Classical Liberty, Renaissance Translation, and the English Civil War," in *Visions of Politics: Regarding Method*, ed. Q. Skinner (Cambridge: Cambridge University Press, 2002), 2:308–343; Quentin Skinner, "Moral Principles and Social Change," in Skinner, *Visions of Politics* (Cambridge: Cambridge University Press, 2002), 1:145–157.

67. Some words, Lewis suggests again, have nothing but a halo, a "mystique by which a whole society lives." Lewis, *Studies in Words*, 282.

68. Quentin Skinner, "Meaning and Understanding in the History of Ideas," *History and Theory* 9, no. 1 (1969): 3–53.

69. Gordon J. Schochet, "Why Should History Matter? Political Theory and the History of Discourse," in *The Varieties of British Political Thought, 1500–1800*, ed. J. G. A. Pocock (Cambridge: Cambridge University Press, 1993), 322, 352.

70. This story is not very different from that of the Enlightenment and its enemies—the *anti-philosophes*—as Darrin McMahon has documented: "Anxiety arose first and foremost from [religion]. Other concerns—civil, political, and economic—flowed from this basic preoccupation." Darrin M. McMahon, *Enemies of the Enlightenment: The French Counter-Enlightenment and the Making of Modernity* (Oxford: Oxford University Press, 2001), 197.

71. Neil Kenny, *The Uses of Curiosity in Early Modern France and Germany* (Oxford: Oxford University Press, 2004); James Engell, *The Creative Imagination: Enlightenment to Romanticism* (Cambridge, MA: Harvard University Press, 1981); Roland Mortier, *L'originalité: une nouvelle catégorie esthétique au siècle des lumières* (Genève: Droz, 1982); Rolf Reichardt, "Révolution," in *Dictionnaire européen des Lumières*, ed. M. Delon (Paris: Presses universitaires de France, 1997), 939–943.

72. Robert Mauzi, *L'idée de bonheur dans la littérature française au XVIIIe siècle* (Paris: Albin Michel, 1979), 9.

73. Reinhart Koselleck, "Begriffsgeschichte and Social History," in *Futures Past: On the Semantics of Historical Time*, ed. R. Koselleck (New York: Columbia University Press, 2004), 75–92.

74. "At certain stages in social evolution, innovation becomes, in turn, its own value." Célestin Bouglé, *Leçons de sociologie sur l'évolution des moeurs* (Paris: Armand Colin, 1922), 113.

75. Brett Bowden, "The Ideal of Civilization: Its Origins and Socio-political Character," *Critical Review of International Science and Political Philosophy* 7, no. 1 (2011): 30.

76. Melvin Lasky suggests that innovation is a precursor term to revolution. I would say instead that innovation (as sudden and violent) simply has connotations of revolution. Melvin S. Lasky, *Utopia and Revolution* (Chicago: University of Chicago Press, 1976), 311.

77. For example, Schumpeter's main concept in the first two editions of *The Economic Theory of Development* (1911 and 1926) is combination—not innovation—combination shifting its characteristics to innovation in the 1934 edition. Schumpeter makes no use of innovation in the German edition of 1911. In the 1926 edition, innovation appears regularly, but as a secondary idea to that of combination. It is novelty of any kind and is used interchangeably in the sense of a "new task," "doing something differently," or simply "something new" and, in one place,

"the function of entrepreneurs." J. A. Schumpeter, *The Theory of Economic Development: An Inquiry into Profits, Capital, Credit, Interest, and the Business Cycle* (Cambridge, MA: Harvard University Press, 1934), 89. This "carrying out of new combinations" is composed of five cases: new good, new method, opening of new market, conquest of a new source of supply, and new organization (66). See also Vilfredo Pareto's instinct of combination in Pareto, *Traité de sociologie générale* (Paris-Genève: Droz, 1968 [1917]).

78. Benoît Godin, "Innovation and Creativity: A Slogan, Nothing but a Slogan," in *Routledge Handbook of the Economics,* ed. Cristiano Antonelli and Albert N. Link (New York: Routledge, 2015).

79. Lewis, *Studies in Words*, 86, 328.

80. John G. A. Pocock, "Languages and Their Implications: The Transformation of the Study of Political Thought," in *Politics, Languages and Time: Essays on Political Thoughts and History* (Chicago: University of Chicago Press, 1989), 3.

10   Failed Inventor Initiatives, from the Franklin Institute to Quirky

Eric S. Hintz

Quirky is a self-proclaimed "social product development" firm based in New York City that turns napkin sketches into finished products. Founded in 2009, Quirky uses the internet to democratize innovation by partnering America's armchair inventors with the wisdom of crowds and the firm's own design expertise. "We started the company to make invention accessible," said Quirky's brash young founder, Ben Kaufman, in 2013. "People come to our site, submit their ideas, and the best new ideas make their way all the way to retail shelves and we do all the heavy lifting in between."[1]

Quirky's best exemplar in this respect is Jake Zien, inventor of Pivot Power, a flexible electrical power strip with outlets that rotate to accommodate large adapters (figure 10.1). In April 2010, Zien, then a college junior, assembled basic sketches and a description of his flexible power strip, joined the Quirky.com community, paid a $99 fee, and submitted his idea online. Within a week, Zien's concept received a favorable evaluation, and Quirky assigned him an "invention ambassador" to shepherd him through the process of developing the power strip into a marketable product.

Over the next several months, Quirky's online community suggested several design and marketing tweaks, including the name Pivot Power. In all, 1,006 "influencers" contributed ideas to improve Zien's concept. Meanwhile, Quirky's staff invested over $1 million in product development—securing the patents, improving the design, arranging for manufacture, and negotiating sales through the retailer Bed Bath & Beyond. Launched in May 2011, Pivot Power quickly became one of Quirky's best-selling products, with 621,589 units sold as of February 2016. In exchange for taking on the risks of development, Quirky kept the lion's share of sales revenues but distributed 10 percent in royalties to its community: 4 percent to

Figure 10.1
Quirky founder Ben Kaufman, left, works with Jake Zien on his new invention, the
Pivot Power flexible power strip, 2011. Photo: Christian Clothier.

Zien, the inventor, with the remaining 6 percent split among the subset of
influencers whose suggested tweaks were actually incorporated into Pivot
Power's design.[2] Zien has made some serious money on the project, banking $827,840 in royalties as of April 2015. He was quick to give credit where
it was due: "I've been very clear that all I had to do was have an idea, and
all the hard work was done by Quirky. I could never have done this myself."
In short, Quirky helped make Zien into a successful innovator.[3]

Stories like Zien's made Quirky the buzzworthy darling of the technical
press, and the company's numbers seemed to back it up. By 2015, Quirky
had attracted approximately $185 million in funding from top-flight Sili-
con Valley venture capitalists (e.g., Kleiner Perkins, Andreessen Horowitz),
partnered with blue-chip product firms (e.g., General Electric, Mattel) and
negotiated sales channels with both big-box stores and e-retailers (e.g., Target, Best Buy, Amazon.com). Quirky had built a virtual community of 1.1
million inventors and influencers, steered by a staff of some 300 employees
from its chic Chelsea offices. In six years, the firm had evaluated 288,733
ideas and launched about 150 products. Like most start-ups, Quirky did not

turn a profit in its first six years, but it earned approximately $100 million in revenues in 2014 alone, while paying out about $10.1 million in royalties since its founding.[4]

Quirky and its advocates promoted the firm as an entirely new and disruptive approach to innovation, the antithesis of traditional corporate R&D, focus groups, and market research. Chief marketing officer Marina Hahn told *Advertising Age* that Quirky was "reinventing invention." Scott Weiss, the Andreessen Horowitz partner who led the firm's $68 million investment, admitted that Quirky's approach to innovation was such an "orthogonal idea" that he wasn't sure how to characterize it. "It has a little of the 'I'm not sure how it works but it does' thing going on."[5]

Quirky's claims of novelty suffer from historical amnesia. Quirky did manage to cleverly build a web-enabled community of independent inventors; in the process, it undoubtedly democratized and streamlined several functions in the process of innovation. However, as this chapter will show, analogs to Quirky's ostensibly novel business practices predated the firm by 185 years. This historical continuity underscores how otherwise independent inventors have always relied on institutions and professional communities to make themselves into commercially successful innovators.

Like many who have failed to remember the past, Quirky was condemned to repeat it. In September 2015, Quirky declared bankruptcy, joining a long list of short-lived independent inventors' associations that never seem to survive. Quirky's rapid demise follows a well-established historical pattern in which dozens of fragile inventors' organizations have emerged and quickly failed, or changed in ways that were detrimental to inventors. Overall, the rise and fall of Quirky (and its predecessors) suggests that American independent inventors have persistently struggled to build and sustain supportive professional communities, a challenge that spans the centuries.

Quirky's Historical Antecedents

In 1907, a mathematician named G. W. Wishard suggested in a letter to *Scientific American* that "some multi-millionaire should found a great institution to aid the worthy inventor." Wishard believed his proposed institution would make inventing a "pleasant and profitable profession for those having inventive genius, for they could devote themselves to pure invention, and leave the manufacturing and business part to those who are better fitted to such purposes." The institution would be well-staffed, with competent "workmen, attorneys, and agents," and crucially, a "board of examiners." An inventor could appear before this board in person or in writing and present his idea. If the concept looked promising, the inventor and the institution would enter into a contract under which the institution would furnish the "tools, machinery, materials, and mechanics to develop the invention." The proposed institution would also "aid the inventor who lacks funds and perhaps business ability by advertising and selling or licensing the rights to persons who will manufacture and sell the goods." In exchange for these "great aids to the inventor," the institution would "reserve a certain share of the revenues," and this would "generally pay it well for its services."[6]

Sound familiar? To give Quirky its due, its impressive combination of web-based crowdsourcing and brick-and-mortar development achieved Wishard's vision much more comprehensively and successfully than any of its predecessors. It was, however, hardly the first to attempt to do so. Starting in the early nineteenth century, a number of inventors' associations provided (or claimed to provide) many of the same services, including professional training, evaluation, patenting, development, manufacturing, and publicity (see box 10.1).[7] Examining these historical associations reveals how organizational mechanisms for achieving those functions reflected their times.

Independent inventors have never lacked for ideas. Rather, they have traditionally stumbled at the point of assembling the necessary investment capital and a community of expertise to develop and commercialize those ideas. In short, inventors past and present have always wondered: Where should I go to figure out what to do with my invention? In 2010, an inventor like Jake Zien could turn to Quirky. In 1910, he might have gone to the Franklin Institute.

The Franklin Institute as an Incubator for Innovators

Philadelphia's Franklin Institute, founded in 1824, remains one of the nation's oldest and most venerable technical institutions. Today, it is largely recognized as a top-flight science museum and honorary society for scientists and engineers. However, during the nineteenth century, the Franklin Institute was among the most important venues in the United States for aiding independent inventors in bringing their ideas to market.[8] The

Box 10.1

Nineteenth- and Early Twentieth-Century Inventors' Organizations

New England Association of Inventors and Patrons of Useful Arts (ca. 1808)

The Franklin Institute of the State of Pennsylvania for the Promotion of the
Mechanic Arts (est. 1824)

The American Institute of the City of New York (1828–ca. 1940)

National Association of Inventors (ca. 1840s)

Inventors' National Institute of Baltimore (ca. 1849)

Inventors', Authors', and Artisans' Grand National Union (ca. 1872)

National Congress of Inventors, with affiliated state organizations (ca. 1874)

National Association of American Inventors (ca. 1878)

Inventors' Institute (ca. 1879)

Inventors' League of the USA (ca. 1889)

American Association of Inventors and Manufacturers (ca. 1891–1902)

International Bureau of Science & Invention (ca. 1900)

Inventors' Guild (ca. 1910–1920)

National Institute of Inventors (ca. 1914–1925)

American Inventors Association (ca. 1928)

National Inventors Congress (ca. 1924–1941)

Inventors' Foundation, Inc. (ca. 1934)

United Inventors and Scientists of America (ca. 1939–1974)

institute provided nineteenth-century analogs of many of the essential services that Quirky now provides for twenty-first-century inventors.

The Franklin Institute was founded at a time when the United States had
enjoyed a half-century of political independence but remained economically dependent on Great Britain for imported technologies and manu-
factured goods. Hoping to stimulate homegrown American industries,
the charter members incorporated the institute for "the Promotion and
Encouragement of Manufactures and the Mechanic and Useful Arts." The
institute proposed to provide public lectures on science and the arts; establish a library and reading room; examine and evaluate new inventions; collect machines, minerals, and other material objects; host exhibitions and
award medals; and publish a journal.[9] The annual dues were modest—eight
dollars in 1895—and attracted a wide spectrum of inventors, mechanics,
industrialists, and scholars, who availed themselves of the institute's many

services.[10] By the institute's centennial in 1924, it boasted 1,378 members, including 563 "nonresident" members from outside the Philadelphia area.[11]

In an era before the widespread establishment of formal universities in the United States, the Franklin Institute provided its members with remarkable opportunities for technical education. Long before the appearance of Quirky's web-based instructional videos (e.g., "Product Design with Quirky—How to Layout Your Sketch"), the institute opened a school of mechanical drawing in 1824, which operated for the next ninety-nine years. Similar to Quirky's weekly Thursday night public evaluations, the Franklin Institute hosted weekly Friday night lectures and new invention demonstrations delivered by members and other invited scientists and inventors (figure 10.2). And just as Quirky has live-streamed and posted its evaluation

Figure 10.2
Inventor and Franklin Institute member Elihu Thomson (1853–1937) and an assistant prepare for an electrical demonstration in the institute's lecture hall, circa 1913. Inventors attended these lectures to stay abreast of the latest technical developments. They also performed demonstrations to attract publicity and financing for their own new inventions. Photo: William N. Jennings. From the Collections of the Franklin Institute, Philadelphia.

videos on its YouTube channel, the Franklin Institute reprinted its lectures and disseminated them widely via the *Journal of the Franklin Institute*.[12]

Philadelphia's local technical community made extensive use of the institute's library, arguably the best technical library in the United States in the nineteenth century. By 1888, the library famously held a nearly complete collection of US and foreign patent reprints and specifications. Unlike Quirky's legal team, the institute did not help inventors secure patents. Nevertheless, as institute secretary William Wahl reported in 1895, "To inventors and manufacturers seeking for information respecting the state of the arts and manufactures, the extensive collection of patent literature … is simply indispensable."[13]

Like Quirky, the Franklin Institute also aided independent inventors by objectively evaluating new inventions. It initially established a five-member Board of Examiners, later known as the Committee on Inventions (1825–1834), then eventually convened a larger, sixty-member Committee on Science and the Arts (or CSA), which continues to this day. The institute's other programs and services—the library, drafting classes, public lectures, collection of models, and exhibitions—had precedents in Europe's mechanics' institutes. However, according to historian A. Michal McMahon, the institute's invention evaluation program was novel and represented the "first substantial attempt by an American organization to direct technological innovation."[14]

Between 1824 and 1900, the CSA formed 2,200 subcommittees to investigate inventions. Approximately one-quarter of these subcommittees were discharged without filing reports, but in some 1,600 cases, inventors received detailed evaluations of their ideas.[15] Often, the CSA's candid reports disappointed inventors. For example, on 16 June 1876, William J. Price of Philadelphia submitted a railroad invention for the CSA's consideration—"an improvement in automatic car couplings"—with an enclosed description and model. A year later, the CSA subcommittee reported that "the car coupling invented by Wm Price … is inferior to many Automatic Car Couplings already in existence," and concluded that "the invention in our opinion cannot be recommended."[16] However, promising ideas received laudatory reports from the CSA, which were subsequently published in the *Journal of the Franklin Institute*. The most meritorious ideas were also awarded cash prizes and medals, serving as an additional endorsement of the invention and further encouragement to the inventor.

In fact, the CSA received applications from several inventors who would go on to achieve high acclaim, including George Westinghouse, Thomas Edison, Elihu Thomson, Nikola Tesla, Elisha Gray, and Rudolf Diesel.[17] According to historian Sydney Wright, many of the CSA's positive evaluations were made "prior to the greatest work of these recipients," so the institute's affirmation provided early encouragement of their inventive work.[18] For example, in 1877, the CSA considered (and eventually endorsed) an electric pen submitted by Thomas Edison.[19] Edison had already achieved a degree of acclaim for his early inventions, including his stock ticker, multiplex telegraph, and an improved carbon telephone transmitter, but by 1877 he had not yet patented his blockbuster phonograph (1878) and incandescent electric lamp (1880).[20] So at an early stage in his career, even America's most decorated and famous inventor found value in submitting his ideas to the CSA for evaluation.

Unlike Quirky, the nonprofit Franklin Institute was not directly engaged in commercializing the new ideas that passed through the CSA. Inventors could nevertheless leverage the CSA's trusted seal of approval and attendant publicity to attract investment capital for their own entrepreneurial ventures. As *Scientific American* noted, if an invention earned one of the CSA's prizes, the "publication of the award [was] made in the journal of the institute, with the result that within a few months the invention [was] brought to the attention of every large corporation in the world maintaining a research laboratory, every organization of scientists, and all the best technical schools and colleges." With this kind of publicity, continued the article, "the greater proportion of the discoveries recognized by [the] Committee on Science and the Arts...have come into wide commercial usage within a relatively short time."[21]

The Franklin Institute also aided inventors by organizing technology fairs and exhibitions to showcase their inventions. Just months after its founding, the institute hosted an exhibition of American manufactures in October 1824, the first of its kind undertaken in the United States. The institute continued to hold fairs either annually or biennially until 1858, and then less frequently through 1899. According to institute historian Thomas Coulson, these exhibitions provided a "medium" connecting "the inventor, the manufacturer, and the consumer," showcasing America's homegrown talent at a time when the institute hoped to stimulate the growth of domestic industries.[22]

At the fairs, the institute's judges evaluated the various technologies and awarded medals and cash prizes to outstanding inventions. Like the CSA's awards, the exhibition prizes and premiums were valuable reputational assets for inventors, who often reproduced renderings of the Franklin Institute medals on their letterhead, product packaging, and printed advertisements.[23] The exhibitions also provided tremendous mass marketing opportunities in an era before national magazines and broadcast media. For example, the 1874 exhibition, marking the institute's fiftieth anniversary, attracted 1,251 exhibitors and 267,638 paying visitors, setting the stage for Philadelphia's enormously successful 1876 Centennial Exhibition. A decade later, the institute's International Electrical Exhibition attracted nearly 300,000 paid visitors to exhibits by some of America's greatest inventors, including Thomas Edison, Alexander Graham Bell, and Frank J. Sprague.[24]

The Franklin Institute Pivots

The Franklin Institute was one of the nation's most esteemed technical societies for its first one hundred years, largely because of the services it offered aspiring inventors. While the institute retains its sterling reputation in the twenty-first century, it operates very differently now. During the first decades of the twentieth century, the institute abandoned many of its technical services, repositioning itself as a museum and purveyor of informal science education for children. These changes saved the Franklin Institute but also deprived inventors of a key professional community and many essential services.

The first change occurred over several decades, as the CSA slowly relinquished its role as an evaluator of emerging technologies. In 1886 and again in 1887, new institute president Charles Banes proposed that the CSA limit itself to evaluating only patented inventions; he feared that the CSA might become embroiled in priority disputes or other damaging litigation. Banes's proposal was initially defeated; however, sometime between 1909 and 1920, the CSA eventually limited itself to examining only patented inventions. While this policy protected the institute from legal entanglements, it also weakened the CSA's influence on the trajectory of emerging technologies. Instead, the CSA adopted a diminished role as a cheerleader and promoter of more established (i.e., patented) technologies.[25]

The proliferation of new endowed prizes and medals also provoked a major shift in the character of the CSA's evaluations. The four medals and certificates that the CSA offered in 1893 gradually ballooned to thirteen by 1958.[26] However, the CSA could no longer rely solely on the volume or quality of outside submissions to annually bestow all of these awards.[27] Over time, the CSA's award criteria shifted away from the merits of an individual technology and toward the recipient's overall record. Eventually, the CSA's awards became something akin to lifetime achievement awards for semiretired inventors. The subcommittees' "investigations" became less concerned with judging the merits of newly submitted inventions and more concerned with nominating appropriate award recipients.[28]

At the turn of the twentieth century, the Franklin Institute also abandoned its long tradition of hosting inventors' fairs and exhibitions. The equipment and infrastructure required to host the exhibitions had become financially burdensome for the institute, which had a history of precarious finances.[29] For example, in consecutive years, the 1884 Electrical Exhibition and the 1885 Novelties Exhibition each ran huge operating deficits (of $6,910 and $9,125, respectively) and nearly bankrupted the institute.[30] By 1895, institute secretary William Wahl had concluded that a local organization such as the Franklin Institute could no longer properly (or profitably) present the breadth and depth of America's inventive talent without some financial support from the government. Instead, Wahl believed that "the great international displays"—government-sponsored exhibitions in Paris (1889, 1900), Chicago (1893), and St. Louis (1904)—were becoming the premier destinations for independent inventors and manufacturers.[31] After mounting some thirty exhibitions since its founding, the institute hosted its final event in 1899, depriving independent inventors of an important showcase.

Instead, the Franklin Institute steadily rebuilt its finances by raising money to construct a new museum, while at the same time recasting itself as a purveyor of popular science education (figure 10.3). On 1 January 1934, the institute moved from its downtown Philadelphia headquarters to a magnificent new museum and planetarium on the Benjamin Franklin Parkway, immediately hailed as "A Wonderland of Science." According to institute historian Sydney Wright, "instead of occasional exhibits," the institute and its museum could now show a "continuous but constantly changing display of the products of industry and invention and of the basic

Figure 10.3
Students visit the Giant Walkthrough Heart exhibit at the Franklin Institute while
a nurse explains heart circulation, 1954. Photo: J. J. Barton. From the Collections of
the Franklin Institute, Philadelphia.

underlying sciences."[32] While the institute had long served an adult population, the new institute was increasingly becoming "a source of science
education for Philadelphia's school children."[33]

While financially wise, the changes in the institute's objectives deprived
inventors of a medium to connect with investors, customers, and fellow
inventors. Granted, the museum did display the work of famous inventors,
such as Linus Yale's pin tumbler lock and Matthias Baldwin's full-sized loco-
motive.[34] However, where the institute's inventors' fairs had been showcases
for cutting-edge, emerging technologies, the museum's permanent exhibits
tended to display technologies so well established that they belonged in
a museum. When the institute ended its long tradition of hosting exhibitions, independent inventors lost an important locus for their professional
activities.

With the slow erosion of inventor-oriented services at the Franklin Institute, commentators such as G. W. Wishard called for new organizations "to aid the worthy inventor," and indeed, several new associations emerged around the turn of the twentieth century (see box 10.1). Some groups, such as the American Association of Inventors and Manufacturers (1891–1902) and the Inventors' Guild (1910–1920), were political organizations that lobbied Congress for changes to the patent laws. Other groups, such as the National Inventors Congress (1924–1941), hosted exhibitions where inventors could meet potential financiers, manufacturers, and customers. Finally, some groups were outright scams. The National Institute of Inventors (1914–1925) ostensibly offered its members impartial evaluations of their new ideas, legal aid for pursuing patents, and financial assistance for marketing their new inventions. But the group's officers simply pocketed the membership dues, embezzling thousands of dollars from America's unsuspecting inventors.[35]

Collectively, these new organizations were extremely short-lived, as each group collapsed within ten to fifteen years of its founding. This situation left inventors frustrated and without the community support and practical services they needed. Moreover, fear and skepticism among inventors probably explains why so few inventors' organizations emerged over the next several decades, and why Quirky seemed so new—and its success so extraordinary—when it burst on the scene in 2009.

Quirky Rises

In 2005, Ben Kaufman was an eighteen-year-old inventor discovering just how difficult it can be to transform an idea into a commercial product. He was an indifferent high school student—"I wanted some way to listen to music that the math teacher wouldn't notice"—so he invented an iPod accessory: retractable headphones concealed in a lanyard. Kaufman persuaded his parents to remortgage their house, taking the $185,000 in seed money to China to find a manufacturer. The "Song Sling" showed well at the 2006 Macworld trade show, and Kaufman's college career took a back seat to building "mophie," his new start-up. However, at Macworld 2007, Kaufman tried something different. Instead of unveiling a new product, he used his booth to host a crowdsourced hackathon, challenging visitors to collectively design and build a new accessory in seventy-two hours. Kaufman distributed pens and paper, hung designs on a clothesline, posted scans of them to the internet, and invited in-person and online visitors to vote on their favorites. The result was the Bevy, an all-in-one case, keychain, and bottle opener for the iPod shuffle.[36]

When mophie's investors forced Kaufman out in August 2007, he took his profits and spent the next year-and-a-half holed up in his Manhattan apartment with a few friends, writing the code for his second start-up, called kluster. Launched in 2008, kluster was a web-based version of his Macworld experiment: a crowdsourcing and collaborative design platform for posting product ideas, voting on concepts, integrating web-based feedback, and dividing up credit among the community. But Kaufman was a bit too early to market. Facebook (2004) and Twitter (2006) were still somewhat new, and end users were still learning how to share, comment on, and "like" ideas on social media. Then, following the 2008 recession, newly unemployed creative people started thinking about how to turn their ideas into new businesses. Three new start-ups launched in 2009, each of which helped democratize the process of invention and new product design: Kickstarter (crowd-funding), MakerBot (3-D printing, rapid prototyping), and Quirky, powered by Kaufman's kluster platform.[37]

In 2013, Kaufman told *Fast Company* that "MakerBot, Kickstarter, and Quirky rose to serve totally different people," each navigating a different branch of the independent inventor's decision tree. According to Kaufman, "MakerBot helps you go from zero to one. Then you're basically left with a choice: 'Do I want to start a business?' If so, raise money; you have Kickstarter. Or if your choice is 'No, I like what I do for a living, I don't want to have to figure out all this crap,' then go to Quirky."[38] Quirky filled a niche by serving independent inventors like Jake Zien who would rather ally with a development partner, license their intellectual property, and collect royalties instead of slugging it out as an entrepreneur.

Quirky Fails

Quirky eventually ran into trouble just six years into its operations, following the pattern of its short-lived, historical predecessors. What happened? By February 2014, Quirky's management team—much like the Franklin Institute's leaders—recognized that the company's core business model had become financially precarious and unsustainable. The firm had spent all but

$50 million of its $185 million in venture capital, while burning through
$5.8 million a month in expenses. In September 2014, Quirky hired Ed
Kremer as its new chief financial officer, and he immediately instituted a
series of cost-cutting measures, including three rounds of layoffs between
November 2014 and February 2015. Kaufman also hosted a live-streamed
town hall meeting for employees and community members to announce
several new strategies. Quirky would stop making so many new products
and instead focus on three categories: connected home, electronics, and
appliances. It would shut down its e-commerce site and sell its products only
through big-box stores and e-retailers. In a new initiative, called Powered by
Quirky, the firm would focus exclusively on partnerships with major brands
such as General Electric and Mattel, using their capital (instead of Quirky's)
to help those firms launch new products. Kaufman also reduced inventors'
and influencers' royalties from 10 percent to between 1.5 and 5 percent,
depending on the retailer. Kaufman believed that making fewer products
for its corporate partners but selling more of them at volume would be a
more sustainable strategy for Quirky and ultimately provide more royalties
for its community. Like the Franklin Institute, Quirky would have to evolve
in order to survive.[39]

Quirky also looked to raise cash by finding new venture investors and
selling off assets—namely, Wink, its smart-home products subsidiary developed in partnership with General Electric. Wink's slate of smartphone-
controllable light switches, thermostats, and air conditioners had generated
$25 million in revenues in its first year. Quirky was hoping to sell the firm
for $30 million when disaster struck. In April 2015, several thousand Wink
home hubs were unable to connect to the internet when a security certificate expired, in what Quirky admitted was a "completely preventable"
error. Quirky was forced to conduct an expensive recall, sending prepaid
mailers to thousands of customers to fix the hubs while simultaneously
removing Wink products from Home Depot's store shelves. More importantly, the recall scared off all potential corporate suitors for Wink as well as
any new venture capital that might have kept Quirky afloat.[40]

By July 2015, Quirky was down to its last $12 million, with Comerica
Bank's $19.9 million line of credit due in October and another $36.8 million in convertible bonds set to mature in December. Quirky suspended
its weekly evaluations, laid off 159 more employees, and searched in vain
for a white knight to purchase the company. Kaufman resigned on 31 July,

and Quirky filed for Chapter 11 bankruptcy protection on 22 September 2015. The firm announced that electronics manufacturer Flextronics would buy Wink for $15 million and that Quirky intended to sell off its crowd-sourcing platform, 3-D printers, tools, and other assets at auction.[41] The announcement generated plenty of animosity. General Electric, for example, complained that the Wink recall and Quirky's poor customer service had damaged its reputation; the firm asked the bankruptcy judge to block the auction of 62,000 Quirky-GE cobranded products in inventory.[42]

Why did Quirky fail? With his usual frankness, Kaufman admitted that Quirky's complex operations—managing a community of 1.1 million inventors, transforming raw ideas into real products, and orchestrating manufacturing and distribution—proved too costly and broke down at scale. Quirky could take calculated risks on a product when it was manufacturing, selling, and shipping only a few hundred units through its e-commerce site. Big-box retailers, however, required Quirky to ship hundreds of thousands of units up front to stock their shelves nationwide; when those products failed to sell, Quirky absorbed huge losses. Quirky also suffered from a sprawling product line, which resulted in brand confusion: it made everything from rubber bands with hooks ($4.99) to web-enabled smart air conditioners ($350). Its emphasis on developing three new products per week meant that Quirky never stopped to refine and improve upon its existing products, many of which garnered poor reviews on Amazon.com and other retail sites. Finally, many of Quirky's products were nonessential at best. Did the world really need an eighty-dollar, web-enabled, smart egg tray?[43]

Quirky's demise dealt a blow to independent inventors worldwide. Quirky had provided a supportive community of like-minded colleagues and a set of essential services for helping move ideas from invention through commercialization. Quirky had also supplied a source of royalty income, now vanished, for independent inventors. Many (former) community members were predictably worried and upset about their ability to recover their intellectual property and owed royalties; an online comment from Mario Riviecchio was typical: "I want my money, I worked long and hard 24/7 [to] influence ideas etc and I'm owed $1400 WHERES MY MONEY AND AM I EVER GOING TO SEE IT."[44]

Other observers hoped that Quirky might somehow be resurrected. For example, in an online forum, a commenter named "Thy" presciently suggested, "Quirky is still a viable business model worth buying. Professional companies will likely understand their failure was not a lack of vision but a lack of planning and improper business management by the staff. This does not mean the ship is a bad one, it simply had a novice crew aboard."[45] Indeed, in December 2015, through a court-approved bankruptcy auction, a Dutch private investment group named Q Holdings LLC paid $4.7 million to acquire Quirky's platform, inventory, and intellectual property. Some observers wondered about Q Holdings' intentions. Had it merely purchased Quirky's IP on the cheap to resell it at a profit? Would it develop the products stuck in Quirky purgatory? Did the firm even intend to engage with Quirky's inventor community?

In a blog post on 8 February 2016, David Hazan, managing partner of Q Holdings, announced his firm's intention "to partner with the community and continue the Quirky brand mission of making invention accessible." The new Quirky, wrote Hazan, would continue to "help talented inventors turn their ideas into commercial success."[46] Quirky's new executives intentionally laid low during the next year and reestablished trust across the community by honoring owed royalty payments. In March 2017, Quirky officially relaunched with a new president, Gina Waldhorn, and a new business model. The firm would cease manufacturing entirely in favor of an "open innovation" model in which established firms could license Quirky's pipeline of crowdsourced product ideas for manufacture and sale under the Quirky name or their own corporate brands.[47]

The business press reserved judgment, but inventors were ecstatic. "Makers rejoice!" wrote Quirky influencer Taron Foxworth. "The return of Quirky is great news for makers and creators." Community member "Eagledancing" simply thanked Quirky for "continuing this dream for us inventors!"[48] Only time will tell if Quirky 2.0 will survive or disappear again like so many of its forerunners.

Conclusion

What can we learn from the stories of Quirky and its predecessors? Can we make some general observations about the nature of communities that cater to independent inventors?

First, by applying the lens of history, we can see that there were nineteenth- and twentieth-century antecedents for most of Quirky's ostensibly novel business practices. When Quirky came onto the scene in 2009, its supporters and a fawning business press characterized the firm as *sui generis*, as "reinventing invention." However, this myopic, presentist view overlooks the history of several organizations that once catered to the needs of independent inventors. Furthermore, a close examination of the Franklin Institute, founded in 1824, demonstrates that this venerable professional society provided nineteenth-century analogs of many of the same services later offered by Quirky (table 10.1). Admittedly, the comparison is not perfect: Quirky is a for-profit consumer products business; the non-profit Franklin Institute supported inventors but stopped short of actually commercializing their inventions. Nevertheless, the Franklin Institute, like Quirky, offered critical services and a supportive community for would-be innovators.

Moreover, examining Quirky in the context of its predecessors underscores the fragile and ephemeral nature of groups founded to support independent inventors. When Quirky declared bankruptcy in 2015 after just six years of operations, its stakeholders—including its investors, business partners, former employees, and unpaid community members—were deeply disappointed, but they should not have been too surprised. Historically, inventors' professional groups founded around the turn of the twentieth century were extremely short-lived.

Among the early inventors' organizations in the United States, only the Franklin Institute remains in operation to this day, and only after evolving in significant ways that were detrimental to independent inventors. If you were to visit the Franklin Institute today, you would not see a technical library, a technology fair, or a drafting class for novice inventors, but only school children visiting a popular science museum and planetarium. While the institute changed in order to survive, inventors were left without the benefit of many essential services. Quirky may yet be reborn in the wake of its bankruptcy, but like the Franklin Institute, it has been forced to evolve and fundamentally restructure its operations.

Finally, the story of Quirky and its predecessors underscores the critical importance of institutions in the success and failure of independent inventors. Successful inventors—both historically and today—do not work alone. They need access to training opportunities and a community of like-minded colleagues to help evaluate, sharpen, and showcase their ideas. Then, in order to move along the spectrum from invention (creating technology) toward innovation (commercializing technology), an inventor

Table 10.1
Services for Independent Inventors at Quirky and the Franklin Institute

	Quirky	Franklin Institute
Lifespan (years active)	6 years (2009–2015, 2017–?)	194 years and counting (1824–present)
Membership	1.1 million (in 2016)	1,378 (in 1924)
Prominent member	Jake Zien	Elihu Thomson
Membership fee	Free to join	$8 annually (in 1895)
Training and professional development	YouTube videos	Technical library, drafting classes, public lectures/demos
Crowdsourcing ideas	Online	In-person, by mail
Idea submission fee	$99, then $10, then free	Free (in 1824) $5 (after 1891)
Evaluation	Online community, Quirky staff, Thursday "Evals"	Committee on Science and the Arts; Exhibition Prize committees
Intellectual property	Staff of attorneys; inventors assign IP to Quirky	Patent law library; IP is inventor's responsibility
Development	In-house by Quirky staff together with the inventor and "influencers"	Responsibility of the inventor
Access to capital, investors	Quirky provides capital from VC, sales revenues, partners	Via exposure to other FI members, through FI exhibitions and the *Journal of the Franklin Institute*
Publicity for the inventor and inventions	Quirky.com; inventor's face on product packaging; press coverage	Public lectures and demos; exhibitions and fairs; *Journal*
Payments, royalties, prizes	Royalties: 4% of revenues for inventors; 6% divided among influencers	Via exhibitions and the CSA: cash prizes and medals for meritorious inventions
End game	Chapter 11 bankruptcy; relaunch	Evolved from professional society for adult inventors into museum for children

needs to work with financiers, patent attorneys, designers, manufacturers, marketers, salesmen, and other business partners. Professional communities such as Quirky and the Franklin Institute serve as crucial intermediaries that connect independent inventors with the people and services required for innovation.[49]

In order to fully understand how innovators are made, we must understand how various professional organizations have crafted cultures for innovators: by providing mechanisms for training inventors, transmitting inventive culture, evaluating ideas, and bringing together the people and resources necessary to commercialize new inventions. Conversely, we must understand the negative impacts when these inventors' communities break down or evolve.[50] Collectively, the struggles of Quirky and its antecedents suggest the persistent difficulty of assembling supportive communities for independent inventors, a challenge that spans the centuries. In the end, we cannot train and encourage innovators unless we can build—and *sustain*—the institutions that support them.

Notes

1. Andy Jordan, "Tech Diary: 'Quirky' Inventions Get a Home Online," *Wall Street Journal*, 24 August 2011, accessed 2 February 2016, http://blogs.wsj.com/digits/2011 /08/24/tech-diary-quirky-inventions-get-a-home-online/; Steve Lohr, "The Invention Mob Brought to You by Quirky," *New York Times*, 14 February 2015, accessed 4 February 2016, http://www.nytimes.com/2015/02/15/technology/quirky-tests-the -crowd-based-creative-process.html; Kaufman quotation in Kevin Chupka, "Quirky Allows Anyone to Become an Inventor," *Yahoo! Finance*, 17 September 2013, accessed 2 February 2016, http://finance.yahoo.com/blogs/breakout/quirky-allows -anyone-become-inventor-160333336.html.

2. Chris Raymond, "How Quirky Turns Ideas into Inventions," *Popular Mechanics*, 7 January 2014, accessed 2 February 2016, http://www.popularmechanics.com /technology/gadgets/a9946/how-quirky-turns-ideas-into-inventions-16344763/; Josh Dean, "Is This the World's Most Creative Manufacturer?" *Inc.*, October 2013, accessed 2 February 2016, http://www.inc.com/magazine/201310/josh-dean/is-quirky -the-worlds-most-creative-manufacturer.html; Lohr, "Invention Mob"; for Pivot Power statistics, see "Pivot Power: Flexible Surge Protector," accessed 17 February 2016, https://invent.quirky.com/invent/243418, login required.

3. Graham Winfrey, "Quirky Makes First Acquisition in Pivot to Serve Corporate Clients," *Inc.*, 13 April 2015, accessed 4 February 2016, http://www.inc.com/graham -winfrey/exclusive-quirky-gets-into-the-acquisition-game-by-purchasing-under

current.html; Raymond, "How Quirky Turns Ideas into Inventions"; Zien quotation in Dean, "Is This the World's Most Creative Manufacturer?"

4. On Quirky's various metrics, see "Quirky: The Invention Platform," accessed 4 February 2016, https://www.quirky.com/; Ruth Simon, "One Week, 3,000 Product Ideas," *Wall Street Journal*, 3 July 2014, accessed 4 February 2016, http://www.wsj .com/articles/one-week-3-000-product-ideas-1404332942; and Jillian D'Onfro, "How a Quirky 28-Year-Old Plowed through $150 Million and Almost Destroyed His Start-Up," *Business Insider*, 29 April 2015, accessed 4 February 2016, http://www.business insider.com/quirky-ben-kaufman-2015-4.

5. John McDermott, "Quirky Wants to Topple CPGs by Turning Ordinary People into Millionaire Inventors," *Advertising Age*, 25 March 2013, accessed 4 February 2016, http://adage.com/article/digital/quirky-topple-cpgs-making-inventors-investors /240486/; Dean, "Is This the World's Most Creative Manufacturer?"

6. G. W. Wishard, "The Need for an Inventors' Aid Institution," letter to the editor, *Scientific American* 97 (6 July 1907): 9.

7. I am grateful to Paul Israel, general editor of the Thomas Edison Papers, for sharing his personal research file on these organizations.

8. For scholarly histories of the Franklin Institute covering the nineteenth century, see Bruce Sinclair, *Philadelphia's Philosopher Mechanics: A History of the Franklin Institute, 1824–1865* (Baltimore: Johns Hopkins University Press, 1974); A. Michael McMahon and Stephanie A. Morris, eds., *Technology in Industrial America: The Committee on Science and the Arts of the Franklin Institute, 1824–1900* (Wilmington, DE: Scholarly Resources, 1977); A. Michael McMahon, "'Bright Science' and the Mechanic Arts: The Franklin Institute and Science in Industrial America, 1824–1876," *Pennsylvania History* 47 (1980): 351–368; Kershaw Burbank, "Noble Ambitions: The Founding of the Franklin Institute," *Pennsylvania Heritage* 18, no. 3 (summer 1992): 32–37.

9. William Wahl, *The Franklin Institute of the State of Pennsylvania for the Promotion of the Mechanic Arts: A Sketch of Its Organization and History* (Philadelphia: Franklin Institute, 1895), 3–7.

10. For the 1895 figure, see "Article IV: Payments," in *Charter and Bylaws of the Franklin Institute*, an appendix of Wahl, *Franklin Institute*, 67.

11. R. W. Lesley, "Report of the Committee on Election and Resignation of Members," 14 January 1925, in *The Franklin Institute Year Book 1924–1925*, 88–89, Franklin Institute Archives, Philadelphia.

12. On the Franklin Institute's drafting school, lectures, and journal, see Wahl, *Franklin Institute*, 19–22, 34–39, 90–91, and Sydney Wright, *The Story of the Franklin Institute* (Lancaster, PA: Lancaster Press, 1938), 16–17. For a sample instructional video, see Quirky with SkillShare, "Product Design with Quirky—How to Layout

Your Sketch," 30 April 2015, accessed 15 July 2016, https://www.youtube.com /watch?v=lpZLEnjnIlY. For several evaluation videos, see Quirky's YouTube channel, accessed 19 February 2016, https://www.youtube.com/user/quirkydotcom.

13. Joseph S. Hepburn, "The Library of the Franklin Institute," *Journal of the Franklin Institute* 269 (March 1960): 221–228; Evelyn S. Paniagua, "American Inventors' Debt to the Institute," *Journal of the Franklin Institute* 247 (January 1949): 1–6; the quotation is from Wahl, *Franklin Institute*, 24–25, 28.

14. A. Michal McMahon, "For the Promotion of Technology: An Historical and Archival Essay on the Franklin Institute's Committee on Science and the Arts," in McMahon and Morris, *Technology in Industrial America*, xiv.

15. Stephanie A. Morris, "A Note on Methodology," in McMahon and Morris, *Technology in Industrial America*, xxxv.

16. CSA case no. 1024, William J. Price, "Railroad Car Couplings," 1876, in Franklin Institute Archives, Philadelphia.

17. See the index of McMahon and Morris, *Technology in Industrial America*, which catalogs all CSA cases from 1824 to 1900.

18. Wright, *Story of the Franklin Institute*, 54.

19. CSA case no. 1085, Thomas A. Edison, "Pen, Electric," 1877, in Franklin Institute Archives. Also, see Thomas A. Edison, "Improvement in Autographic Printing," US patent 180,857, filed 13 March 1876, issued 8 August 1876.

20. Paul Israel, *Edison: A Life of Invention* (New York: Wiley, 1998).

21. William A. McGarry, "First Aid for Inventors: The Franklin Institute and the Committee through Which It Assists Meritorious Patents," *Scientific American*, 13 November 1920, 502.

22. For an overview of the institute's exhibitions, see Wahl, *Franklin Institute*, 40–52, and Wright, *Story of the Franklin Institute*, 21–28, 49–51. The quotations are from Thomas Coulson, *A Short History of the Franklin Institute* (Lancaster, PA: Lancaster Press, 1957), 6–7.

23. Petra Moser and Tom Nicholas, "Prizes, Publicity, and Patents: Non-Monetary Awards as a Mechanism to Encourage Innovation," *Journal of Industrial Economics* 61, no. 3 (September 2013): 763–788.

24. Exhibitor and attendance figures from the 1874 and 1884 exhibitions are from the chairman's reports of those exhibits, as quoted in Wahl, *Franklin Institute*, 40–52. For a list of inventor-exhibitors to the 1884 exhibit, see *Official Catalogue of the International Electrical Exhibition*, 1884, Franklin Institute Archives. For reminiscences by several of the attendees, see "Celebration of the Thirtieth Anniversary of the International Electrical Exhibition, Held in Philadelphia in 1884, under the Auspices of

the Franklin Institute," *Journal of the Franklin Institute* 178 (August 1914): 195–220. For an overview, see Jane Mork Gibson, "The International Electrical Exhibition of 1884: A Landmark for the Electrical Engineer," *IEEE Transactions of Education* E-20 (August 1980): 172–173.

25. On Banes, see McMahon, "For the Promotion of Technology," xxvii–xxx.

26. *Regulations for the Government of the Committee on Science and the Arts*, 1893, 18–30, pamphlet, Franklin Institute Archives. On the proliferation of medals, see *Awards of the Franklin Institute*, 1958, pamphlet, Franklin Institute Archives.

27. In 1909 the CSA established a standing Subcommittee on New Subjects and Publicity to internally suggest new inventions worthy of study. See Article III, section 4, "Regulations of the Committee on Science and the Arts of the Franklin Institute," adopted 15 December 1909, reprinted in the *Journal of the Franklin Institute* 170 (August 1910): 142–150.

28. This change is reflected in looking across two CSA case files devoted to inventor, Franklin Institute member, and General Electric cofounder Elihu Thomson. In 1900–1901, the CSA investigated case no. 2141 concerning Thomson's "Constant Current Arc Light Transformer." The case file included measurements and performance graphs from tests conducted on the transformer, as well as endorsement letters from various cities that had successfully employed the transformer for their municipal lighting. Thomson was duly awarded the institute's John Scott Legacy Medal. In contrast, the subject of Thomson's CSA case no. 2577 in 1912 was "Distinguished Achievement." The subcommittee's report was not technical; it merely awarded the Elliott Cresson Medal for Thomson's "leading and distinguished work in the industrial applications of electricity." See CSA case no. 2141, Elihu Thomson, "Constant-Current Arc Light Transformer," 1901, and CSA case no. 2577, Elihu Thomson, "Industrial Applications of Electricity," 1912, in Franklin Institute Archives, Philadelphia.

29. "The wants of the Institute are vital.... Yet it is embarrassed in every direction.... Every year its accounts present a deficiency. This cannot go on forever." See "Annual Report of Board of Managers for the Year 1887," *Journal of the Franklin Institute* 125 (February 1888): 169–70.

30. Gibson, "International Electrical Exhibition of 1884," 173; "Proceedings of the Annual Meeting, held Wednesday, January 20, 1886," *Journal of the Franklin Institute* 121 (February 1886): 156–159.

31. Wahl, *Franklin Institute*, 51–52; Paul Greenhalgh, *Ephemeral Vistas: The Expositions Universelles, Great Exhibitions and World's Fairs, 1851–1939* (Manchester, UK: Manchester University Press, 2000).

32. Wright, *Story of the Franklin Institute*, 51.

33. Michelle Tucker, "A Partnership for Public Science Education: Reinventing the Franklin Institute, 1925–1934," *Penn History Review* 8, no. 1 (spring 2000): 10.

34. Wright, *Story of the Franklin Institute*, 54, 69, 104.

35. See Eric S. Hintz, "The Post-Heroic Generation: American Independent Inventors, 1900–1950," PhD diss., University of Pennsylvania, 2010, especially chapter 2: "The Professional Lives of American Independent Inventors."

36. Jennifer Wang, "Quirky: The Solution to the Innovator's Dilemma," *Entrepreneur*, 26 July 2011, accessed 8 February 2016, http://www.entrepreneur.com /article/220045; Alice Truong, "Invention Takes a Quirky Turn," *Yahoo! News*, 20 June 2012, accessed 8 February 2016, http://news.yahoo.com/ben-kaufman-s--crazy -journey--to-success.html.

37. J. J. Colao, "Can a Crowdsourcing Invention Company Become 'The Best Retailer in the World?,'" *Forbes*, 27 May 2013, accessed 8 February 2016; http://www.forbes .com/sites/jjcolao/2013/05/09/can-a-crowdsourcing-invention-company-become -the-best-retailer-in-the-world/; Lohr, "Invention Mob Brought to You by Quirky."

38. "The Crowd Takes Over," *Fast Company*, October 2013, 60.

39. D'Onfro, "How a Quirky 28-Year-Old Plowed through $150 Million"; Jillian D'Onfro, "Fresh Funding and More Departures at Quirky, the New York Start-up That Burned through $150 Million," *Business Insider*, 13 June 2015, accessed 22 February 2016, http://www.businessinsider.com/quirky-funding-and-changes-2015-6.

40. Julie Bort, "Ben Kaufman Is Surprisingly Open about Why Quirky Struggled and What He's Doing about It," *Business Insider*, 16 July 2015, accessed 22 February 2016, http://www.businessinsider.com/ben-kaufman-on-why-quirky-failed-2015-7; Julie Jacobson, "Quirky 'Terribly Embarrassed' over Wink Home Automation Hub Recall (Updated)," *CEPro*, 20 April 2015, accessed 22 February 2016, http://www.cepro.com /article/quirky_terribly_embarrassed_over_wink_home_automation_hub_recall/.

41. Stacey Higginbotham, "Quirky Has $12 Million in Cash and Here's What It Wants to Do Next," *Fortune*, 15 July 2015, accessed 22 February 2016, http://for tune.com/2015/07/15/quirky-has-12-million-in-cash-and-heres-what-it-wants-to-do -next/; Stephanie Gleason and Ted Mann, "Invention Startup Quirky Files for Bankruptcy," *Wall Street Journal*, 22 September 2015, accessed 22 February 2016, http:// www.wsj.com/articles/invention-startup-quirky-files-for-bankruptcy-1442938458.

42. Stephanie Gleason and Ted Mann, "GE Says Quirky Has Hurt Its Reputation," *Wall Street Journal*, 3 December 2015, accessed 22 February 2016, http://www.wsj .com/articles/ge-says-quirky-has-hurt-its-reputation-1449179311.

43. Higginbotham, "Quirky Has $12 Million in Cash"; Steve Lohr, "Quirky, an Invention Start-Up, Files for Bankruptcy," *New York Times*, 22 September 2015, accessed 22 February 2016, http://www.nytimes.com/2015/09/23/business/the

-invention-start-up-quirky-files-for-bankruptcy.html; Ruth Simon, "Invention Isn't Easy: Quirky's Hits and Misses," *Wall Street Journal*, 2 July 2014, accessed 22 February 2016, http://blogs.wsj.com/corporate-intelligence/2014/07/02/invention-isnt-easy -quirkys-hits-and-misses/.

44. Stephanie Gleason, "Quirky Inventors Seek Path to Reclaim Products," *Wall Street Journal*, 20 October 2015, accessed 22 February 2016, http://blogs.wsj.com /bankruptcy/2015/10/20/quirky-inventors-seek-path-to-reclaim-products/; online comment by Mario Riviecchio, 17 November 2015, in response to Michael Reed, "As a Quirky Community Member, What Are Your Future Plans?" *QuirkyQuits.com*, 27 September 2015, accessed 23 February 2016, archived at https://web.archive.org /web/20160227180814/http://quirkyquits.com/2015/09/27/as-a-quirky-community -member-what-are-your-future-plans/.

45. Online poster Thy, commenting within the thread "Quirky Sold," 4 November 2015, accessed 23 February 2016, http://quirky.freeforums.net/post/7407, login required.

46. Lindsey Ellis, "Quirky Assets OK'd for Sale," *Schenectady Times-Union*, 11 December 2015, accessed 23 February 2016, http://www.timesunion.com/business /article/Quirky-assets-OK-d-for-sale-6692437.php; David Hazan, "An Introduction," *A Quirky Blog*, 8 February 2016, accessed 23 February 2016, archived at https:// web.archive.org/web/20160223133649/http://aquirkyblog.squarespace.com/home /2016/2/8/quirky-20.

47. Jack Anzarouth, "Quirky Appoints New President, Is Primed for Relaunch," *PRWeb*, 16 March 2017, accessed 11 August 2017, http://www.prweb.com/releases /2017/03/prweb14155691.htm; "June Update: What's Going On," *A Quirky Blog*, 12 June 2017, accessed 19 July 2017, archived at https://web.archive.org/web /20170719113258/http://aquirkyblog.squarespace.com:80/home/2017/6/june -update-whats-going-on.

48. For news coverage, see Stephanie Gleason, "Invention Platform Quirky Re-launches," *Wall Street Journal*, 17 March 2017, accessed 11 August 2017, https://www .wsj.com/articles/invention-platform-quirky-relaunches-1489787519. For inventors' reactions, see Taron Foxworth, "What the Return of Quirky Means for Us," *Medium*, 7 April 2017, accessed 11 August 2017, https://medium.com/stay-connected/what -the-return-of-quirky-means-for-us-42c31d6210d5; Eagledancing, comment from 7 April 2017, on Gina Waldhorn, "A Letter from Our President, Gina Waldhorn," 21 March 2017, accessed 11 August 2017, https://shop.quirky.com/blogs/news /a-letter-from-our-president-gina-waldhorn.

49. Peter Whalley, "The Social Practice of Independent Inventing," *Science, Technology, and Human Values* 16, no. 2 (spring 1991): 208–232; Naomi R. Lamoreaux and Kenneth L. Sokoloff, "Intermediaries in the US Market for Technology, 1870–1920," in *Finance, Intermediaries, and Economic Development*, ed. Stanley L. Engerman, Phillip

T. Hoffman, Jean-Laurent Rosenthal, and Kenneth L. Sokoloff (New York: Cambridge University Press, 2003).

50. To paraphrase historian Tim Lenoir, "I advocate a careful look at the conditions of the production of scientific [and inventive] work in question and the social relations [and institutions] that support it." After all, "the effort to construct disciplines"—or in this case, a community of otherwise independent inventors—"is simultaneously an effort to inscribe supportive structures that sustain a culture." See Lenoir, *Instituting Science: The Cultural Production of Scientific Disciplines* (Stanford, CA: Stanford University Press, 1997), 14, 21.

11 Building Global Innovation Hubs: The MIT Model in Three Start-Up Universities

Sebastian Pfotenhauer[1]

Innovation has become a policy obsession—as have the individuals and institutions responsible for it.[2] Hardly a week passes without a government announcing an innovation strategy for a city, region, or country, or an organization or company branding itself as an innovation leader. From climate change, urban mobility, and energy efficiency to public health, aging populations, poverty, or hunger—innovation is heralded as the solution for virtually every identified challenge.[3] In fact, it has become virtually impossible to talk about economic development or social progress without invoking the need for innovation and innovators in one way or another.[4]

Universities play a central role in this ubiquitous innovation discourse. For one, universities are considered principal sources of the scientific knowledge and technological invention needed to spur innovation. For another, universities are the breeding grounds for innovators and imbue the latter with the necessary knowledge, skills, connections, and mindset to fulfill their destinies as agents of social change and economic competitiveness.[5] Whether in the form of celebrity dropouts such as Bill Gates, Mark Zuckerberg, or Steve Jobs, or as successful academic serial entrepreneurs such as Robert Langer, Leroy Hood, or Michael Bristow, academic institutions have become renowned for creating innovators, which in turn has become an explicit mission of many research universities today. Finally, universities are frequently at the heart of "innovation ecosystems," where they engage with other innovation actors on research and development and provide critical expertise in technological domains. Increasingly, universities have themselves become hotbeds of innovation, whether through dedicated technology transfer units or linkages to industry and government.

Unsurprisingly, therefore, innovative universities have become objects of political desire in their own right, and many innovation strategies around the globe center on universities in one form or another.[6] Yet university-centered innovation strategies are frequently accompanied by the sobering observation that only a handful of universities and regions around the world are able to live up to this promise as innovation engines. As a consequence, policymakers and institutional leaders around the globe increasingly look to those presumed innovation leaders—MIT, Stanford, Technion, and the like—for orientation and authoritative "best practices."[7] Underlying this trend is the belief that the innovation expertise and practices encountered at these institutions are transferable – that is, that they can be distilled, codified, shipped, and grafted onto other institutions elsewhere.[8] Almost without exception, "best-practice transfer" is conceptualized here as a one-way street between a well-defined, successful model-practice on the one end, and an underperforming adopter in need of help on the other.[9]

In this chapter, I challenge this common understanding of "best-practice transfer" in innovation. I explore how three start-up universities in three different countries have sought to emulate best practices from one prominent innovative university—MIT—despite being vastly different contexts: the Singapore University of Technology and Design (SUTD), Masdar Institute of Science and Technology in Abu Dhabi (henceforth "Masdar Institute"), and Skolkovo Institute of Technology ("Skoltech") in Russia. Contrary to what the idea of best-practice transfer commonly suggests, my research finds that these three emulations look extremely different—to the extent that it would be hard to identify a common MIT core among them. While some activities and organizational routines at these start-up universities indeed follow current MIT practice in straightforward ways, many do not. All of them are chosen to fit local constraints and visions, and some do not even exist at MIT or directly oppose MIT practice. Yet all three start-up institutions claim that they follow the "MIT model" and were, in fact, established in collaboration with MIT.

Conventional wisdom would suggest that these start-up universities have failed to emulate the model adequately. I offer a different explanation. A more productive way to understand this apparent contradiction between common identity and divergent practice is to treat the MIT model not as a fixed, well-defined—let alone codified—set of practices, but rather as a flexible *boundary object* that fulfills many other social functions besides providing a template for how to design a university. A *boundary object* straddles different social contexts and accommodates different meanings and imaginaries, while still maintaining a recognizable common identity across these settings.[10] On the one hand, the MIT model is robust enough to serve as a marker of common identity that effectively renders diverse initiatives part of the same model—including, for example, a common language about innovation and its institutional forms, common reference points for success, common aspirations, and a (partly) common network. On the other hand, it is plastic enough to appeal to different sites and accommodate local needs and constraints. At each site, different understandings prevail as to what innovation is, why it is needed, how it ought to be implemented, and how it relates to the existing institutional landscape. At each site, therefore, the model is understood to entail different understandings of "best practice" based on different understandings of what is lacking and different expectations towards a possible solution, which gives the model a local meaning that makes sense in the specific context in which it is being applied. At the same time, the MIT name carries authority within the global political economy of innovation, and its political mobilization enables initiatives that might otherwise not have been possible. In fact, as we shall see, the legitimacy and ability to "do something different" from existing institutional practices is one of the few things that is shared across places and precisely what many of the actors are after. Together, this leads to markedly different emergent institutional forms that nevertheless all see themselves as based on the MIT model.[11]

The lack of standardization of the MIT model and the resulting divergence in "best practices" adopted by the recipient countries are not a weakness of the model, however; rather, they are a strength. The MIT model travels well and can be effective in various contexts *precisely because* it combines a common identity and quasi-universal authority as a model with weak codification and adaptability, achieving concrete local relevance.

This view of the MIT model as a *boundary object* raises a different set of policy questions than commonly associated with "best practice transfer," however—questions that have more to do with articulation of needs, constrains, and legitimacy than with efficacy, similitude, and emulation success. It also changes how policymakers and institutional leaders should think about best-practice transfer as a policy instrument: instead of attempting to define or capture the models or best practices in advance, they should pay attention to the specific ways by which the model becomes locally meaningful. What is more, my research suggests that the success of a "transfer" effort, and of innovation initiatives more generally, cannot be seriously measured in a comparative, one-size-fits-all fashion across different sites. Rather, it needs to be weighed against the idiosyncratic expectations, visions, and constraints encountered in these various societies. This perspective provides a counterpoint to some of the other contributions to this volume, which aim to distill essential characteristics of successful innovative places or people.

Capturing the MIT Model: Similar Desires, Divergent Imaginations

On the face of it, the three start-up universities discussed in this chapter—SUTD, Masdar Institute, Skoltech—have a lot in common in the way they approach best-practice transfer. All three are relatively recent, large-scale initiatives, launched between 2006 and 2013 as national flagship initiatives by their national governments "in collaboration with MIT." Each initiative involves hundreds of people, has a budget of tens to hundreds of millions of dollars, covers multiple scientific areas, and involves a mix of educational, scientific, entrepreneurial, and institution-building activities. Moreover, each initiative represents a limited-time, contractual, "capacity-building" agreement. That is, they are *not* MIT branch campuses and will, in the long run, not be operated by MIT. Rather, MIT is advising and supporting local governments and institutional leaders in the design, implementation, and staffing of these institutions and helping jump-start their operations according to certain predefined goals.[12] At the time of this writing, SUTD completed its initial seven-year engagement with MIT in June 2017 and is currently phasing out collaborative education activities, though research collaboration will continue. Skoltech was initially based on a three-year contract between MIT and the Russian government signed in 2011, which was later stretched to four and a half years and renewed for another three years in 2016, though with reduced intensity and funding. Masdar Institute is currently completing its second five-year phase of collaborative institution-building activities, though the nature of the collaboration has changed considerably as the institution has matured.

So what exactly do these three initiatives seek when striving to build new universities with the help of MIT? In common parlance, MIT has come to epitomize the key role that excellent technical universities can play in innovation and for regional economies when they are embedded in facilitating ecosystems. In the innovation policy literature, MIT is frequently cited as living proof of the central role of universities in innovation and their emergence as heavyweight economic actors that increasingly engage in the creation of proprietary knowledge and research commercialization through spin-offs or licensing. The MIT model has repeatedly been the target of scholarly attempts to abstract it into a theoretical model, most notably in the "entrepreneurial university" and "triple helix" models by Etzkowitz and colleagues.[13] MIT's "impact of innovation" has been documented in regular publications by the Kauffman Foundation, Bank Boston, and MIT itself, which find, for example, that "MIT alumni of both undergraduate and graduate programs have been among the founders of at least 30,000 currently active companies. [MIT] estimate that these enterprises employ 4.6 million individuals and generate annual global revenues of $1.9 trillion, which is roughly equivalent to the GDP of the world's 10th largest economy as of 2014."[14]

Actors at the three start-up universities are keenly aware of these achievements and reference both the tangible impact and the model character (both as a paragon institution and an abstract best-practices model) in their justification for why they turn to MIT. Yet their focus remains highly selective. Much less attention is being paid to the historical circumstances under which MIT came to be the landmark American institution it is today. For example, references to MIT usually show little appreciation of the fact that MIT came out of a particular American land grant college tradition ushered in by the Morrill Act of 1862, or to MIT's strong tradition of basic science that has laid the foundation for much of its innovation success.[15] Instead, all three start-up universities focus primarily on applied and "translatable" engineering research. Likewise, there is little reference to MIT's longstanding engagement with defense research dating back to the development of innovative World War II military technology, which set the stage for a long and successful history of capturing government contract research and establishing close ties to industry. While emissaries from MIT working at the start-up institutions as part of the collaborations might know about these trajectories, the general focus—especially among local stakeholders—tends to be on a set of abstract desirable characteristics around innovation, including aspects of applicability, problem-solving, economic impact, entrepreneurship, "changing the world," and creativity.

As will become evident in the case studies below, this reference to an ostensibly well-understood, well-codified "MIT model" by the three start-up universities does not withstand scrutiny. Actors in the three countries focus on, and import, different practices (and with it different bits of history) from MIT, reflecting different understandings of what "best" means for each context. Perhaps more surprisingly, however, there is also no uniform understanding at MIT itself regarding the key components of its supposed success model. When prompted, MIT faculty and institutional leaders point to a range of different characteristics, from rigid training in math and science; to specific training in innovation and entrepreneurship; to sufficient space for the unfolding of creativity; to a can-do entrepreneurial spirit, aggressive collaboration, high competition, and selectivity; to an engineering tradition founded in basic science; to extreme decentralization; to a multifaceted innovation ecosystem; to close affinity to industry and government.[16] What is more, all MIT faculty and senior administrators interviewed for this research emphasize that there is no formal blueprint for the MIT model, nor is there a prepackaged set of institutional practices ready to be deployed. Consequently, each time MIT is approached by a potential partner, senior administrators and faculty take stock and make different selections based largely on both their own view on MIT and the input of the foreign partner, resulting in utterly different partnership architectures.[17]

Likewise, opinions as to why MIT should (or should not) get involved in these institution-building efforts inevitably vary. Some faculty emphasize the privileged access to research sites and questions that overseas partnerships might provide. Others value the preferential access to some of the best students and researchers in the partner country, hoping that they can attract some to work with them at MIT. Others are attracted by the quasi-experimental character of these initiatives, which allows them to try out new organizational forms or educational approaches that would be difficult to implement at MIT. Some actors emphasize how the partnerships allow MIT to raise substantial additional research funds for both individual faculty projects and institutewide initiatives. Finally, there is a general (though not necessarily explicit) sense among both faculty and administrators that MIT is essentially global in its mission and impact, and that it should aim to "educate students in science, technology, and other areas of scholarship that will best serve the nation and the world" and "bring this knowledge to bear on the world's great challenges."[18]

These divergent stances on the essence of MIT and the goals and risks of institution-building underscore the fact that convergence on a single MIT model is not required for the best-practice transfer logic to work. As I will argue, these divergences only increase when looking beyond MIT to the three different sites.

Singapore University of Technology and Design

In 2010, MIT and Singapore signed an agreement to jointly establish the fourth autonomous university in the city-state: the Singapore University of Technology and Design. SUTD opened its doors in 2012 and, by the time of this writing, has since recruited seven student cohorts of gradually increasing size (from 340 in 2012 to 457 in 2018), in addition to about 160 faculty and teaching staff.

From the outset, SUTD was explicitly conceived to be "something different from the existing institutions," according to Singapore's prime minister, Lee Hsien Loong. The university would provide

> not just an academic education, but one which is going to stimulate students to go beyond the book knowledge, to apply it to solving problems. It will teach students to be creative, not just in technology and design, but also to be creative in bringing ideas out of the academic environment into the real world, into the business arena, into the economy and make a difference to the world.[19]

The prime minister's aspirational message is instructive in that it invokes what seem to be essential characteristics of innovators—the primacy of applied knowledge, problem-solving, creativity, business-oriented ideas, and an ambition to change the world. These terms closely echo the description of MIT given by SUTD's founding president, Tom Magnanti, a former MIT dean of engineering, as

> an organization that bubbles with enthusiasm and has a passion to literally change the world. MIT does this through first-class scholarship, the development of big and important ideas, a deep commitment to educating the most talented students to be found anywhere, and an unwavering commitment to sustaining a culture of innovation, leadership and entrepreneurship.

Magnanti blends this description of MIT into a vision for SUTD:

> Simply put, SUTD's aspirations are no less.... Through creative research and education anchored on technology and design, SUTD aims to create a new type of technically grounded leader and inventor, one fully equipped to address the challenges and issues of today and tomorrow [to achieve its dream of creating] the same kind of magic that MIT and several other of the world's great universities

have achieved in scholarship, innovation, and social and economic impact. My dream is that SUTD will do for Singapore what MIT and Stanford have done for Massachusetts and Silicon Valley, as well as for the U.S. and the World—it will become an intellectual hub and an engine of growth for Singapore and the world.[20]

In this logic, SUTD is intended to reproduce the success of MIT because it emulates essential characteristics of the latter and imbues students with an MIT-like set of attributes.

How does this aspiration to mimic MIT hold up in practice? On the face of it, the institutional differences between SUTD and MIT seem to outweigh the similarities: SUTD's primary focus is on undergraduate education (though this is gradually changing), with a strong emphasis on innovation and design in four thematic "pillars": engineering systems and design, information systems technology and design, engineering product development, and architecture and sustainable design (figure 11.1). This focus on "Big D" design, as SUTD calls it, is seen as the key to producing "a new breed of the brightest technical minds that understands form to design the new innovations of tomorrow."[21] In contrast, MIT has considerably more graduate students than undergraduates, and most of its undergraduate programs are aligned with traditional disciplines, including physics, mathematics, and "classical" engineering domains. Design and systems, while important to MIT's undergraduate education, do not feature as the headline for entire degree programs. In fact, MIT recently discontinued its own engineering systems division as an independent institutional unit. What is more, in contrast to MIT's term and curricular structures, SUTD offers trimesters, a combined "freshmore" year, integrated internship periods, and capstone projects. It merges courses on science, engineering, design, leadership, entrepreneurship, humanities, and art to a degree that goes considerably beyond the humanities requirements at MIT. While some of these features exist in one form or another at MIT, their central role in the identity of SUTD sets the latter apart from the former, raising the question of where, exactly, the MIT model is located in the educational domain.

Closer inspection reveals that both the focus on design and the integrated curriculum reflect key recommendations made by an MIT internal review committee (the Task Force on the Undergraduate Educational Commons, known as the Silbey Committee) and a previous report by the MIT Engineering Council ("From Useful Abstractions to Useful Design") on

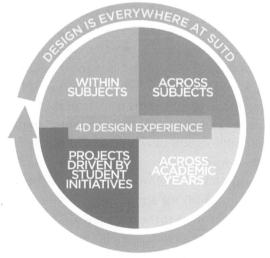

Figure 11.1
Fostering creativity and design-based engineering education through "4D Pedagogy" at SUTD. Graphic from "Towards a Better World by Design" (https://www.sutd.edu .sg/cmsresource/brochure/undergraduate_brochure.pdf). Courtesy of the Singapore University of Technology and Design.

how MIT ought to restructure its own undergraduate education, published shortly before SUTD's inception. These recommendations, however, were never fully pursued and implemented at MIT. According to a senior faculty member at MIT familiar with the process, a large fraction of the faculty at MIT argued for a cautious, conservative stance on the basic structure of the undergraduate curriculum, which in the view of many had proven very successful. Yet the institutionwide initiative for an educational overhaul, and especially the focus on design, had sparked significant interest among some faculty members, some of whom were involved in the ongoing planning processes surrounding Singapore. In fact, many of the faculty engaged in the partnership emphasized that they were interested in SUTD *precisely* as an opportunity to try out some of the practices originally contemplated for MIT. This suggests that SUTD could be best understood as a form of MIT-led experimentation, further complicating the idea of a straightforward best-practice transfer and a unified vision of what an innovation-oriented education curriculum ought to look like.

The prime minister expressed a second aspiration—the need to be "different from existing institutions." Given the small size of Singapore's educational landscape, this mandate can be read to mean "different primarily from the two large public universities," namely, the National University of Singapore (NUS) and Nanyang Technological University (NTU), which until relatively recently represented the entirety of the Singaporean higher education landscape. Both NUS and NTU were central to the rise of Singapore as an economically, scientifically, and technologically highly advanced nation in the 1980s, when the city-state experienced the limits of foreign direct investment-driven growth. Until then, low-wage manufacturing had expanded rapidly but had left the Singaporean workforce poorly educated. As a result, the government made the consequential decision "to phase out its labor-intensive industry and focus on skills-intensive, high-value-added, technology-intensive industries such as electronics manufacturing, data storage, and petrochemicals."[22] Since the mid-2000s, the policy discourse in Singapore has further shifted "from efficiency-driven growth to innovation-driven growth," accompanied by a growing belief among Singapore's leadership that the key factor hampering growth was a lack of creativity, not engineering capability.[23] SUTD fits well within this trajectory, as does the call for "something new."

However, a more complicated picture emerges when we take into account the history of MIT's engagement in Singapore. Following a number of smaller initiatives in the early 1990s, MIT entered into a series of three major engagements with the city-state, starting in 1999 with the Singapore-MIT Alliance (SMA). SMA was designed primarily as an educational collaboration between MIT, NUS, and NTU that would implement graduate programs in Singapore modeled after practices at MIT. SMA utilized such distance-learning models as video lectures, student mobility periods to MIT, a dual degree option, and guided student research with advisors from both MIT and Singapore. SMA was succeeded by SMA2, which marked a shift toward the life sciences, reflecting a broader change in Singapore's self-imagination from "intelligent island to biopolis" at that time.[24] SMA and SMA2 were in turn followed by the Singapore-MIT Alliance for Research and Technology (SMART) in 2007, conceived primarily as a research enterprise. SMART's philosophy, appropriately summarized in a self-description as a "collaboratory," is to bring world-class researchers to Singapore for longer periods to conduct research likely to provide applications for Singaporean priority areas—for instance, future mobility in the city-state or infectious diseases. SMART also adopted MIT organizational practices around innovation, including an innovation center modeled after MIT's Deshpande Center, a translational unit designed to shepherd nascent faculty research into commercialization.

This predecessor history of SUTD casts the mandate "to be different" in a new light: MIT—now enrolled in setting up SUTD—was equally involved in the construction of the very landscape from which SUTD is supposed to deviate. Part of this succession arguably reflects MIT's own evolution over the last two decades, which has produced many new institutional initiatives around innovation. However, this explanation takes us only so far: as we have seen, most of what is distinctive about SUTD—the curriculum structure, the design focus, and so on—does not exist at MIT itself to a similar extent.

A more satisfactory explanation can be found by looking at Singapore's changing *needs* rather than MIT's changing practices. SMA corresponded to the extraordinarily important role that engineering education had been playing in Singapore, and SMART responded to the insight that the small city-state's "small domestic talent group" could not fulfill its research demands alone. SUTD, finally, reflects a new vision of innovation as creativity and design. In this sense, SUTD is but the latest iteration of a process in which Singapore has, time and again, built the MIT model in a way that was useful in the Singaporean context, based on the city-state's changing needs and visions for development at any given time.[25]

Skolkovo Institute of Technology (Skoltech)

On 25 October 2011, MIT and the government-run Skolkovo Foundation in Russia signed a multiyear agreement to establish a new graduate research university on the outskirts of Moscow, to be known as the Skolkovo Institute of Technology, or Skoltech. A flagship initiative of the Medvedev government, Skoltech was soon heralded by politicians and the media as the heart of Russia's "Would-Be Silicon Valley," envisioned "to create, on the European scale, something similar to MIT, prestigious and well designed."[26] Soon, however, Skoltech was also embroiled in political turbulences that followed on the heels of changes in government, which affected funding and public support for the initiative.[27]

Skoltech opened its doors in fall 2012 to an inaugural cohort of twenty students, and it has since scaled up to 450 students, who come from about twenty different countries. At present, the university has about seventy-five faculty members. Like SUTD, Skoltech was conceived as a distinctive counterpoint to the existing university system. It would focus on innovation and entrepreneurship geared toward technology and application, it would be international, and, in contrast to the traditional institutional separation between the Russian Academy of Science and the teaching universities, it would combine both research *and* education. At the same time, Skoltech did not break with the existing system entirely, but extended an olive branch to its national peers. According to Skoltech's mission, it aims to import "international research and educational models [to] integrate the best Russian scientific traditions with twenty-first-century entrepreneurship and innovation" in the form of a "modern, international university."[28]

Skoltech's mission reflects a different set of institutional premises and constraints than those operating at SUTD, pointing to an inherent tension. On the one hand, Skoltech is being established precisely because Russia lacks a modern, international university with a focus on innovation and entrepreneurship. What is more, Russia's proud tradition of science and engineering has undergone two-and-a-half decades of continual decline, and many of the attempts to revive the system or spur innovation have had very limited success.[29] On the other hand, Skoltech is meant to affirm Russian tradition and incorporate what is good about the Russian system. More than that, Skoltech should, according to one senior administrator at the Skolkovo Initiative, ideally also "support other institutions" and act as a "catalyst for reform" in the Russian system. Another administrator interviewed for this research suggests that, in the long run, Skoltech will likely be unable to "survive entirely without the support from other universities and the academies," not even as a presidential initiative.

Skoltech thus has to walk a fine line between institutional differentiation and integration, straddling opposing needs without being usurped by the system. This has several implications for its organizational architecture and how it takes up MIT's practices. For example, Skoltech's research is organized in the form of fifteen integrative Centers for Research, Education, and Innovation (CREIs) with "thematic research missions," such as space, energy systems, hydrocarbon recovery, and data-intensive biomedicine and biotechnology (figure 11.2).[30] These CREIs are reminiscent of one of MIT's hallmark institutional features—mission-oriented research centers. At MIT, this tradition goes back to the Cold War era, when mission-oriented centers with an often decidedly military orientation (including the Research Laboratory for Electronics, the Draper Laboratory, and the Lincoln Lab) received considerable public support. More recently, various integrative, collaborative research centers—such as the Whitehead Institute for Biomedical Research, the Broad Institute, or the Koch Center for Integrative Cancer Research—have been founded in the life sciences.

At Skoltech, the CREIs also provide an interface to the rest of the Russian system. Although Skoltech—a private university with special legal and administrative status that is funded through the designated public "Skolkovo Foundation"—is very explicitly located *outside* this existing system, all CREIs host "distributed collaborative research programs between Skoltech, international, *and Russian institutions.*"[31] That is, they include at least one national research partner, which also receives funding from the Skolkovo Foundation, alongside an international partner (not necessarily MIT). This nationally collaborative model allows Skoltech to avoid explicit confrontations with the existing state universities and academies, and to jump-start research in the form of sponsored "research

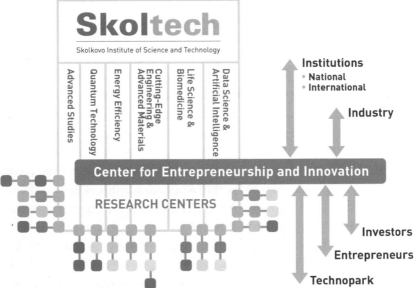

Figure 11.2

Skoltech's integrative Centers for Research, Education, and Innovation (CREIs) are designed to stimulate applied science, entrepreneurship, and institutional change in Russia. On the top, Fyodor Antonov, director of Anisoprint, is a resident of Skolkovo's nuclear technologies cluster. From Shura Collinson, "Skolkovo Residents Seek to Print Satellites Directly in Space," *Sk Skolkovo*, 13 May 2016, http://sk.ru/news/b/news /archive/2016/05/13/skolkovo-residents-seek-to-print-satellites-directly-in-space.aspx. On the bottom is the vision for CEI@Skoltech and its integration into the university. From MIT Skoltech Initiative, "Skoltech CEI Concept," http://75.119.204.15/sktech /sktech-program/entrepreneurship-innovation/concept.html. Courtesy of Skoltech.

collaborations ... between the CREI partner institutions" that allow researchers to make use of partner laboratories while "the Skoltech campus and research laboratories are constructed and equipped," as described by one senior administrator. This national dimension of the CREI concept has no counterpart in SUTD, which, however, features several research centers with international partners.

Another layer of the "fine line" is Skoltech's research orientation. The university chose (or rather, was mandated to pursue) five "presidential" priority areas, most of which correspond to established research fields in Russia: energy, space, nuclear technology, information technology, and biomedicine. These sectors are broad enough to accommodate a host of different subtopics, but they also hint at a desire to continue Russia's traditional areas of research strength, rooted in basic physical sciences and mathematics as well as current sectors of economic strength and competitiveness. Russia remains a nuclear power and space heavyweight. It has entered into technology transfer agreements with several emerging nuclear nations, and its space program has regained ground with the ongoing downsizing of NASA. This partial orientation toward fundamental science is in line with MIT's approach to innovation, which remains thoroughly grounded in a strong commitment to fundamental science. SUTD and Masdar Institute do not share this emphasis. At the same time, it is difficult to imagine a flourishing entrepreneurial scene of young innovators seeking to establish tech start-ups in the nuclear or space fields—not least because of the considerable infrastructure requirements and lead times involved. Information technology, biotech, and energy (at least in the sense of sustainable energy) seem to afford better opportunities. In either case, the contrast to SUTD's vision of the MIT model, which emphasizes creative undergraduates rallying around questions of product manufacturing and design, with little traditional engineering and science, is stark.

Education at Skoltech has a decidedly hands-on focus, modeled after the "Conceive, Design, Implement, Operate" (CDIO) approach originally developed by Skoltech president and former MIT professor Ed Crawley. By offering specialized graduate education programs with a strong research focus, Skoltech in part manages to address one of the biggest challenges of Russian science: the organizational separation of research and higher education between the Academy of Science and teaching universities.[32] But by organizing *all* research primarily in centers (as opposed to departments),

Skoltech arguably also reproduces part of the Academy model. This stands in contrast to MIT's principal organization as a comprehensive research university; at MIT, all centers come *on top of* an otherwise traditional land-grant college structure with five schools and departmental substructures. What is more, at MIT, the centers draw upon, and feed into, a wide range of general and specialized educational programs, both undergraduate and graduate—neither of which is available at Skoltech.

Another characteristic feature of Skoltech is its Center for Entrepreneurship and Innovation (CEI), conceived as the heart for all innovation activity at Skoltech, including innovation education, research, and institutional services (such as tech transfer). In fact, CEI was explicitly designed to combine MIT's various innovation-related units under one roof. CEI draws directly on MIT practice, but it nevertheless stands apart in one important regard: at MIT, innovation activity is essentially distributed across the institution and is not owned or managed by one particular entity, and no single unit has a similar convening authority around innovation. CEI's more centralized, orchestrated approach arguably makes more sense for a small start-up institution; it is also reminiscent of the Russian government's earlier attempts in the late 1990s and early 2000s to boost innovation by creating top-down tech transfer offices, economic zones, technoparks, and the like, many of which continue to struggle because of their prescriptive, technocratic, nonorganic character. In contrast, many MIT faculty consider the "highly decentralized character" and "redundancy" of "many rather loosely coordinated entities" to be one of MIT's strengths. As one faculty member suggests, at times faculty or students who use one of these services are not even aware that other services exist. In this regard, Skoltech also differs from both SUTD and, as discussed below, Masdar Institute. Particularly at SUTD, innovation is more subliminally tied into the notion of design rather than being the explicit target of research and education.

Masdar Institute of Science and Technology (Masdar Institute)

Like other wealthy Middle-Eastern nations, Abu Dhabi and the United Arab Emirates (UAE) have gained a reputation for institutional imports, which include (among others) college education, fine arts, architecture, sports, and international organizations.[33] Some analysts have suggested that the acquisition of these building blocks and trappings of a modern state has happened, in the style of expensive accessories, through direct purchase from prestigious, high-quality providers. The choice of MIT as a partner to build Masdar Institute seems to fit this pattern. As a member of the Masdar faculty suggests,

> Well, the best way to transform an economy, what you do is just buy a company, build a factory. What do you do… if it is a knowledge-based economy that you'd like to build? Then you need to build a nucleus for generating knowledge which is innovative.… This nucleus of this Masdar Initiative is the Masdar Institute, the university, … and they went to the best in the business, which is MIT, to build a technology-based university that would be the nucleus of this transformation.

However, to think of Masdar Institute as a mere turnkey import modeled solely after MIT would do injustice to the initiative and its idiosyncrasies. Understanding the role of the MIT model in creating the Masdar Institute requires a detour into the rationale for the much larger national development project of which it is part: Masdar, also known as the Abu Dhabi Future Energy Company. One of several high-profile science and innovation initiatives currently underway on the Arabian Peninsula, Masdar was launched in 2006 to spearhead Abu Dhabi's evolution as "a leader in global energy," driving the transition in the UAE "from a provider of fossil fuels to a developer of alternative energy and clean technologies" and acting as "a catalyst for the economic diversification of the emirate, … driving new sources of income for the emirate and strengthening Abu Dhabi's knowledge-based economic sectors."[34] Since the 1960s, oil and gas exports have fueled the breakneck pace of development in the UAE. This dependence on the fossil fuel economy has made the country highly vulnerable to the boom-and-bust cycles of the hydrocarbon commodity markets and has led to a range of critical sustainability challenges—both economically and environmentally. Besides the looming threat of limited oil and gas resources, the country also features a food sector in which most food is imported and water is desalinated on a daily basis. It produces one of the highest carbon footprints per capita in the world due to challenges in agriculture, housing, and transportation in the desert, and a high economic dependency on foreign companies, labor, and technical expertise.

Masdar is the government's principal response to this set of interrelated sustainability challenges. At an operational level, the Masdar Initiative has four prongs: First, at its center sits Masdar Institute, the first graduate-level research university in the UAE, intended as a training ground and magnet for local and global talent, as well as a producer and accumulator of expertise around sustainability technologies (figure 11.3). Surrounding the institute, second, is Masdar City, an urban development project intended as a testing ground and living laboratory for innovative technologies emanating from Masdar Institute. Masdar City is imagined as a carfree place fully reliant on renewable energy sources, with a low carbon and waste footprint, home to some 50,000 highly skilled scientists, engineers, and entrepreneurs as well as hundreds of firms specializing in energy and sustainability technology, manufacturing, and investment.[35] The third prong, Masdar Clean Energy, is a major developer of large-scale sustainable energy projects in the region and is an intended major customer of energy technology emerging from MIT. Finally, Masdar Capital is a globally active (though regionally focused) venture capital fund focused on technology commercialization, primarily in clean technologies. Together, these four components represent a massive research-to-demonstration-to-valorization-to-habitation pipeline—an innovation-economy-in-a-box, if you will—that is at once an ambitious institution-building project and a touchstone for the socioeconomic and environmental future of an entire region.

What existed in 2006, then, was an extremely ambitious vision of Masdar as the key to the economic, technological, environmental, and social future of the UAE. What did not exist, however, were any R&D infrastructure, scientific or technological expertise, or the necessary human resources to produce the very technologies that would drive this envisioned high-tech economy. As an MIT faculty member laconically noted, "Masdar Institute was meant to solve that."

Masdar Institute entails a specific Emirati vision markedly different from that of SUTD and Skoltech. Instead of creating something different from an existing university landscape, it represents the country's first attempt at building an advanced research university. Instead of overcoming established educational and scientific practices, it was to introduce them for the first time. Instead of training innovators in a broad range of scientific fields and with a generic purview, it exclusively targets advanced energy and sustainability technology.

This focus on a specific domain is visible in Masdar Institute's education portfolio. The institute offers nine master's degree programs and, as of 2014, an interdisciplinary doctoral program. Although all of the education tracks are classical engineering degrees (e.g., chemical engineering, computer and

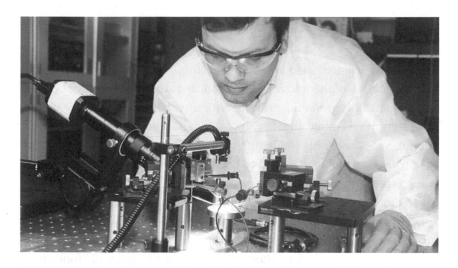

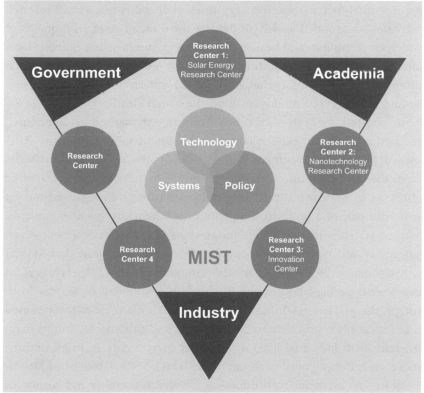

Figure 11.3

On the top is Jaime Viegas, assistant professor of microsystems engineering at Masdar Institute, working on his joint project with MIT in February 2017 to use sunlight to turn water into fuel in the form of hydrogen. Photo: Tahra Al Hammadi, Masdar Institute News. On the bottom is the triple helix of Masdar Institute, from the 2012–2017 strategic plan. Diagram courtesy of Masdar Institute.

information science, mechanical engineering, water and environmental engineering), all of them have been customized around the theme of energy and sustainability, deviating from the MIT sister programs that were used as blueprints. For example, the learning goals of the chemical engineering program include "an ability to identify and address current and future chemical engineering problems related to energy sources, generation, conversion, and green chemical production within a broader framework of sustainable development," and "an ability to apply a multidisciplinary approach to conceive, plan, design, and implement solutions to chemical engineering problems in the field of energy and sustainability."

In terms of research, too, Masdar Institute focuses primarily (one could argue, entirely) on advanced energy and sustainability matters. Its research activities are grouped loosely under the three broad research domains of water, environment, and health; future energy systems; and microsystems and advanced materials. Both education and research are therefore much more focused thematically than at SUTD or Skoltech (let alone MIT), representing a highly mission-driven, applied research ideal.

In contrast to both SUTD and Skoltech, Masdar's mission-oriented research is tied to an explicit corporate agenda. A subsidiary of the state-owned Mubadala Investment and Development Company, Masdar has from the start been an economic development project—not a scientific effort per se. Masdar Institute is officially a not-for-profit organization, yet it is "still part of a profit-making enterprise," a faculty member explains. Mission statements that include phrases like "catalyst for economic diversification" and "source of income for the emirate" suggests that Masdar will ultimately be evaluated by its economic returns.[36] The corporatist character is perhaps best exemplified by the enrollment of Masdar Institute in the MIT Energy Initiative (MITEI), MIT's main cross-departmental unit for energy research (both conventional and sustainable) and its main interface with industrial R&D interests on energy matters. In an unusual move, and after extended negotiations with MIT, Mubadala enlisted Masdar Institute—an academic institution—as a *corporate* member and cosponsor of MITEI, where it stands out among the other multinational firms, such as Shell, BP, or Saudi Aramco, that are affiliated with MITEI. Yet this role is consistent with Mubadala's corporate understanding of Masdar, which officially calls Masdar Institute its "research arm."[37] In other words, the government of Abu Dhabi funds and uses an academic institution (Masdar

Institute) as the research arm of a state-owned company (Mubadala) to drive and diversify the economy and advance technological development for urban and energy infrastructure projects.

Masdar Institute thus seems to be less concerned with reproducing a comprehensive MIT model than with leading a very targeted, foundational effort in capacity-building for national development. In this way, Masdar is reminiscent of MIT's global engagements from the 1960s to the 1990s in such locations as India, Iran, Egypt, Saudi Arabia, Thailand, or Argentina, although without the overt political agenda of a Cold War context.[38] Moreover, MIT was also brought into Abu Dhabi to build local expertise in designing, implementing, ramping up, and operating research facilities and associated policy infrastructure that had no precedent in the country (or elsewhere under similar conditions, for that matter).[39] Masdar Institute has become a vehicle for the government to develop certain policy frameworks and regulations required for advanced research that simply had not been relevant in the country before. For instance, Masdar Institute's activities helped shape environmental, occupational health, and safety regulations and develop local institutional review board procedures. In this sense, despite the narrow focus on advanced energy and sustainability, Masdar plays a much broader, pioneering role in Abu Dhabi's development than MIT could possibly play in the diverse and highly developed institutional landscape of the United States.

Understanding Best Practices: Expertise, Legitimacy, and Identity

The above case studies illustrate how three start-up universities sought to build innovation capacity and train innovators according to the MIT model in entirely different ways. All three did so by teaming up with MIT through a multiyear, institution-building partnership, with the explicit goal of transferring best practices with the help of MIT. Yet comparison of the cases reveals that the three start-up universities bear little resemblance to one another—or to MIT, for that matter. Although all three universities did incorporate *some* MIT practices—including educational curricula, research projects, and organizational arrangements—these practices were typically selectively chosen, modified, and implemented in disparate institutional architectures. What is more, they were usually complemented by activities that did not exist at MIT at all.

The three start-up universities speak to a fragmented and incongruent understanding of the MIT model, putting into question the usefulness of the notion of best practice as it is commonly understood—that is, as a set of universally valid and desirable practices that try to reproduce the activities from a supposedly well-defined, uniform base model. However, the cases also underscore that all three initiatives nonetheless put the MIT model front and center in conceiving, implementing, and justifying their activities.

This apparent contradiction between a common reference to the MIT model and differences in local "best" practices becomes intelligible when looking at the MIT model as a *boundary object*. At each site, a range of actors came together with different interests to establish a new institution according to the model: local government leaders sought credit and credibility in the pursuit of a politically risky and expensive policy shift (e.g., the need for "something different" at SUTD). Regional societal stakeholders (like local student populations or existing institutions) shaped key ideas about what was deemed desirable or acceptable in a local social context (e.g., the inclusion of other Russian universities in Skoltech's CREIs). MIT faculty were interested in exciting new research opportunities and funding (e.g., clean-tech research in Abu Dhabi), as well as in trying out new educational models (e.g., around technology and design in SUTD). MIT administrators sought to strengthen MIT's global footprint and revenue streams. Institutional leaders at the new universities were aiming to jump-start new activities by adopting organizational templates and presumed best practices (e.g., the transfer of educational curricula). Finally, national and international students were looking for quality educational opportunities that were considered to be lacking in this specific context. These different actors may fundamentally disagree on what the MIT model is and which parts of it are relevant; in fact, they may not be interested in the specifics of the model at all. Still, they agree that this new institution is important for the pursuit of their objectives.[40] Importantly, the conceptual stability and interpretive flexibility around the MIT model holds true even as the model travels to different social contexts.

Three interrelated dimensions appear particularly pertinent for understanding the role of boundary objects in shaping and stabilizing best-practice transfer activities in innovation policy: expertise, legitimacy, and identity. *Expertise*, first, is closest to the traditional notion of best-practice transfer as a supposedly straightforward, unidirectional emulation of activities that have proven successful in one place. All three countries sought a certain kind of innovation expertise as embodied by the MIT model. All three locations also imported *some* practices more or less directly from MIT. The strong emphasis on creating *something different* from what existed before in all three cases—a countermodel to established universities in the case of SUTD and Skoltech and Abu Dhabi's first research university in the case of Masdar—suggests that the local institutional landscape and society at large were seen as deficient with regard to innovation.[41] It also provides a clue as to why the desired expertise had to come from an outside expert entity: policymakers believed that critical innovation expertise was unavailable in their countries at that time. However, this focus on the *source* of expertise only takes us so far in understanding the role of best practices. As we have seen, MIT itself lacks consensus on what exactly MIT's expertise is and how it is institutionally embodied. Moreover, the focus on expertise does not explain, for instance, the specific decisions made to determine which practices from MIT were deemed useful and which other practices needed to complement it, or how countries arrived at the point where a (new) kind of innovation expertise was needed in the first place.

Legitimacy, second, points to a complementary function of how the MIT model becomes relevant in different contexts. Questions of expertise are inextricably tied to questions of legitimacy with regard to who counts as an expert and why.[42] Expertise serves as a powerful political resource because of its ability to resolve disputes through epistemic authority. In our cases, reference to the success of the MIT model justified organizational changes and allowed stakeholders to shore up support for reforms that might otherwise have received little domestic backing. For example, actors involved in the early stages of the Skoltech initiative underscore that it would have been unlikely, if not impossible, to receive presidential support or launch an institution-building initiative outside the powerful Russian science establishment without the MIT brand attached to it. In contrast, MIT's legitimacy in Singapore was rooted not only in its global reputation but also in an "incremental building of trust and building of understanding" over several decades between MIT and Singapore, according to a senior MIT administrator involved in the partnership negotiations.

The legitimacy of the MIT brand also played a key role for all three universities in attracting top faculty and students who saw MIT's involvement as a sign of credibility. For faculty at all three institutions, MIT's involvement was a key factor in their decision to join the start-up university. The same argument applies to students, who value the MIT connection and potential opportunities to visit the United States. For example, as a result of its deliberate use of the MIT brand, Masdar Institute was able to choose 89 students from 1,200 applicants for its initial class in 2009, with an average quantitative GRE score of 765 (comparable to the MIT average). Legitimacy, therefore, seems particularly important in the context of start-up institutions without an existing track record, which face the challenge of attracting world-class researchers and students to a place that does not yet exist—to the proverbial middle of the desert, if need be.

This leads us to our final ingredient of best-practice transfer: *identity*. Differences among the three start-up institutions can be understood as expressions of local identity, expressed in the particular ways they elicit and mobilize foreign expertise and legitimacy. Squeezed between the two poles of expertise and legitimacy, markers of identity shape which authoritative MIT practices get picked, how they get re-embedded, and how they are justified and reconciled with local ways of reasoning through policy change. Singapore primarily sought a break with existing engineering education practices and focused on cutting-edge reform efforts currently underway at MIT itself, which matched evolving government priorities in the city-state and helped supplant earlier versions of MIT best practices in innovation. Skoltech was conceived as a countermodel to Russian universities but at the same time provided a bridge to the dignified existing institutional landscape. Its research priorities largely resemble traditional areas of strength rooted in basic science. Masdar Institute was launched primarily as a vehicle for socioeconomic development, addressing various sustainability challenges in the context of a corporate superstructure. Masdar Institute focused on foundational capacity and institution-building efforts, not unlike MIT's early partnerships in India and elsewhere. These different visions bring home the point that what counts as useful, acceptable, or valid knowledge is always coproduced within local social, political, and institutional orders.[43]

In light of this analysis, then, can an innovative university be (ready-)made? Specifically, can it be modeled after best practices from a leading institution such as MIT? My research indicates that the concept of best-practice transfer is far from straightforward—and not just because of complex historical path-dependencies at MIT itself. Instead, my findings suggest that the concept of an innovative university has no universal meaning but depends on the vastly different social functions and constraints it faces in different contexts. The above case studies show that no agreement on the purpose of innovation and best practices exists, and that transfer efforts will invariably lead to vastly different outcomes across different social contexts *even if implemented with the help of the source of those practices.*

These invariably unique imaginaries of innovation do not reduce the potential utility of best-practice transfer, but they force us to adjust the way we think and talk about it. As we have seen, key features in the design, implementation, and performance of these start-up universities remain unexplained if treated as mere variations on a common MIT model. The very idea that there is an MIT model suggests that MIT's innovation success has some abstract, scientific quality with quasi-universal applicability, much in line with other innovation models like the linear model, the chain link model, or the triple helix model. Unlike these abstract ideal types, however, the MIT model does in fact have one concrete exemplar—MIT in Cambridge, Massachusetts—which points to the slippery slope when referring to best practices. In practice, the boundaries between the concrete and the ideal MIT model that international leaders seek to emulate and that MIT occasionally markets are not easily drawn and are often deliberately blurred.

Making innovative universities thus requires, first and foremost, unpacking the social and political role of innovation (and universities) in a given context, and tailoring institutional practices in a way that is cognizant of these unique imaginaries of innovation. An adequate account of best-practice transfer must consider how innovation expertise travels alongside political legitimacy and how both are inflected by elements of local identity. In this light, the diversity and interpretive flexibility surrounding the MIT model is not a weakness but a necessity and a strength that can be exploited. It explains how an imagined institutional ideal of innovation can indeed be relevant in locations as diverse as Abu Dhabi, Singapore, and Russia, beyond mere branding. And through that, it can help anticipate and explain why certain models for cultivating innovation succeed or fail.

Notes

1. This work was partly supported by the NSF Science of Science and Innovation Policy Program under the collaborative grant "Technology, Collaboration, Learning: Modeling Complex International Innovation Partnerships" (#1262263), the MIT

Leading Technology Policy Initiative, and the MIT-Portugal Program. It benefited greatly from collaborative engagement with the project "Traveling Imaginaries of Innovation: The 'Practice Turn' and Its Transnational Implementation," sponsored by the NSF Science, Technology, and Society Program. I would like to thank Sheila Jasanoff, Dava Newman, Dan Roos, Fred Moavenzadeh, Artyom Morozov, Oleg Alekseev, Danielle Wood, Josh Jacobs, Lori Breslow, Mac Hird, the members of the MIT International Advisory Committee, and all interview participants at MIT and across the three countries for their support, generous sharing of thoughts, and information.

2. Sebastian Pfotenhauer, Joakim Juhl, and Erik Aarden, "Challenging the 'Deficit Model' of Innovation: Framing Policy Issues under the Innovation Imperative," *Research Policy* (forthcoming).

3. Sebastian Pfotenhauer and Sheila Jasanoff, "Panacea or Diagnosis? Imaginaries of Innovation and the 'MIT Model' in Three Political Cultures," *Social Studies of Science* 47, no. 6 (2017): 783–810.

4. Sebastian M. Pfotenhauer and Joakim Juhl, "Innovation and the Political State: Beyond the Myth of Technologies and Markets," in *Critical Studies of Innovation: Alternative Approaches to the Pro-Innovation Bias*, ed. Benoît Godin and Dominique Vinck (Northampton, MA: Edward Elgar, 2017), 68–94.

5. See the contributions of Wisnioski (chapter 1), McManus and MacDonald (chapter 4), Gustetic (chapter 7), and Russell and Vinsel (chapter 13) in this volume.

6. Edwin Mansfield and Jeong-Yeon Lee, "The Modern University: Contributor to Industrial Innovation and Recipient of Industrial R&D Support," *Research Policy* 25 (October 1996): 1047–1058; Attila Varge, *University Research and Regional Innovation* (Boston: Kluwer Academic Publishing, 1998); H. Holden Thorp and Buck Goldstein, *Engines of Innovation: The Entrepreneurial University in the Twenty-First Century* (Chapel Hill: University of North Carolina Press, 2010); Georg Winckler, "Innovation Strategies of European Universities in the Triangle of Education, Research, and Innovation," in *University Research for Innovation*, ed. Luc E. Weber and James J. Duderstadt, Glion Colloquium (London: Economica, 2010), 107–124.

7. Sebastian M. Pfotenhauer and Sheila Jasanoff, "Traveling Imaginaries: The 'Practice Turn' in Innovation Policy and the Global Circulation of Innovation Models," in *The Routledge Handbook of the Political Economy of Science*, ed. David Tyfield et al. (London: Routledge, 2017), 416–428.

8. Pfotenhauer and Jasanoff, "Panacea or Diagnosis?"

9. See, e.g., Eugene Bardach, "Comment: The Problem of 'Best Practice' Research," *Journal of Policy Analysis and Management* 13, no. 2 (1994): 260–268, doi:10.2307/3325011; Derek L. Ransley and Jay L. Rogers, "A Consensus on Best R&D Practices," *Research*

Technology Management 37, no. 2 (3 April 1994): 19; OECD, *Policy Evaluation in Innovation and Technology: Towards Best Practices* (Paris: OECD, 1997), http://www.oecd.org/document/23/0,3746,en_2649_34451_1822359_119681_1_1_1,00.html; Patricia Keehley, *Benchmarking for Best Practices in the Public Sector: Achieving Performance Breakthroughs in Federal, State, and Local Agencies* (San Francisco: Jossey-Bass, 1997); Booz Allen Hamilton, "Best Practice Transfer: Unleashing the Value Within," 2001, http://www.boozallen.com/media/file/BestPrac.pdf; Julio A. Pertuze et al., "Best Practices for Industry-University Collaboration," *MIT Sloan Management Review*, 26 June 2010; Charles W. Wessner, ed., *Best Practices in State and Regional Innovation Initiatives: Competing in the 21st Century* (Washington, DC: National Academies Press, 2013); David N. Resende, David Gibson, and James Jarrett, "BTP—Best Transfer Practices; A Tool for Qualitative Analysis of Tech-Transfer Offices: A Cross Cultural Analysis," *Technovation* 33, no. 1 (January 2013): 2–12.

10. The concept of boundary objects was first introduced by the science and technology studies scholars Susan Leigh Star and James Griesemer (Star and Griesemer, "Institutional Ecology, 'Translations,' and Boundary Objects: Amateurs and Professionals in Berkeley's Museum of Vertebrate Zoology, 1907–39," *Social Studies of Science* 19 [1989]: 387–420). Their groundbreaking study showed how groups of stakeholders came together and agreed on the importance and creation of a new scientific institution, Berkeley's Museum of Vertebrate Zoology, despite different understandings of what this institution represented and why it was needed. The flexible meaning of the museum and its exhibit made it possible for individuals who did not agree on a common set of goals to obtain the necessary momentum and backing to create and stabilize the institution. In contrast to Star and Griesemer, my focus is on an institutional *model*, not a single institution.

11. Pfotenhauer and Jasanoff, "Panacea or Diagnosis?"

12. MIT faculty and institutional leaders offer two reasons for this limited-time capacity-building approach—an approach that MIT has pursued since the establishment of the Indian Institutes of Technology in the 1960s (see, e.g., Stuart W. Leslie and Robert Kargon, "Exporting MIT: Science, Technology, and Nation-Building in India and Iran," *Osiris* 21 [2006]: 110–130). On the one hand, it ensures eventual ownership of the project by local stakeholders who know that they cannot rely on MIT for operation forever and hence work toward truly local capacity. On the other hand, it is a mechanism to safeguard the unique brand of the MIT in Cambridge, Massachusetts, and limit the risk and exposure in case of potential failure.

13. Henry Etzkowitz, *MIT and the Rise of Entrepreneurial Science* (New York: Routledge, 2002); Henry Etzkowitz, *The Triple Helix: Industry, University, and Government in Innovation* (New York: Routledge, 2008).

14. Edward B. Roberts, Fiona Murray, and J. Daniel Kim, "Entrepreneurship and Innovation at MIT: Continuing Global Growth and Impact," *Martin Trust Center for MIT Entrepreneurship*, 2015, https://innovation.mit.edu/assets/Entrepreneurship InnovationMIT-8Dec2015-final.pdf, 6–7.

15. Some adjustments have been made over time. For example, Masdar Institute gradually added more capacity in some fundamental research domains (e.g., material science), and Skoltech has from the start tried to tap into traditional Russian research strengths in nuclear and space engineering. Yet the central focus remains firmly on applied research.

16. This research draws upon more than sixty semistructured interviews, additional informal conversations at each university, and document research (e.g., public and program-internal reports). The author also wishes to acknowledge his close ties to MIT. He joined MIT as a graduate student and was later employed as a research scientist and lecturer. He worked as a part-time research and project manager for one of MIT's major international partnerships and at one point served as a policy advisor to the Skolkovo Foundation. This involvement with MIT's international community facilitated access and provided opportunities for participant observation.

17. Sebastian M. Pfotenhauer et al., "Architecting Complex International Science, Technology, and Innovation Partnerships (CISTIPs): A Study of Four Global MIT Collaborations," *Technological Forecasting and Social Change* 104 (2016): 38–56.

18. "MIT Facts 2016: Mission," 2016, http://web.mit.edu/facts/mission.html.

19. "Towards a Better World by Design," SUTD undergraduate brochure, 2010, https://www.sutd.edu.sg/cmsresource/brochure/undergraduate_brochure.pdf.

20. "SUTD President's Message," 2012, http://www.sutd.edu.sg/presidents_message .aspx.

21. "The Big-D," 2012, http://www.sutd.edu.sg/thebigd.aspx.

22. Lim Chuan Poh, "Singapore: Betting on Biomedical Sciences," *Issues in Science and Technology* 26, no. 3 (spring 2010): 69–74.

23. Kim-Song Tan and Sock-Yong Phang, *From Efficiency-Driven to Innovation-Driven Economic Growth: Perspectives from Singapore* (Washington, DC: World Bank Publications, 2005).

24. Chuan Poh, "Singapore: Betting on Biomedical Sciences."

25. Pfotenhauer and Jasanoff, "Panacea or Diagnosis?"

26. Peter Savodnik, "Skolkovo, Russia's Would-Be Silicon Valley," *Bloomberg Businessweek*, 1 September 2011, http://www.businessweek.com/magazine/skolkovo -russias-wouldbe-silicon-valley-09012011.html; Mary Carmichael, "MIT and Russian

Foundation Join to Open Tech Research Hub Meant to Be a Silicon Valley Outside Moscow," *Boston Globe*, 26 October 2011, https://www.bostonglobe.com/metro /2011/10/26/mit-russian-foundation-open-tech-research-hub/mBnvoEgsWPpKoKB Deh3I3M/story.html.

27. Alec Luhn, "Not Just Oil and Oligarchs," *Slate*, 9 December 2013, http://www .slate.com/articles/technology/the_next_silicon_valley/2013/12/russia_s_innova-tion_city_skolkovo_plagued_by_doubts_but_it_continues_to.html; Monty Munford, "Kremlin Intrigue Threatens Russia's Silicon Valley," *BloombergView*, 17 September 2013, http://www.bloomberg.com/bw/articles/2013-07-18/kremlin-intrigue-threatens -russias-silicon-valley.

28. "About Skoltech," 2013, http://www.skoltech.ru/en/about/.

29. Loren R. Graham and Irina Dezhina, *Science in the New Russia: Crisis, Aid, Reform* (Bloomington: Indiana University Press, 2008).

30. "Skoltech CEI Concept," 2012, http://web.mit.edu/sktech/sktech-program/entre preneurship-innovation/concept.html.

31. "About Skoltech" (emphasis added).

32. Graham and Dezhina, *Science in the New Russia*.

33. Sebastian M. Pfotenhauer, "Co-Producing Emirati Science and Society at Masdar Institute of Science and Technology," in *Accelerating Science and Technology Development in the Middle East: Unleashing the Potential of Near Ties*, ed. Afreen Siddiqi and Laura Diaz-Anadon (Berlin: Gerlach Press, 2016), 89–113.

34. "Delivering Sustainability: Sustainability Report," 2012, Masdar Institute, http:// www.masdar.ae/assets/downloads/content/669/masdar_2012_sustainability_report .pdf; "Who We Are," 2014, Masdar Corporate, http://www.masdar.ae/.

35. David Hopwood, "Abu Dhabi's Masdar Plan Takes Shape," *Renewable Energy Focus* 11, no. 1 (January 2010): 18–23; Sam Nader, "Paths to a Low-Carbon Economy—The Masdar Example," *Energy Procedia* 1 (February 2009): 3951–3958.

36. "Who We Are," 2013, Masdar Corporate, http://www.masdar.ae/en/masdar/our -story.

37. "Our Research Arm," 2014, Masdar Corporate, http://www.masdar.ae.

38. Leslie and Kargon, "Exporting MIT."

39. Pfotenhauer, "Co-Producing Emirati Science and Society."

40. As Star and Griesemer emphasize, "Consensus is not necessary for cooperation nor for the successful conduct of work." Star and Griesemer, "Institutional Ecology, 'Translations' and Boundary Objects."

41. Pfotenhauer and Jasanoff, "Panacea or Diagnosis?"

42. Sheila Jasanoff, "Judgment under Siege: The Three-Body Problem of Expert Legitimacy," in *Democratization of Expertise?*, ed. Sabine Maasen and Peter Weingart (Berlin/Heidelberg: Springer-Verlag, 2005), 24:209–224; H. M. Collins and Robert Evans, *Rethinking Expertise* (Chicago: University of Chicago Press, 2007).

43. Sheila Jasanoff, ed., *States of Knowledge: The Co-Production of Science and the Social Order* (New York: Routledge, 2004).

12 The Innovation Gap in Pink and Black

Lisa D. Cook

From a number of perspectives, innovation is a good thing for the economy.[1] Economists have long recognized how the generation and implementation of ideas drives economic growth.[2] Historians also have demonstrated the positive relationship between the commercialization of invention, industrialization, and economic activity in studies documenting early American inventor-entrepreneurs and the creation of the patent system.[3] Statisticians provide further evidence of the innovation economy's importance to the nation; for example, from 1960 to 2013, the number of workers in innovation jobs grew 3 percent annually, compared to 2 percent for the broader workforce.[4]

The benefits of the innovation economy, however, have not been evenly distributed. Despite numerous initiatives to train and cultivate innovators, women and African Americans continue to participate at each stage of the innovation process—from education to patent activity to start-ups—at lower rates than their male and white counterparts. As a consequence, women and African Americans have not enjoyed their proportionate share of innovation's benefits.[5]

For women and African Americans throughout the history of the United States, this pathway to success has been curtailed because of entrenched gender discrimination and racial segregation. Women and African Americans have had less than equal access to education, especially advanced technical training. Discriminatory laws and policies once forbade enslaved blacks from earning patents and married women from owning them. Meanwhile, women and African Americans were for decades systematically excluded from the professional scientific and engineering societies. Likewise, women and African Americans had limited access to wealthy backers and mainstream banks and so were forced to develop different sources for the start-up capital necessary for commercializing their inventions.[6]

Despite major gains since the 1970s, women and African Americans remain underrepresented in the innovation economy.[7] Empirical evidence gathered over decades reveals an even more pervasive underrepresentation. Women and African Americans earn fewer advanced degrees in the STEM disciplines (science, technology, engineering, and mathematics) than the population figures would suggest. Likewise, women and African Americans earn fewer US patents than would be expected and are less likely to commercialize them.

This *innovation gap* represents a lost opportunity, a discriminatory drag on our economy, and further structural evidence of the wide income and wealth gaps in the United States. More seriously, the underrepresentation of women and African American innovators is a failure to deliver on the American ideals of equality and equal opportunity for all. In this chapter, I draw on observations from educational surveys, employment and income figures, and US patent data to describe inequalities at all stages of innovation. I will explain how these inequalities emerged, how they persist, why they matter, and what we can do to close the gaps.

Participation in the Innovation Economy

Fundamentally, economists and the public care about innovation, because it is a critical factor in economic growth, wealth generation, and higher living standards. If we consider the components of economic growth—labor, capital, and total factor productivity—innovation can substantially affect each one. In 1957, economist Robert Solow demonstrated that aggregate economic growth owed more to innovation, or technical change, than additional inputs of labor or capital. Solow believed that innovation had an economywide effect and that its impact on the other factors—labor and capital—would be neutral.[8] For decades macroeconomists agreed, and technological progress was viewed as benefiting all workers equally. More recently, theory and evidence suggest that technical change is skill-biased and that its economic benefits favor highly skilled workers. The inequality in wages that arises from this process is a core feature of the more general debate about inequality today.[9]

Many of the key measures of the innovation economy track the participation of *innovators*. The National Science Foundation (NSF), for example, defines the "science and engineering (S&E) workforce" by the number of participants in science and engineering occupations, by the number of holders of science and engineering degrees, and by the use of technical expertise on the job.[10] In addition to educational, occupational, and income metrics, we can measure participation in the innovation economy via patent holders. Data on patents recorded and disseminated by the US Patent and Trademark Office (USPTO) are available from 1790 to the present and provide a relatively consistent historical metric.[11] While the NSF collects demographic data such as gender, race, and ethnicity, such factors are not recorded in patent data. However, my colleagues and I have developed or taken advantage of new methods for inferring which historical and contemporary patents were granted to women and African Americans from 1966 to 2014.[12]

As economists measure innovation's contribution to the economy with increasing precision, it is clear that innovation's importance is growing.[13] In 2013, the NSF calculated that the innovation economy comprised roughly 6 to 21 million workers.[14] These innovation workers earn substantially more than the median income for all workers. In 2014, the median innovation worker earned $81,000, compared to $36,000 for all workers. Innovation economy jobs also are growing faster than in other sectors, and unemployment rates are lower. In February 2013, the unemployment rate for scientists and engineers was 3.8 percent, compared to 4.3 percent for all college graduates and 8.1 percent for the United States overall.[15] During the Great Recession (2007–2009), moreover, the US workforce contracted; however, the innovation workforce was less affected by the overall economic contraction.[16] Amid the recession, the income gap between innovation workers and the general labor force also widened. In 2012, innovation economy earnings were double those of other workers; by 2014, the median innovation worker earned an additional 25 percent more than the median worker in the general labor force.[17] Across a number of measures, the science-based innovation workforce provides a tremendous boost to the overall economy.

Within the innovation economy, however, both participation and salaries vary greatly by gender, race, and ethnicity. In what follows, I examine how the racial and gender gaps are manifest throughout different stages of the innovation process. I provide longitudinal, quantitative evidence to outline the nature and scope of these gaps over time. I then complement this statistical picture with historical and contemporary examples from individual women and African American innovators who were impacted by racial and gender discrimination during the innovation process. This analysis across different scales illuminates both the macroeconomic impact and lived experience of the innovation gap in pink and black.

Participation Gaps throughout the Innovation Ecosystem

At the risk of oversimplifying a complicated, nonlinear process, it helps to imagine that an individual participates in the innovation economy by passing through three stages. First, innovation typically begins with formal education or an apprenticeship. An innovator needs to master the specialized canon of knowledge in his or her chosen technical field, increasingly through the acquisition of an advanced degree in a STEM field. Second, workers in the innovation economy participate in actual invention in corporate research facilities, university laboratories, government agencies, or sometimes in garages or other informal workspaces. Finally, innovation, or the commercialization of invention, occurs when an inventor sells or licenses her patent, or launches a new start-up or business unit to profit directly from the development of the invention. The innovation gap in pink and black is present, to varying degrees, in all three of these stages.

The Preparation and Education Gap

The first stage in participating in the innovation economy is obtaining a formal education or an apprenticeship. Historically, women and African Americans have been denied equal access to training in STEM fields. When the first agricultural and mechanical colleges appeared in the 1870s, women were discouraged from enrolling and were prevented from joining professional engineering societies. Women first began to enter the coeducational technical universities and technical job training programs during World War II's manpower shortage; the Society for Women Engineers was founded shortly thereafter in 1950. Persistent sexism in engineering education and internships still dissuades many women from pursuing engineering degrees, while workplace discrimination encourages many women engineers to eventually exit the field. As we will see, women have increasingly pursued undergraduate and advanced technical degrees, but their numbers are nowhere near gender parity.[18]

Similarly, African Americans have historically been denied equal access to educational opportunities. Before the Civil War, most enslaved African Americans were systematically denied simple literacy for fear that education would lead to rebellion. Even with the postbellum establishment of black colleges and technical schools—such as Virginia's Hampton Institute (1868) and Alabama's Tuskegee Institute (1881)—freed men and women generally had fewer opportunities for formal education and technical training. Aspiring black inventors were not welcome at the mainstream technical associations; they were also denied apprenticeships by white tradesmen and generally barred from entering the "shop culture" of machinists and telegraphers that was a crucial training ground for many inventors. The civil rights movement precipitated desegregation of public universities in the 1950s and 1960s, and the National Society of Black Engineers emerged soon afterward during the 1970s. However, persistent racial bias in admissions criteria has continued to work against the equal participation of African Americans in the STEM fields.[19]

Women and African Americans have enjoyed significantly improved access to technical training over the last few decades, but an education gap remains. Women and African Americans have increasingly been involved at the beginning of the innovative process, which is at the stage of doing basic research that undergirds changes in the stock, flow, and direction of knowledge (figure 12.1). Between 1970 and 2014, the share of PhDs in S&E fields awarded to women grew substantially, from just 9 percent to 41.6 percent. Over the same period, the share of S&E PhDs awarded to African Americans, though small, more than tripled, from 1 percent to 3.5 percent. We see similar trends for master's and bachelor's degrees.[20]

Increases among women and African Americans, however, have not been uniform across fields of study. In particular, women and African American doctoral recipients have tended to gravitate toward psychology and the life sciences, and to avoid engineering. For example, in 2014, women accounted for 73.5 percent of psychology doctorate degrees. Alternatively, women have traditionally received the lowest share of doctoral degrees in engineering; that figure was just 22.8 percent in 2014. Similarly, among STEM fields, the highest share of African American doctorates was in psychology

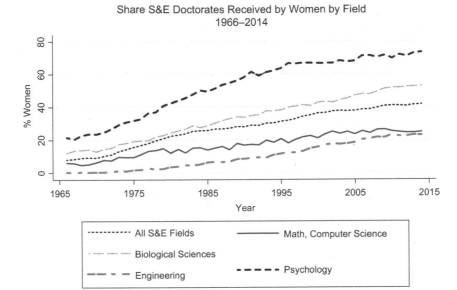

Source: NSF NCSES; NCES data
Note: Earth sciences include atmospheric and ocean sciences; biological sciences include agricultural sciences.

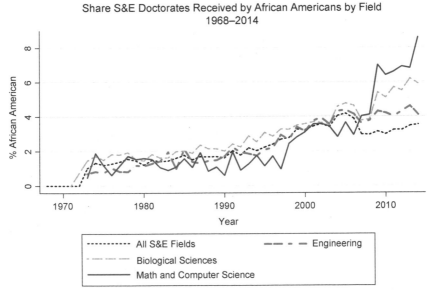

Source: NSF NCSES, Survey of Earned Doctorates

Figure 12.1
Share of S&E doctorates received by women and African Americans by field.

(7.9 percent), and the lowest was in engineering (1.7 percent). These disciplinary differences are important, because engineering is the discipline most closely associated with patenting.

Clearly, there are persistent barriers to women and African Americans pursuing advanced degrees in STEM fields in general and in engineering in particular. But what does this *education gap* look like in practice? Jennifer Selvidge, a senior honors student in materials engineering at MIT, captured the frustrations of many women and African Americans in her 2014 article "Pushing Women and People of Color out of Science before We Go In." She reported being told "hundreds of times" that, as a woman, she did not deserve to be at MIT and that metallurgy was a "man's field." She endured sexual harassment by her male teaching assistants and witnessed male professors attempting to publicly humiliate the few women professors in her department. She observed students of color being actively advised to change majors and leave her department; she also heard a teaching assistant argue that "black Americans are genetically inferior due to slavery era breeding practices." And this was at the leading engineering school in the country, if not the world.[21]

America will not fully realize its scientific potential and ever higher economic growth and living standards unless we encourage more women, African Americans, and other underrepresented groups to earn degrees in STEM fields and train for STEM careers. Indeed, the education gap in pink and black—that is, the limited pool of technically trained women and African Americans—helps explain the gaps in the second and third stages (invention, commercialization) of the innovation process.

The Invention Gap

Invention is the second stage of participation in the innovation economy. As with the education gap, there are historical and structural barriers underlying the invention gap. For centuries, individual women and African Americans were unwelcome in the white, male culture of the corporate R&D labs and were barred from joining professional scientific and engineering societies. Discrimination thus deprived them of the social capital and connections required to advance their careers and develop their inventions.[22]

Legal access to the US patent system offered greater but still limited opportunities for women and African Americans. There was no language in the original Patent Act of 1790 limiting patentees based on gender, race, age, or religion; decades before emancipation and universal suffrage, women and (free) blacks could, and did, invent and earn US patents.[23] Still, women and African Americans did not have equal protection under the patent laws. Although free blacks were allowed to obtain patents, the Patent Office refused to grant patents to enslaved blacks. Moreover, laws in many states assigned all marital property rights to husbands and prohibited married women from owning or controlling patents in their own names.[24]

Contemporary measures of inventive activity among women and African Americans reveal evidence of increased participation, but also continued barriers to access. We can measure the relative participation of women and African Americans in invention through data on employment, salaries, and patents.

In the realm of technical employment, women's participation in the invention stage has grown modestly. Between 1993 and 2015, women in S&E occupations rose from 23 percent to 28 percent.[25] Still, there were significant intersectional differences. White women constituted 18 percent of the total, while African American women accounted for just 2 percent.[26]

Digging into the data of what women and African American doctorate holders actually do on the job raises additional concerns about their participation as innovators (figure 12.2). First, the majority of such graduates are not employed in science and engineering occupations at all. Additionally, while more than half the people in S&E-related occupations are women scientists and engineers, they tend to be in supporting roles, such as technicians and precollege teaching, rather than inventing roles. Moreover, differences in fields of educational training are compounded in the career trajectories of women and African Americans with S&E degrees. More than two-thirds of psychologists are women, and women are more concentrated in life sciences relative to men but less concentrated in computer and mathematical sciences.[27]

Similarly, African American scientists and engineers make up just 4.8 percent of S&E occupations. African American scientists and engineers also are more concentrated among social and related scientists and computer systems analysts than in other occupations. Among S&E-related occupations, African American scientists and engineers are more concentrated in health-related occupations and in precollege teaching than in other occupations. Almost twice as many African American scientists and engineers are in non-S&E occupations as are in S&E occupations.

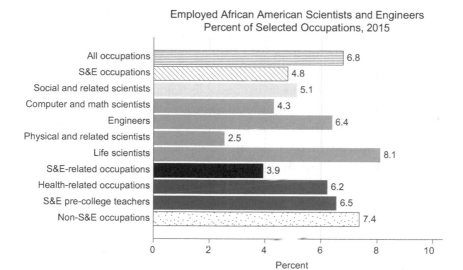

Figure 12.2
Employed women and African Americans in S&E fields, 2015.
Source: NSF WMPD 2017 Digest.

Problems of retention also plague the innovation economy. Women leave for various reasons, including childcare, family leave policies, and workplace environment.[28] Such departures have implications for the earnings of women innovators. On average, women's wages will be lower outside the innovation economy, so the departures of technically trained women tend to exacerbate the income inequality that already exists between the innovation and noninnovation economy.

While employment rates are increasing among women and underrepresented minority scientists and engineers, unemployment rates vary significantly by gender and by racial and ethnic group (figure 12.3). The unemployment rate for African American women is higher than the unemployment rate overall, nearly double that of all scientists and engineers and more than double that of white women scientists and engineers. Unemployment for underrepresented minority men, at just above 4 percent, is higher than for white and Asian men and higher than the average for all scientists and engineers.[29] Similar to the data on occupations, scientists and engineers with greater experiences of unemployment will likely be poorer and less able to accumulate wealth than their counterparts with lesser experiences of unemployment.

Yet another gap related to invention is based on the kinds of institutions in which women and African Americans are employed (figure 12.4). Most scientists and engineers are employed in industry. Apart from underrepresented minority men, the second and third sectors of employment are education and government. On average, government and education salaries are lower than those in industry, further deepening the income inequality among S&E workers. Most importantly, while many workers in government laboratories work hard at patenting, they have binding constraints relative to their private sector peers and are less likely to commercialize their inventions. This can have even greater implications for wealth inequality in the innovation stage of technology commercialization.

Salary and income data indicate other markers of inequality in innovation jobs. The earnings or income gap between workers in the innovation economy and the overall economy is substantial. A worker in the innovation economy earned 63 percent more than the average American worker in 2014.[30] To be sure, this divergence in income is consistent with and related to overall income inequality in the United States.

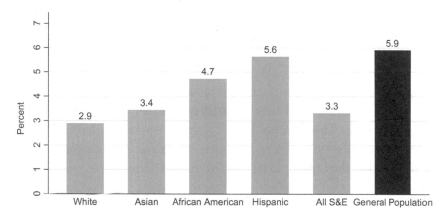

Figure 12.3
Unemployment rates among scientists and engineers, 2015.
Source: NSF WMPD 2017 Digest. Note: The general population consists of the U.S. civilian noninstitutional population of 16 years and over. Unemployment rates based on individuals actively seeking employment.

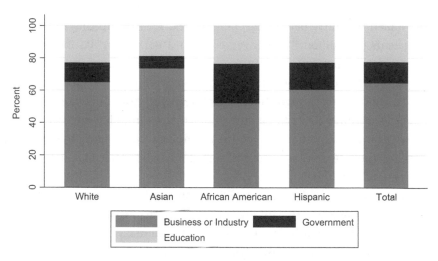

Figure 12.4
Employment sector of scientists and engineers by race and ethnicity, 2015.
Source: NSF WMPD 2017 Digest.

Just as incomes vary between innovation and the rest of the economy, they also vary among those within the innovation economy, particularly by gender and race. While the median salary for men in the innovation economy in 2010 was $80,000, it was only $53,000 for women. The gap between the median salary for African Americans and whites is not as large as it is between men and women. In 2010, the median salary for whites was $72,000, and for African Americans it was $56,000, or 78 percent of the median salary for whites.[31] Overall, there are still significant gender and racial gaps, but they appear to be closing. In 2015, for example, the median salary was $87,000 for men and $62,000 for women, or 71 percent of the median male salary.[32] In 2015, this share had moved only slightly to 79 percent. For S&E occupations, this share narrows to 92 percent.

Patent data provide yet another means of measuring inequality in inventive activity.[33] In earlier research, my colleagues and I demonstrated that women and African Americans lag far behind other US inventors with respect to patent activity. Using USPTO data from 1970 to 2006, we calculated that patent output for all US inventors is 235 patents per million; for women it is forty patents per million; and for African Americans it is six patents per million.[34] Moreover, economist Raj Chetty has found that a propensity to patent is closely associated with prior exposure to inventive activity and multigenerational income and wealth disparities. In other words, children from high-income families who grow up around other inventors are more likely to patent, while children from low-income families with limited exposure to emerging technology are less likely to patent.[35] Together, these two findings suggest a perpetual and intractable invention gap that is difficult to close.

Overall, women and African Americans are less likely to work in science and engineering jobs, are paid less for doing those jobs, and are less likely to earn US patents. But how is this discrimination manifested in the workplace? The discrimination underlying the invention gap was on public display in 2017 when Google engineer James Damore wrote an internal memo that leaked and went viral arguing that women were underrepresented in technology careers because of "inherent psychological differences."[36] A few weeks later, a former Google software engineer, Kelly Ellis, and two other women sued Google, alleging discrimination in both pay and promotion. These incidents followed a spring 2017 report in which the US Department of Labor investigated systemic discrimination against women at Google.[37]

Although the courts have yet to rule on the case, one could hypothesize that unequal salaries contribute to women's anemic pursuit of S&E degrees and careers, thereby perpetuating their underrepresentation. Consequently, men like Damore—who, because of the gaps, have never worked on diversified teams—may continue to perpetuate the stereotype that women and minorities simply do not belong in the high-tech professions. It is a vicious cycle.

The Innovation Gap

The commercialization of inventions is the third and final stage of participation in the innovation economy. This stage is also where diverse groups—women and underrepresented minorities—are most scarce. The preceding section showed that women and African Americans are less likely to have a patent and to work in the innovation economy. Given these gaps, it is reasonable to assume that the commercialization gap is wide from the start. Indeed, women and African Americans are less likely to found a firm based on a patented invention or to be a venture capitalist or tech investor in early-stage start-ups. Therefore, they are less likely to profit from the commercialization stage.

Commercialization requires drawing on financial and social capital to introduce the invention into society. Historically, women and African Americans have had diminished access to these resources and so have developed alternative strategies.[38] For example, women and African American inventors once purposefully obscured their identities in order to sidestep discrimination and profit from their inventions. Garrett Morgan, an African American inventor based in Cleveland, correctly surmised that white consumers would be skeptical of buying products from a black inventor. At the turn of the twentieth century, Morgan hired white actors to portray him in order to sell his gas mask, traffic signal, and other products.[39] Discrimination also forced women and African Americans into bad deals or to forego commercialization entirely. For example, in 1888, Ellen Eglin of Washington, D.C., sold her patented clothes wringer to a patent agent for a mere eighteen dollars rather than build a business around the patent.[40]

Social conditions have improved enormously over the last century, but today's women and African American inventors continue to struggle with commercialization. Contemporary inventors have four methods of generating income from an invention: (1) engage in entrepreneurship and start a new firm (or business unit) to develop, manufacture, and sell the invention; (2) assign (i.e., sell) the patented invention for a lump sum; (3) license the patented invention to another manufacturer and collect royalties until the patent expires; or (4) sue patent infringers to collect owed royalties and damages as a so-called patent troll. Economists can access data on certain of these approaches but not others. Consequently, I measure the innovation gap by focusing on data regarding entrepreneurship, firm ownership, patent assignments, and wealth accumulation from assets developed in the innovation economy.

Commercialization, particularly entrepreneurship and equity ownership of high-tech firms, is the stage of the innovative process where entrepreneurs find the largest pecuniary gains. Those who own equity stakes in high-tech companies—for example, angel and venture capital investors, founders, and employees with stock options—stand to profit greatly from initial public offerings, acquisitions, mergers and acquisitions, stock splits, and other liquidity events. Among the *Forbes* list of richest people in the world, five of the top ten derive their wealth primarily from the innovation economy. However, all ten are men, and none are black.[41]

The high market valuations and sheer size of transactions among high-tech firms mean that those with equity ownership stakes stand to profit handsomely. For example, Apple's market capitalization recently hit the $1 trillion market cap, which is greater than the size of the economy (GDP) of a number of the world's richest countries, including Argentina, Sweden, Switzerland, and Turkey.[42] The nine tech firms with initial public offerings (IPOs) in 2017 are collectively valued at roughly $36 billion.[43]

However, it is difficult to find women and African Americans among the ranks of venture capitalists, entrepreneurs, and (senior) management teams. For example, according to a study by the National Venture Capital Association (NVCA) in 2015, only 11 percent of venture capitalists were women, and only 2 percent were African Americans.[44] The NVCA also reported that in 2014, less than 7 percent of US venture capital funds were invested in businesses founded by women, and less than 1 percent were invested in businesses founded by African American women.[45] A different study by First Round Capital provided similarly bleak numbers: female CEOs receive only 2.7 percent of all venture funding, while women of color get virtually none—0.2 percent.[46]

This homogeneity extends from the investors to high-tech entrepreneurs and executives. Women and African Americans often work in the legal and marketing departments of high-tech firms, but there are few women in senior technical roles, in executive positions, or on boards of directors. For example, in 2014, *Fortune* ranked the relative diversity of several large tech firms based on recently released demographic data. With respect to women executives, Indiegogo was ranked highest, with women constituting 43 percent of leadership roles. Cisco and Pinterest were ranked lowest, with only 19 percent women in these roles. Women constituted just 18.7 percent of boards of S&P 500 firms in 2014, which was up from 16.3 percent in 2011.[47]

This lack of gender, racial, and ethnic diversity in the innovation economy is often attributed to a lack of mentors and social networks, implicit bias, and, to a lesser extent, the feeble pipeline of potential entrepreneurs, executives, board members, and funders.[48] Again, we see a vicious cycle: there are few women and African Americans in high tech, so there are fewer mentors, social networks, and colleagues helping to shepherd a more diverse set of investors and entrepreneurs into the innovation economy.

With regard to commercialization, these findings are troubling because a large and growing literature suggests that more diverse teams produce better outcomes.[49] For example, First Round Capital reports that founding teams that include a woman outperform their all-male peers by 63 percent.[50] This homogeneity among VCs, and the resulting suboptimal financing of projects, is clearly a bad outcome for women and African American founders and entrepreneurs, but also for the overall economy, which depends on the commercialization of new ideas to raise incomes and living standards.

Patent assignment is another simple measure of potential commercialization recorded in USPTO data. Patents are typically sold by independent or employee inventors to entities such as corporations, government agencies, universities, and research institutions. For example, venture capitalists often prefer investing in founders with patents and patents pending, and they typically require them to assign the patents to the start-up as a condition of investment. In addition, as part of their employment contracts, corporate, university, and government employees who produce inventions on the job are contractually obligated to immediately assign their patents to their employers (usually for one dollar) once they are issued by the USPTO. The assignees (the buyers) may or may not choose to commercialize these inventions, so assignments are an admittedly imperfect proxy for commercialization.[51]

Economists have particularly good data on corporate and government patent assignments as a measure of the commercialization gap. The previous section showed that women and African Americans are less likely to earn patents. But they are also less likely to sell those patents. Women inventors' odds of assigning a patent at issue to a public firm are 51 percent lower than men's odds; similarly, African American inventors' assignment odds are 46 percent lower than other US inventors' odds.[52]

In addition, the high concentration of African American scientists and engineers in the government sector (see figure 12.4) has implications for wealth accumulation and inequality. The Bayh-Dole Act of 1980 permitted universities and corporations to own and commercialize any patents earned as a result of federal research funding. However, the US government still owns any patents earned by federal employees.[53] In general, commercialization is more difficult in the government (versus the private) sector due to strict ethics policies and is less likely to occur due to lack of incentives and risk aversion among government employees and contractors.[54] Moreover, African Americans who begin assigning their patents to government entities are more likely to continue to assign to government entities rather than corporate entities, unlike their white coinventors on the original government-assigned patents.[55] Therefore, African American scientists and engineers, who tend to be more concentrated in government service, are less likely to realize financial returns from their patents.

Overall, women and African American inventors are significantly less likely than their US inventor counterparts to obtain and commercialize a patent. Women and African Americans are also underrepresented among venture capitalists, are less likely to receive start-up capital and launch new firms, and thus are less likely to own equity stakes in high-tech firms as a founder, senior manager, executive, or board member. In short, women and African Americans face significant challenges when it comes to commercializing their inventive activity. But how is this manifest in the real world? What does this innovation gap look like in practice?

This innovation gap was in the public eye in 2012 when Ellen Pao sued her employer, the noted venture capital firm Kleiner Perkins Caufield & Byers, for gender discrimination. Pao alleged that the firm's male partners excluded her from key meetings and harassed her when she spoke out; she was ultimately passed over for promotion. In the midst of the lawsuit, she became the CEO of Reddit and faced additional gender-related harassment and threats of violence when she called for the removal of objectionable content from the site's message boards. Pao ultimately lost her discrimination lawsuit, but in the process she has become an advocate for diversity and inclusion, and a symbol of the difficulties women face in the male-dominated high-tech and venture capital industries.[56] In the end, systemic gender and racial discrimination in the high-tech and venture capital sectors drives a wedge between good ideas, capital, and the commercialization of those ideas.

Why Does It Matter?

Women and African Americans are underrepresented in the innovation economy. At each stage of the innovation process—including technical training, employment, patenting, financing, entrepreneurship, and commercialization—women and African Americans participate at lower rates than their counterparts.

Why does it matter? First, because of these gaps, our economy is not as strong as it should be. These findings suggest a misallocation of resources that could contribute to suboptimal rates of economic growth. For example, we know that coed patent teams are more productive than single-sex teams with respect to the most valuable patents; likewise, start-up teams that include a woman outperform their all-male peers.[57] There is an opportunity cost to gender and racial discrimination, so patent teams, start-up firms, and the overall economy will continue to underperform without more emphasis on diversity and inclusion.

Second, because of these gaps, women and Africans Americans have not been able to enjoy their fair share of innovation's ample rewards. As we have seen, relative to other sectors, the innovation economy generates high incomes and considerable wealth. In addition, the leaders of the high-tech sector—including Bill Gates, Elon Musk, Mark Zuckerberg, and Jeff Bezos—wield considerable social, cultural, and political influence. However, if women and African Americans are disproportionately absent from the sector, they are deprived of their fair share of opportunity, wealth, and influence. Indeed, inequality in the innovation sector undermines the American ideals of equal opportunity and a shared responsibility to achieve shared prosperity.

Third, these gender and racial gaps raise fundamental concerns related to growing income and wealth inequality. In the economics literature, there is mounting evidence of increases in both types of inequality. For example, income-tax data suggests that levels of income inequality are higher now than they were during the Gilded Age.[58] In just the period from 1993 to 2011, real income growth was nearly ten times higher for the top 1 percent compared to the bottom 99 percent—57.5 percent compared to 5.8 percent.[59]

With its high salaries, stock options, and huge IPOs, the innovation economy has been a driver of this overall income and wealth inequality. However, there are also huge income and wealth inequalities *within* the innovation economy. The financial industry notwithstanding, the innovation sector likely offers the starkest examples depicting that, over time, the rate of return to financial capital (dividends, interest, and especially capital gains) strictly exceeds the rate of return to human capital (wages).[60] We can see this income and wealth divergence among the sector's highly skilled (but salaried) software programmers, marketers, and lawyers, and the billionaire founders and venture capitalists who own sizable equity stakes in publicly traded firms. Women and African Americans have a harder time even entering the innovation sector, and those who do tend to be salaried employees versus owner-capitalists. Clearly, the income and wealth gap within, and relative to, the innovation economy is also related to overall income and wealth inequality in the United States.

Why would economists and the public care about these distributional issues? First, with respect to well-being, individuals assess their incomes, or economic well-being, in relative rather than absolute terms. Large and sustained divergence in incomes within the innovation economy may result in additional discontent and turmoil within that sector. In addition, workers in other sectors of the economy may become increasingly demoralized as they witness the huge income and wealth gains—and disproportionate political influence—of the top tier of the innovation sector. It is rare for economic policymaking bodies and leaders—such as the IMF, former Council of Economic Advisers chair Alan Krueger, and former Federal Reserve chair Janet Yellen—to discuss income and wealth inequality and its implications publicly, but recently they have done so and are doing so increasingly.[61] Three consequences are of particular concern to economic policymaking bodies: a fall in intergenerational mobility, families borrowing beyond their means (which could lead to another financial crisis), and political decisions that are likely to result in policies that lead to even lower growth. Any of these frustrations could eventually boil over into more social unrest, which in turn may lead to lower growth rates.

Conclusion: What Is to Be Done?

How might we address the underrepresentation of women and African Americans in the innovation economy? How might we intervene at each stage of the innovation process? Is there a role for policy?

Before suggesting any policies, researchers should first calculate and quantify the opportunity cost associated with current practice. Recent research finds that women's underrepresentation in engineering, development, and design jobs explains much of the patent gap between men and women and that closing the gap could increase US GDP per capita by 2.7 percent.[62] In an analysis of the NSF's Survey of Earned Doctorates, I have also suggested that the inclusion of more women and African Americans in the initial stages of the innovation process could increase GDP per capita by 0.64 to 3.3 percent.[63] Although this research focuses on BA and PhD holders, the policy prescription is similar: address barriers that keep women and African Americans from engaging more fully in the innovation economy.

Another modest policy proposal involves data. It is still difficult to obtain demographic data on patentees, so there should be a sustained collaborative effort by government agencies to make such data available to researchers. Currently, there is an effort to match patent data to census data at secure federal Research Data Centers (RDCs). Similarly, the UMETRICS initiative tracks federal spending on science and technology research projects, with some additional demographic data available via the RDCs. These data initiatives should be sustained and extended.

Finally, policymakers should specifically focus on addressing diversity and inclusion in patenting. Since most patenting occurs within firms, such policy prescriptions will likely be developed and implemented by the firms themselves. Fortunately, there is a compelling business case for reforms. Diverse, coed patent teams are more productive than homogeneous, single-sex teams, so enlightened workplace inclusion policies are good for high-tech workers but also good for business.

Economists need to collect more data and conduct more research to address the innovation gap adequately, recommend appropriate changes, and reallocate resources better. The broader inclusion of women and African Americans in the innovation economy should help us realize our innovation potential and, as a result, higher living standards and greater shared prosperity. There is much to be gained from this effort and much to be lost if we fail to act. If women and African Americans continue to be underrepresented in the innovation economy, we may face a future with lower rates of economic growth, widespread income and wealth inequality, a political process influenced only by top earners, and ultimately, social unrest.

Notes

1. An earlier version of this essay was presented at the Institute for New Economic Thinking annual conference as a keynote address in April 2014 in Toronto. The author is grateful for thoughtful comments from participants in the conference. The author also thanks Jan Gerson, Alex Ginsberg, and Madelaine Irwin.

2. Paul M. Romer, "Increasing Returns and Long-Run Growth," *Journal of Political Economy* 94, no. 5 (October 1986): 1002–1037. Prior to Romer, economists thought of innovation as an exogenous phenomenon that was generated outside of economic models. For example, see Robert M. Solow, "Technical Change and the Aggregate Production Function," *Review of Economics and Statistics* 39, no. 3 (August 1957): 312–320; Zvi Griliches, "Hybrid Corn: An Exploration in the Economics of Technological Change," *Econometrica* 25 (October 1957): 501–522.

3. Zorina Khan and Kenneth Sokoloff, "Schemes of Practical Utility: Entrepreneurship and Innovation among 'Great Inventors' in the United States, 1790–1865," *Journal of Economic History* 53, no. 2 (June 1993): 289–307; Joel Mokyr, "Long-Term Economic Growth and the History of Technology," chapter 17 in *Handbook of Economic Growth*, vol. 1, part B, ed. Philippe Aghion and Steven Durlauf (Amsterdam: Elsevier, 2005), 1113–1180; Ross Thomson, *Structures of Change in the Mechanical Age: Technological Innovation in the United States, 1790–1865* (Baltimore: Johns Hopkins University Press, 2009).

4. "Science and Engineering Indicators 2016, Chapter 3," National Science Foundation, 2016, accessed 13 September 2017. In this chapter, the terms "innovation economy," "science and engineering (S&E) economy," and "S&E occupations" will be used interchangeably.

5. To be sure, other forms of discrimination in the American innovation economy exist, including discrimination against various ethnic groups, religious identities,

and sexual orientations. However, this chapter focuses exclusively on the under-representation of women and African Americans.

6. There is an abundant historical literature documenting both the triumphs and struggles of women and African Americans engaged in technological innovation. For example, see Anne L. Macdonald, *Feminine Ingenuity: Women and Invention in America* (New York: Ballantine Books, 1992); Autumn Stanley, *Mothers and Daughters of Invention: Notes for a Revised History of Technology* (Metuchen, NJ: Scarecrow Press, 1993); Rayvon Fouché, *Black Inventors in the Age of Segregation: Granville T. Woods, Lewis H. Latimer, and Shelby J. Davidson* (Baltimore: Johns Hopkins University Press, 2003); Patricia Carter Sluby, *The Inventive Spirit of African Americans: Patented Ingenuity* (Westport, CT: Praeger, 2004).

7. News outlets regularly report on the lack of diversity in the venture capital industry and the paucity of women and African Americans who serve as executives and board members for high-tech companies. For example, see Anna Wiener, "Why Can't Silicon Valley Solve Its Diversity Problem?" *New Yorker*, 26 November 2016, accessed 17 July 2017, http://www.newyorker.com/business/currency/why-cant-silicon-valley-solve-its-diversity-problem. Also see the U.S. Equal Employment Opportunity Commission, "Diversity in High Tech," May 2016, accessed 17 July 2017, https://www.eeoc.gov/eeoc/statistics/reports/hightech/upload/diversity-in-high-tech-report.pdf.

8. Solow, "Technical Change and the Aggregate Production Function."

9. For a good summary of this debate in the growth literature, see Giovanni L. Violante, "Skill-Biased Technical Change," in *The New Palgrave Dictionary of Economics*, 2nd ed., ed. Steven N. Durlauf and Lawrence E. Blume (London: Palgrave Macmillan, 2008).

10. NSF collects data on S&E students, graduates, and workers using a variety of surveys and sources, including the NSF Survey of Earned Doctorates (SED) and the National Center for Education Statistics (NCES) Integrated Postsecondary Education Data System Completions Survey. In addition to collecting data on fields of study, I have assembled NSF data on doctoral degrees earned by women (1966 to 2014) and African Americans (1968 to 2014).

11. This chapter focuses on utility (versus design) patents, which constitute the largest category of issued patents. A utility patent is issued for any new and useful process, machine, manufacture, composition of matter, or any new and useful improvement thereof.

12. Lisa D. Cook, *The African American and Women Inventors and Patents Data Set* (Stanford University, August 2003), last revised in November 2008, extended to women inventors, August 2009.The original data set did not include a separate, distinct file of women inventors. Women inventors were separately identified in 2009, and the term "women" was added to the data set. Lisa D. Cook and Chaleampong

Kongcharoen, "The Idea Gap in Pink and Black," NBER working paper no. 16331, September 2010.

13. See, e.g., Erik Brynjolfsson and Andrew McAfee, *Race against the Machine: How the Digital Revolution Is Accelerating Innovation, Driving Productivity, and Irreversibly Transforming Employment and the Economy* (Lexington, MA: Digital Frontier Press, 2011); Stephen D. Oliner and Daniel E. Sichel, "The Resurgence of Growth in the Late 1990s: Is Information Technology the Story?" *Journal of Economic Perspectives* 14, no. 4 (2000): 3–22; Stephen D. Oliner, Daniel E. Sichel, and Kevin J. Stiroh, "Explaining a Productive Decade," FEDS Working Paper no. 2007–63 (January 2007), https://ssrn.com/abstract=1160248 or http://dx.doi.org/10.2139/ssrn.1160248.

14. The definition varies based on the three ways NSF measures the S&E workforce. In the United States in 2013, roughly 5.74 million college graduates were employed in S&E occupations, and roughly 21.1 million college graduates had a bachelor's or higher-level degree in an S&E field. "Science and Engineering Indicators 2016, Chapter 3."

15. Ibid.

16. Ibid.

17. Ibid.; "Science and Engineering Indicators 2014, Chapter 3," National Science Foundation, 2014, accessed 13 September 2017, https://www.nsf.gov/statistics/seind 14/index.cfm/chapter-3/c3h.htm.

18. On gender discrimination in technical education and the professions, see Ruth Oldenziel, *Making Technology Masculine: Men, Women, and Modern Machines in America, 1870–1945* (Amsterdam: Amsterdam University Press, 1999); Amy Sue Bix, *Girls Coming to Tech! A History of American Engineering Education for Women* (Cambridge, MA: MIT Press, 2014); and Carroll Seron, Susan S. Silbey, Erin Cech, and Brian Rubineau, "Persistence Is Cultural: Professional Socialization and the Reproduction of Sex Segregation," *Work and Occupations* 43, no. 2 (December 2016): 178–214.

19. On racial discrimination in technical education, see David E. Wharton, *A Struggle Worthy of Note: The Engineering and Technological Education of Black Americans* (Westport, CT: Greenwood Press, 1992); Bruce Sinclair, ed., *Technology and the African-American Experience: Needs and Opportunities for Study* (Cambridge, MA: MIT Press, 2004); Mabel O. Wilson, *Negro Building: Black Americans in the World of Fairs and Museums* (Berkeley: University of California Press, 2012); and Amy E. Slaton, *Race, Rigor, and Selectivity in U.S. Engineering: The History of an Occupational Color Line* (Cambridge, MA: Harvard University Press, 2010).

20. While the focus of this section is on S&E doctorates, most commercialized inventions originate from those with bachelor's degrees and master's degrees. "Doctorate Recipients from U.S. Universities," National Science Foundation, 2017, accessed 13 September 2017, https://www.nsf.gov/statistics/2017/nsf17306/static

/report/nsf17306.pdf; "Doctorate Recipients from United States Universities: Summary Report 2000," National Science Foundation, 2001, accessed 13 September 2017, http://files.eric.ed.gov/fulltext/ED459639.pdf.

21. Jennifer Selvidge, "Pushing Women and People of Color Out of Science before We Go In," *Huffington Post*, 18 September 2014, accessed 18 September 2017, http://www.huffingtonpost.com/jennifer-selvidge/pushing-women-and-people-_b _5840392.html.

22. Lisa D. Cook, "Inventing Social Capital: Evidence from African American Inventors, 1843–1930," *Explorations in Economic History* 48, no. 4 (December 2011): 507–518.

23. For example, in 1809, inventor Mary Kies was the first woman to receive a US patent, for an improved method of weaving straw with silk thread to make hats. Similarly, Thomas L. Jennings was the first free person of color to receive a US patent in 1821, for a dry-cleaning process. On Kies, see United States Patent Office, *Women Inventors to Whom Patents Have Been Granted by the United States Government, 1790 to July 1, 1888* (Washington, DC: Government Printing Office, 1888). On Jennings, see Sluby, *Inventive Spirit of African Americans*, 15–17. On the egalitarian nature of the US patent system, see B. Zorina Khan, *The Democratization of Invention: Patents and Copyrights in American Economic Development, 1790–1920* (New York: Cambridge University Press, 2005).

24. On discriminatory patent laws and policies, see Matilda Joslyn Gage, "Woman as an Inventor," *North American Review* 136, no. 318 (May 1883): 478–489, especially 488; Henry Baker, "The Negro as an Inventor," in *Twentieth Century Negro Literature*, ed. D. W. Culp (Naperville, IL: J. L. Nichols, 1902): 399–413, especially 400; Carroll Pursell, "Women Inventors in America," *Technology and Culture* 22, no. 3 (July 1981): 545–549, especially 546; Steven Lubar, "The Transformation of Antebellum Patent Law," *Technology and Culture* 32, no. 4 (October 1991): 932–959.

25. "Science & Engineering Indicators 2014, Chapter 3."

26. "Women, Minorities, and Persons with Disabilities in Science and Engineering," National Science Foundation, 2017, accessed 14 September 2017, https://www.nsf .gov/statistics/wmpd (author's calculations).

27. "Science & Engineering Indicators 2014, Chapter 3;" "Women, Minorities, and Persons with Disabilities in Science and Engineering."

28. A growing number of researchers are examining why women leave S&E occupations. For example, see "Women, Minorities, and Persons with Disabilities in Science and Engineering"; Julianne Pepitone, "Silicon Valley Fights to Keep Its Diversity Data Secret," *CNN Money*, 9 November 2011, http://money.cnn.com/2011/11/09 /technology/diversity_silicon_valley.

29. Underrepresented minorities include scientists and engineers who are black, Hispanic, and American Indian or Alaska Native. While the disaggregated data are not available, the unemployment rates in the innovation economy for these groups are somewhat similar. Data on gender by race and ethnicity are reported in "Women, Minorities, and Persons with Disabilities in Science and Engineering," but the accompanying data do not allow this calculation to be made.

30. "Science & Engineering Indicators 2016, Chapter 3."

31. Salary data for 2010 are from "Science & Engineering Indicators 2014, Chapter 3," and are for full-time workers with the highest degree in an S&E field. If using the measure of S&E occupations, men's median salary is 19 percent higher than women's. Salary data for 2015 are from "Women, Minorities, and Persons with Disabilities in Science and Engineering."

32. "Women, Minorities, and Persons with Disabilities in Science and Engineering," author's calculations. If considering only S&E occupations, the share of female-to-male median salary narrows to 81 percent and ranges from 77 percent for ages 29 and younger to 85 percent for ages 50 to 75. The share of female-to-male median salary is slightly higher in S&E-related occupations, 73 percent, and slightly lower for non-S&E occupations, 69 percent. "Mathematical scientist" is the only occupation in which the median female salary exceeds the male median salary, and the ratio of female-to-male median salary is 1.13.

33. For a variety of reasons, patent data is an imperfect proxy for measuring inventive activity. First, not all inventions are legally protected. Second, the mechanisms for legal protection vary widely, including patents, copyrights, trademarks, trade secrets, or some combination thereof. Finally, many patents are not economically viable; these include vanity patents, defensive patents (patents obtained not to be developed but to prevent a competitor from inventing in a complementary area), and inventions whose commercialization may be cost-prohibitive. On the methodological possibilities and limitations of using patent data, see Jacob Schmookler, *Invention and Economic Growth* (Cambridge, MA: Harvard University Press, 1966); Zvi Griliches, "Patent Statistics as Economic Indicators: A Survey," *Journal of Economic Literature* 28 (1990): 1661–1707; and Adam Jaffe and Manuel Trajtenberg, eds., *Patents, Citations, and Innovations: A Window on the Knowledge Economy* (Cambridge, MA: MIT Press, 2002).

34. Lisa D. Cook, "Inventing Social Networks: Evidence from African American 'Great Inventors,'" working paper, Michigan State University, 2007; Cook and Kongchareon, "Idea Gap in Pink and Black"; Lisa D. Cook, "The Innovation Economy in Pink and Black," working paper, 2014.

35. Alex Bell, Raj Chetty, Xavier Jaravel, Neviana Petkova, and John van Reenan, "The Lifecycle of Inventors," working paper, 13 June 2016, accessed 21 July 2017, http://www.rajchetty.com/chettyfiles/lifecycle_inventors.pdf.

36. Mark Bergen and Ellen Huet, "Google Fires Author of Divisive Memo on Gender Differences," *Bloomberg*, 7 August 2017, accessed 29 September 2017, https://www.bloomberg.com/news/articles/2017-08-08/google-fires-employee-behind-controversial-diversity-memo; Michael Cernovich, "Full James Damore Memo—Uncensored Memo with Charts and Cites," accessed 29 September 2017, https://medium.com/@Cernovich/full-james-damore-memo-uncensored-memo-with-charts-and-cites-339f3d2d05f.

37. Bourree Lam, "The Department of Labor Accuses Google of Gender Pay Discrimination," *Atlantic Monthly*, accessed 29 September 2017, https://www.theatlantic.com/business/archive/2017/04/dol-google-pay-discrimination/522411/.

38. Cook, "Inventing Social Networks."

39. Lisa D. Cook, "Overcoming Discrimination by Consumers during the Age of Segregation: The Example of Garrett Morgan," *Business History Review* 86, no. 2 (summer 2012): 211–234, especially 227–229.

40. Eglin's 1888 letter quoted in Charlotte Smith, "Colored Woman Inventor," *Woman Inventor* 1, no. 1 (April 1891): 3.

41. Luisa Kroll and Kerry Dolan, "Meet the Members of the Three-Comma Club," *Forbes*, 6 March 2018, https://www.forbes.com/billionaires/#26f4dea7251c.

42. Google Finance, *Google.com*, https://finance.google.com/finance; Google public data, world development indicators, *Google.com*, https://www.google.com/publicdata/explore?ds=wb-wdi.

43. Alex Wilhelm, "2017 Tech IPOs Are on a Tear Compared to Last Year," *Techcrunch.com*, 1 May 2017, https://techcrunch.com/2017/05/01/2017-tech-ipos-are-on-a-tear-compared-to-last-year.

44. "Building a More Inclusive Entrepreneurial System," National Venture Capital Association, accessed 29 September 2017, https://nvca.org/wp-content/uploads/delightful-downloads/2016/07/NVCA-2016-Diversity-Report.pdf.

45. "Building a More Inclusive Entrepreneurial System."

46. First Round Capital, cited in Kimberly Weisul, "Venture Capital Is Broken. These Women Are Trying to Fix It," *Inc.*, November 2016, accessed 29 September 2017, https://www.inc.com/magazine/201611/kimberly-weisul/new-face-of-funding.html.

47. Tom Huddleston Jr., "Boardroom Breakthrough: Gender Diversity Is Flourishing among Board Nominees," *Fortune*, 25 September 2014, accessed 29 September 2017, http://fortune.com/2014/09/25/boardroom-breakthrough-gender-diversity-is-flourishing-among-board-nominees.

48. For more information on gender bias, specifically implicit bias, see Sanders and Ashcraft (chapter 17) in this volume.

49. David Rock and Heidi Grant, "Why Diverse Teams Are Smarter," *Harvard Business Review,* 4 November 2016, accessed 29 September 2017, https://hbr.org/2016/11/why-diverse-teams-are-smarter.

50. Weisul, "Venture Capital Is Broken."

51. Joan Farre-Mensa, Deepak Hegde, Alexander Ljungqvist, "What Is a Patent Worth? Evidence from the U.S. Patent 'Lottery,'" 14 March 2017, NBER working paper no. 23268, accessed 8 October 2017, http://www.nber.org/papers/w23268; Catherine L. Fisk, *Working Knowledge: Employee Innovation and the Rise of Corporate Intellectual Property, 1800–1930* (Chapel Hill: University of North Carolina Press, 2009); Philip Mirowski, *Science-Mart: Privatizing American Science* (Cambridge, MA: Harvard University Press, 2011).

52. Cook and Kongchareon, "Idea Gap in Pink and Black"; odds are calculated relative to assignment to an individual.

53. David C. Mowery, Richard R. Nelson, Bhaven N. Sampat, and Arvids A. Ziedonis, *Ivory Tower and Industrial Innovation: University-Industry Technology Transfer before and after the Bayh-Dole Act* (Stanford, CA: Stanford Business Books, 2004).

54. Beyond the greater barriers to commercialization in the government sector relative to other sectors there is also an issue of selection. Government jobs are often quite stable, and government agencies have traditionally been risk averse; for example, see Gustetic (chapter 7) in this volume.

55. Cook, *African American and Women Inventors.*

56. Ellen Pao, *Reset: My Fight for Inclusion and Lasting Change* (New York: Spiegel and Grau, 2017).

57. Weisul, "Venture Capital Is Broken"; Cook and Kongchareon, "Idea Gap in Pink and Black."

58. Thomas Piketty and Emmanuel Saez, "Income Inequality in the United States, 1913–1998," *Quarterly Journal of Economics* 118, no. 1 (2003): 1–39; Anthony B. Atkinson, Thomas Piketty, and Emmanuel Saez, "Top Incomes in the Long Run of History," *Journal of Economic Literature* 49, no. 1 (2011): 3–71.

59. Emmanuel Saez, "Striking It Richer: The Evolution of Top Incomes in the United States (updated with 2012 preliminary estimates)," University of California, Berkeley, 3 September 2013, https://eml.berkeley.edu//~saez/saez-UStopincomes-2012.pdf.

60. Thomas Piketty, *Capital in the Twenty-First Century*, trans. Arthur Goldhammer (Cambridge, MA: Belknap Press, 2013).

61. See Alan Krueger, "The Rise and Consequences of Inequality in the United States," Council of Economic Advisers, The White House, 12 January 2012, https://obamawhitehouse.archives.gov/sites/default/files/krueger_cap_speech_final_remarks.pdf;

Dabla-Norris et al., "Causes and Consequences of Income Inequality: A Global Perspective," IMF Staff Discussion Note, SDN/15/13, October 2015, https://www.imf.org/external/pubs/ft/sdn/2015/sdn1513.pdf; Janet Yellen, "Perspectives on Inequality and Opportunity from the Survey of Consumer Finances," speech delivered at the Conference on Economic Opportunity and Inequality, Federal Reserve Bank of Boston, 17 October 2014, https://www.federalreserve.gov/newsevents/speech/yellen20141017a.htm.

62. Jennifer Hunt, Jean-Philippe Garant, Hannah Herman, and David J. Munroe, "Why Don't Women Patent?" NBER working paper no. 17888, National Bureau of Economic Research, March 2012, http://www.nber.org/papers/w17888.pdf.

63. Lisa D. Cook and Yanyan Yang, "The Commercialization Gap in Pink and Black," Michigan State University, mimeo, 2017.

13   Make Maintainers: Engineering Education and an Ethics of Care

Andrew L. Russell and Lee Vinsel

We start with a crucial distinction: the difference between innovation and the way people talk about it. On the one hand, there are the various *acts* that we can refer to as innovation. On the other hand, there is all the *talk* about innovation, a public discourse that we refer to as "innovation-speak." These things are related but conceptually distinct; indeed, for over ten years now we have seen prominent professionals complain about innovation-speak as a way to defend the act of innovation. For example, in 2005 designer and writer Michael Bierut bemoaned the "cult of innovation," complained that innovation was a "euphemism," a "bandwagon," and a "fad," and reminded his readers of a warning from the legendary designer Charles Eames: "Innovate as a last resort: More horrors are done in the name of innovation than any other."[1]

Bierut's skepticism places him in a distinct minority. Innovation-speak flourished over the next decade, despite warnings from cheerleaders of business and technology that innovation had become "the most overused word in America" (*Wired*) and that the term had "begun to lose meaning" (*Wall Street Journal*).[2] Academics also began to wonder what the appealing term was obscuring. In 2008 historian Benoît Godin began a critical history of the idea and concept of innovation; by early 2014, we were regularly pointing out the overuse of the term "innovation" in our classrooms, at conferences, and in online discussions.[3] We published our views in the online magazine *Aeon* in a 2016 essay titled "Hail the Maintainers," which laid out a critique of innovation-speak and proposed an alternative vision of technology-in-society with maintenance at the center.

The starting point of our critique is a simple idea: innovation-speak does not adequately capture the essence of human life with technology. It is true that our culture's recent obsession with innovation has generated a deeper and more meaningful understanding of where innovation and innovators come from. Innovation is important. It has played an essential role in economic growth and improved quality of life.

But this focus on innovation has an unfortunate side effect, which has been to obscure so many other aspects of technology and its social consequences. More troubling, innovation is often treated as value-in-itself and as a panacea: technological change will save us without our ever having to enter into human dialogue. At its most extreme, innovation-speak actively devalues the work of most humans, including most college graduates, and could actually harm the self-conceptions of students who end up in completely essential but noninnovative careers. To put it another way, chronicles of various acts of creation and innovation are not one and the same as the totality of human experience with technology. Indeed, when we reflect on human life with technology, we conclude that most human effort around technology involves maintenance, repair, upkeep, and mundane labor.

Our purpose in this essay is to offer a holistic picture of human life with technology and to give suggestions for how education might be aligned with this picture. We focus on engineering education, because innovation-speak is particularly rampant in that domain. We argue in the end that reorienting engineering education around an ethics of care provides a new and refreshing vision that liberates us from the constraints of innovation-speak. In turn, this creates space for both innovative and noninnovative work and provides a more accurate and grounded vision of technology and society. The entirety of the subjects we are engaging—from innovation-speak, to the social roles of the maintainers, to the ethics of care—are rooted in the stories we tell each other about the world. If we are correct in arguing that we would be better off once we move past our societal obsession with innovation, our first steps should be to change the tales we tell about technology and society.

Innovation-Speak and the Transformation of American Universities

Use of the word "innovation" has increased greatly since World War II and even more intensely since the 1990s, but this shift builds on a much longer history of technology and culture. Recent work by historians such as Deirdre McCloskey and Joel Mokyr suggests that one important source of the

British Industrial Revolution was a cultural revaluation of work, technical knowledge, and material novelty.[4] As invention was accorded increasing social status, more bright and capable individuals went into the business of invention and the exploitation of nature. In the United States, these cultural developments were associated with heroes such as Benjamin Franklin. More explicitly, by the 1850s and 1860s, popular authors such as Samuel Smiles celebrated engineers both as idols who brought material improvements to the lives of many and as paragons of Smiles's moral ideal of "self-help." By the late nineteenth century, a "cult of invention" had developed around popular figures such as Thomas Edison and Alexander Graham Bell.

In the early twentieth century, corporations started building R&D labs to institutionalize the method of invention and to build corporate strategies around continuous and predictable patterns of innovation.[5] The pioneers of industrial R&D figured out how to harness the imagery of invention for the purposes of marketing and self-promotion. Two prominent examples were General Electric's "House of Magic" and General Motors' annual model changes, auto shows, and industrial musical films.

This corporatization of invention—both as a material reality and as corporate imagery to hawk on the market—often went hand-in-hand with a deeper cultural reliance on material progress. Scholars refer to this reliance as the "technological fix," which is a fundamental faith that deep social problems can be resolved simply though technical change rather than through a political rearrangement of social structures. In the post–World War II period, this worldview in the United States increasingly became tied to an anti-communist celebration of free enterprise, such as that in the Kitchen Debate in Moscow and at Disney World's EPCOT Center (the Experimental Prototype Community of Tomorrow), which received support from a number of corporations.

The discourse of innovation-speak developed in this context, and it stemmed from multiple sources. One of the most important for our purposes is the rise of the economics of innovation, or more broadly, "innovation studies." In the late 1950s and 1960s, economists including Robert Solow and Kenneth Arrow hypothesized that technological change, or innovation, was a significant factor in economic growth. Within a few years, this hypothesis had hardened into orthodoxy within some schools of economic thought. The notion of innovation increasingly became tied to technology, and the term "technological innovation" took off in the

1960s.[6] A turn to making innovation a normative aspiration—something we *should* do, rather than something that just happens in the world—was significantly enhanced by the rise of "innovation policy" in the late 1970s, which asserted that government activity could and should increase innovation.[7]

The connections between innovation and fear in economic policy were supported by American foreign policy. In the two world wars and throughout the Cold War, American policymakers agreed that military superiority depended on scientific and technological superiority. This consensus drove substantial investments in conventional weapons; more lethal chemical and nuclear weapons; new approaches to naval, aerial, and space vessels; and basic science investments in solid-state components and computing devices.

Even after the end of the Cold War and the emergence of the United States as the world's sole military superpower, fear continued to be a powerful motivator for American innovation policy at home. In the United States, the turn to innovation policy was directly tied to a cultural fear of Japan, particularly economic competition from that nation but also a worry that the Japanese would take over US institutions and push their cultural practices on American workers. From that time forward, innovation-speak has been a discourse of fear. Rust Belt towns that were falling behind sought to make themselves the next Silicon Valley. Businesses paid oodles to professor and consultant Clayton Christensen, who coined the term "disruptive innovation" in the hopes that he could help them avert the possibility of their companies being overthrown by outsider, upstart firms. Corporate executives, university presidents, and science policy gurus increasingly told stories about how the American education system was falling behind, especially when it came to science and technology, and about how young people would be cast adrift unless they received degrees in so-called STEM fields.

Moreover, since the 1980s, American universities have increasingly been re-created in the corporate image, and most of the changes have been made in the name of innovation. As Philip Mirowski and others have detailed, new laws and other institutional changes have been aimed at turning universities into patent factories.[8] The Bayh-Dole Act of 1980, for instance, allowed researchers to patent inventions created through federal funding, something that had previously been forbidden when the running assumption was that federal money should benefit public rather than private goods. As corporations have scaled back expenditures on R&D, the National Science Foundation and other funders have increasingly become focused on knowledge that is exploitable in the short term, rather than on long-run basic science. In their grant proposals, scientists and engineers have to claim that they are doing something novel and innovative rather than advancing fundamental scientific knowledge. University business models have become more and more dependent on the "overhead" from sponsored research, and university administrators have come to measure the value of faculty members by how much grant money they can pull down.

Universities have also come to accept the idea that it is their core mission to create *innovators*. Often this impetus goes hand in hand with a celebration of STEM education, with scientific and engineering knowledge being seen as the key to innovative activity, but the current focus on innovation and entrepreneurship in higher education goes well beyond the bounds of STEM.[9] The University Innovation Fellows (UIF) program, is a good example of this wider cultural trend.[10] Initially funded by the National Science Foundation, the UIF is a training program and social network for students at all levels of university education. UIF encourages students to imagine themselves as "change agents" who must disrupt the stodgy ways of their universities and introduce innovations. Armed with sticky notes, whiteboards, and a "fail fast" mentality, the students are "empowered" to value discontinuity, novelty, and change rather than continuity, tradition, and care. They go down this path of disruption with little reflection on what ends such changes are meant to accomplish. Innovation is assumed to be a value in itself: UIF's website is filled with words such as "change," "innovation," "creativity," and "entrepreneurship," with minimal reflection on what changes are desirable or what ends are hoped to be reached. While it is unclear what values motivate the UIF—beyond the nonvalue of change for its own sake—it seems certain that the fellows get a deep education in creating hype.

Taking all of these recent historical developments together, it is clear that the innovation idea is more than an overused business slogan. It has come to form the basis of a thoroughgoing reform of basic cultural institutions including but not limited to schools and universities. Innovation has become the yardstick in universities both for the outcomes of faculty research, that is, patents and grant money, and for the outcomes of undergraduate education—STEM majors winning high-paying jobs in tech sectors.

Innovation-Speak and the Training of Engineers

Engineering schools have become particularly fertile grounds for innovation-speak. Many engineering students now take required courses in entrepreneurship and design sequences focused on innovation. These students are rarely told that the narrative of innovation-speak, particularly versions having to do with STEM education, also serve the economic interests of these schools. The rhetoric of "innovation" and high-paying engineering jobs becomes a natural and almost effortless form of marketing in today's culture of uncertainty and anxiety, including the very real concerns about the cost of education, student debt, and return on investment.

The innovation focus in university engineering schools builds on long traditions within engineering education and the engineering profession, which typically center on invention and design. Engineering degrees commonly end with capstone or "senior design" projects that involve the creation of new things and not with more mundane (and realistic) engineering undertakings. For example, engineering students often help build robots or electric cars, create a computer program, or design remotely piloted drones. Yet most engineers do not take part in design activities once on the job. Professional engineering societies reinforce this focus on design and novelty in several key ways. They hand out awards and fellowships primarily to engineers who have created new technologies rather than to engineering leaders who have played fundamental roles in keeping systems and enterprises running smoothly. The IEEE's highest award, the Medal of Honor, for instance, has as its evaluation criteria "substantial significance of achievement, originality, impact on society, impact on profession, publications, and patents related to achievement." Lists of recent winners make clear that these criteria are understood in terms of invention and innovation. Similarly, the National Academy of Engineering's Draper Prize—its highest prize—typically rewards new inventions.[11] Moreover, in 2008, the National Academy of Engineering put out its Grand Challenges for Engineering, which are almost wholly described in terms of creating new things to solve deep social problems—the technological fix writ large.

Some proposals to reform engineering education go further, arguing that it should be remade in the image (and language) of Silicon Valley. One clear example of this today is the so-called Big Beacon Movement in engineering education, which, borrowing from the "revolutionary" language of innovation-speak, aims to show "how all stakeholders can collaborate to disrupt the status quo."[12] Unsurprisingly, Big Beacon receives a laudatory shout out on the homepage of the UIF, which itself views universities as backward, bureaucratic organizations in need of revolutionary change. In their book *A Whole New Engineer: The Coming Revolution in Engineering Education*, Big Beacon cofounders David Goldberg and Mark Somerville put forward a vision that badly misrepresents the nature of technology and society.[13] The entire book is conceived in destructive and fearful terms as detailed in the epilogue, "Invitation to Collaborative Disruption: Will Disruption Shape Us, or Will We Shape It?" On top of chapters full of buzzwords ("Changing How We Change: From Bureaucracy to Change Management"), four of the book's nine chapters contain the phrase "whole new." According to the authors, we need "whole new" engineers, learners, professors, even a "whole new" culture. But is there really nothing in our culture worth preserving? Is it really true that the technologies around us are entirely new, or should be? Is it even imaginable that engineers will deal only with the "whole new" rather than having to learn how to wisely manage and maintain the old?

Engineering Is Maintenance

If you adopt even a modestly critical point of view, you will quickly conclude that the rhetoric in works such as *The Whole New Engineer* is simply out of touch with ordinary life. If you look at the room around you, you will see many mundane technologies—including tables, chairs, light bulbs, bookshelves, books, electric fans—that have gone through long processes of incremental change but have been largely unaltered for decades, even centuries. Just behind the walls are other technologies—water and waste pipes, HVAC ducts, electric wiring—that are similarly old and unremarkable. If you commuted today, you likely crossed roads, bridges, railroad beds, or subway systems that would not have looked surprising or foreign to someone living in the 1920s. Many of the technologies that you have used to live today—electric or gas stoves for cooking your breakfast, running water for washing your dirty body, toilets for sending your waste away—are not "whole new," are not revolutionary, are not innovative in any significant way, and yet they are totally necessary. Moreover, the vast quantity of human labor is aimed at keeping these fundamental systems running, rather than at introducing wholly new technologies, and human society relies on these systems to keep itself going (for instance, prepping food to keep us from starving). According to one study, over 70 percent of engineers work on maintaining or overseeing existing systems rather than designing new ones.[14] Furthermore, there are many technological systems—such as electricity, water, phone, and internet services—that we do not want to see "disrupted"; rather, we value reliable, continuous, high-quality service.

Unfortunately, it is not only engineering education that misses the fundamental importance and ubiquitous nature of maintenance. Much of the scholarly literature about technology fails to reckon with these basic facts of ordinary life with technology. Because historians and others who study the social dimensions of technology grew up in a culture that celebrated and centered on invention and innovation, their work also has been focused on these phenomena. While a few classic works emphasize the centrality of maintenance and repair for sustaining and conserving society, in most technology studies, maintenance, repair, and upkeep are largely ignored, rendered invisible.[15] The scholarly focus on invention and innovation has greater consequences than simply creating "gaps in the literature." After all, how are engineering professors and other educators to learn and teach about the broad history of their fields' technologies if the available literature focuses so narrowly on invention?

Mercifully, a growing body of literature has started to improve this situation. Ruth Schwartz Cowan, for instance, in her classic study *More Work for Mother* examined how women's housework, much of it maintenance-focused, perpetuated and sustained family life.[16] Another touchstone book in maintenance studies, David Edgerton's *The Shock of the Old*, emphasizes that most basic technologies around us are old rather than new, ordinary rather than novel. Edgerton points out that one reason it is difficult to talk about maintenance as a social process is that it often is not counted in economic metrics. Canada did ask about maintenance costs for many years in an economic survey. For those years, maintenance accounted for between 11 and 21 percent of GDP, a vastly higher number than innovation-centric expenditures such as spending on research and development (R&D), which only comprises about 2 percent of GDP in OECD countries today.[17]

Moreover, the study of maintenance and repair has greatly expanded in the last decade.[18]

The real shame of the matter is this: a more holistic, sober, and accurate picture of human life with technology has been around for decades, and some of the authors who have put it forward, such as Cowan, are relatively well-known beyond the boundaries of the small field of technology studies. The evangelists of innovation who buy too wholly into the rhetoric of "whole new" are acting irresponsibly by ignoring diligent research that has actionable insights.

Once a more grounded vision is established, it is easy to see that most engineering work will always be dedicated to maintaining and conserving existing technological systems and using those systems for production, not in introducing new systems. Because of the way that industrial societies have developed, it could not be any other way. Most civil engineers work on keeping up existing physical infrastructures, such as roads and bridges. Even in "cutting edge" fields such as software, about 70 percent of budgets go into maintenance and upkeep, whereas only about 8 percent of budgets go into new design, as historian Nathan Ensmenger has noted.[19] Moreover, the structure of the engineering workforce means that most engineers work with large-scale technological systems, where companies create value through quality of service. These engineers know that radical or revolutionary changes usually do little more than irritate customers—and these customers tend to complain to regulators and their elected representatives.

To summarize, most engineers are going to be maintainers, and if we include our perspective to include all workers, not just engineers, the percentage of maintainers will be even higher. Yet innovation-speak actively devalues this essential work, which will never be radical, revolutionary, or "whole new." As a discourse that is shoved down the throats of young people, innovation-speak has the potential to generate in them false self-images as innovators that turn out to be harmful when they end up in jobs that are essential but basically noninnovative. This can lead to real disillusionment, not only with society at large but with specific authority figures, with students feeling they have been lied to by their university, their professors, and maybe even their parents, who encouraged them to pursue engineering. We have heard several anecdotes from leaders in business and education that acknowledge the crux of the problem: prevailing rhetoric encourages *everyone* to be entrepreneurial innovators who come up with big ideas, but all organizations need many more people who can maintain and execute—in other words, who can simply get things done. Given the moral hazards of innovation-speak, is there a better way of thinking and telling stories about the role of technology in society that can offer a holistic vision of maintenance *and* innovation? We believe so, and we think it is rooted in an ethics of care.

An Ethics of Care

In the opening sections, we described a trend: American culture is saturated with the ideology of innovation-speak, and that ideology's celebrated concepts of entrepreneurship and disruption have seeped into engineering education. This trend is troubling because it misrepresents the character of the work that actual engineers do. We believe these students—and the communities they serve—will be better off if they replace notions of innovation and disruption with an ethics of care. The ethics of care arose as part of feminist theory in the late twentieth century, most famously in Carol Gilligan's 1982 book *In a Different Voice*.[20] The starting point for the ethics of care was a fundamental critique of existing ethical paradigms. Gilligan and others believed these paradigms were overly abstract and intellectual and, therefore, did not reflect how ethical decisions were actually made in ordinary, everyday life.

The ethics of care is rooted in a few basic ideas. First, we are fundamentally dependent on one another—a conceptual departure from classical liberal theory, which cast us as basically independent and autonomous. Here, the authors' background in technology studies compels us to add that one way we humans depend on each other is through technologies and infrastructures, which require massive collaborative and coordinated efforts to sustain. Second, our decision-making must first attend to the marginal and vulnerable. Such a perspective is often left out of innovation-speak, which brackets how technological change affects people. Silicon Valley, the kingdom of the innovation-mouthed, is a horribly unequal place, where multiple poor families pack into small ranch houses just to make ends meet.[21] Third, rather than being rooted in abstract principles, our moral choices should attend and respond to the immediate conditions of our context. Indeed, the ethics of care can be thought of as an ethics of responsiveness.

The authors find the ethics of care to be a helpful way of thinking about all education, including engineering education, particularly because the ethics of care reorients us to thinking about ends rather than means. For instance, for many people, the goals of a just society are to provide a high quality of life to all in an environmentally sustainable manner. Obviously, there are many different ideas about how best to reach these goals, and often discussions about these issues are founded on traditional divisions. Some individuals believe that the "free market" provides the optimal society and that government intervention can only interfere with and degrade these processes, while others assert that the state has an active role to play in improving life for all.

Ultimately, then, the ethics of care pushes us to have explicit conversations about values—or put another way, what we each value. Clearly, there is no unanimity or even rough consensus around the values our society holds dearest, as we live in a diverse social world with many different individuals and groups, who hold many different, sometimes conflicting, values. To make matters more complicated, the United States has increasingly become a partisan society: members of different political parties do not like each other. Yet when we help students to reflect on their actual cares and values, what they say often flies in the face of the ideology of innovation and entrepreneurship. For instance, a colleague noted that one of his engineering students—a young man who emigrated with his family from India—found innovation-speak wholly alienating. The student was interested in finding a good job that would allow him to provide for his parents, siblings, and his eventual wife and children. In other words, his actual values were oriented toward interconnection and care. Our point is that his ultimate work as an engineer would likely be similarly oriented. If he came to work as a power systems engineer for an electric utility, the reliable electricity he would work to produce would help run medical devices and other technologies that keep people alive. This work is critical, even if it has nothing to do with innovation.

To put the point directly: maintenance *is* caring. In some cases, individuals perform maintenance as an expression of care directed at particular objects, such as when they oil bicycle chains or replace air filters. In other cases, this expression of care is directed at people or groups, such as when individuals participate in birthday parties or visit nurses, doctors, or therapists. All of these activities are maintenance activities, and they all involve care; as such, they invite us to ask: What values and interests are cared for when maintenance work is performed? Applied to engineering work and engineering education, this question raises an opportunity for reflection, not merely on the instrumental value of engineering but also on the deeper human values that engineering can support.

We know that engineers are more than capable of reflecting on the fundamental values that their work engages. Engineers often conduct such reflections through the vehicle of their professional societies' codes of ethics. Let us consider briefly the Code of Ethics of the American Society of Mechanical Engineers (ASME). As with many other engineering societies, ASME's code focuses primarily on the need for engineers to be objective, fair, and honest in their business dealings.[22] In other words, the code largely relates to ensuring and increasing the social status and prestige of engineers and toward supporting the healthy functioning of capitalism by avoiding crime and corruption. But some aspects of the code go beyond such professional matters. ASME's code is built on three fundamental principles. The first holds that engineers should use "their knowledge and skill for the enhancement of human welfare." This notion is further elaborated in two of the eight "fundamental canons," which build on the fundamental principles. Canon 1 asserts that "engineers shall hold paramount the safety, health, and welfare of the public in the performance of their professional duties." And canon 8 reads, "Engineers shall consider environmental impact in the performance of their professional duties." While these principles and canons are fairly vague and certainly leave a great deal of leeway for interpretation, they can be used to start deeper conversations about values. At a bare minimum, they should remind us that engineering goes well beyond innovation.

Put another way, much of modern life depends on well-functioning technological systems, and the vast majority of human work will always be aimed at maintaining them—that is, the labor is oriented toward taking care of the world and its inhabitants. This work is essential, and we should value it. Yet care also involves change. If we find a better method of caring for the world, we should adopt it, but not in ways that degenerate the quality of life for others.

For engineering education, this means that we must strike a balance between pedagogies that value maintenance and innovation. Innovation is important, and it should be part of engineering programs. We know that some reliable factors hamper innovation processes and lead to innovation "valleys of death," and we should teach our students how to surmount these barriers if they can. We also need to ignite the imaginations of young people, to nurture their creativity, and to teach them that they should resist the arbitrary exercise of authority. But such lessons need to sit in a more expansive context and broader moral compass. Engineering is fundamentally about caring for technological systems, the humans that rely on them, and the natural environments that surround them. Innovation is but a small part of that overall process of stewardship.

For sure, we see precursors to the ethics of care in long-running engineering traditions. For instance, during the 1920s and 1930s, the high moment of engineering progressivism, Herbert Hoover and other influential figures worshipped at the altar of "efficiency."[23] Increasing efficiency often involved the introduction of new technologies and processes—"innovation" in today's language—but it was carried out in the name of *conserving* resources, both financial and natural, and reducing *waste*, an important moral term of that period. In other words, efficiency was more focused on ends than means.

Although it has been around since the 1980s, the ethics of care framework and examples focused on operations and maintenance have made little headway in engineering education and the ethics courses and modules that make up engineering curricula. To give one example, Gail Baura's textbook *Engineering Ethics: An Industrial Perspective* (2006), in many ways a strong work, contains thirteen case studies of ethical problems.[24] Yet of these, nine are wholly or mostly focused on the early stages of technology—design, research, and development. As we have seen, roughly 70 percent of engineers actually spend their work time focusing on maintenance and the oversight of existing technologies. In this way, most existing engineering ethics texts do not reflect the actual work that engineers will do, in part because they buy into the ideological self-image of engineers as *creators*.

Moreover, as mentioned earlier, these texts do not fit engineering students' own moral self-understandings. In an interesting study, engineer Angela Bielefeldt introduced sixty-four engineering students to five standard frameworks for thinking about ethics (rights ethics, duty ethics, utilitarianism, virtue ethics, and ethics of care) and asked them which theory was closest to their own moral worldview.[25] The largest number, eighteen, chose the ethics of care, a view usually not even covered in such courses.

Furthermore, this choice had strong gender and racial/ethnic components. About 40 percent of women in the course chose the ethics of care (as opposed to 23 percent of their male counterparts), and a staggering 57 percent of Hispanic American students made the same choice. These findings suggest that standard engineering education may actually alienate women and minorities by limiting them to moral frameworks that do not accord with their actual beliefs and experiences. As Bielefeldt suggests, "Teaching engineering ethics through the ethics of care may be helpful to retain women and minority students," a constant, well-known problem in the engineering field.

When it comes to teaching the ethics of care, engineering has much to learn from other fields and disciplines. The healthcare and K-12 education fields have made the approach a central feature of their training programs for decades.[26] Put another way, engineering educators do not need to invent anything from scratch, but they can effectively adapt thinking and teaching tools from other fields. For instance, political scientist Joan Tronto divides caring into four phases:

1. Caring about, in which the caregiver realizes that there is a potential problem
2. Taking care of, in which the caregiver decides the proper course of action
3. Caregiving, in which the action is carried out
4. Care receiving, in which the caregiver assesses the success of the action[27]

This simple ethics of care schema alone—and there are several others of its type—has broad applicability in engineering practice and is something not covered in most engineering education. Moreover, it touches on real and serious moral lapses in the history of engineering—from the spectacular, such as Enron using fake maintenance to spike energy prices and induce blackouts, to the mundane, such as the fact that maintenance workers are often the most frequently injured and killed in industrial operations—that could act as cases for this approach.

Taking the ethics of care seriously means that, in addition to courses on innovation and entrepreneurship, engineering programs need to introduce more courses, experiential opportunities, and capstone projects that focus on conservation, maintenance, and upkeep. There are multiple ways to introduce engineering students to themes of maintenance. First, students should learn how corporations and engineering professionals manage maintenance regimes. Melinda Hodkiewicz, an engineering professor at the University of Western Australia and a member of the Maintainers network, regularly teaches basic maintenance theories and concepts in her classes.[28] She believes that, at a minimum, students should become familiar with reliability-centered maintenance, a formal and standardized process for managing system maintenance. But her own teaching goes far beyond these basics, and she has some evidence of success. One of her students started an internship with a petroleum company and wrote to her soon afterward to thank her: "Everybody in the team was really impressed that I had previous exposure" to basic maintenance theories and concepts.[29] "Pretty much everything I learned in [Hodkiewicz's class] is what I'm using in practice. It saved them a lot of time when they were explaining the scopes I'd be responsible for and also made me look great the first week I started." In other words, introducing students to maintenance is important because often it is what they will be doing on the job.

Second, even as students learn about innovation, they should do so with an emphasis on its inherent relationship to maintenance. For example, students should also understand the notion of designing for easy and efficient maintainability. Here, ethical and political topics are unavoidable. Since corporations introduced practices of planned obsolescence in the 1920s, they have designed for the opposite of maintainability, particularly when it comes to consumer products. Some firms go even further, creating what some call "forced obsolescence." For example, Apple stops supporting and updating its iPhones after putting out a certain number of new products and system upgrades. Even when older phones are still fundamentally sound, they become basically unusable. Given that cell phones involve many environmentally unfriendly and politically problematic parts and materials, forced obsolescence raises serious moral questions. Designing for maintainability involves certain established practices, but it is also a rich opportunity to involve students in ethical discussions about what they owe other humans in their professional lives.

Third, maintenance and upkeep can and should form the basis of capstone projects. Such projects could take many forms. Of course, this could be as simple as apprising students of how the university's facilities and physical plant staff keep the school going. Other options include having students maintain university depositories of student and faculty publications, having computer science students work with updating and altering back-end legacy code, and working with local transport and infrastructure organizations, whether private or public, to manage and update systems maintenance routines, particularly if the organization's practices are inadequate or out-of-date. We think that environmental engineering and sustainability management provides a particularly rich way to explore these issues, however, and that they should be required of more engineering students. Achieving a more sustainable future that greatly reduces the amount of greenhouse gases being emitted will of course require innovation. But it will also involve rethinking how we use our resources, maintain our physical infrastructures, and take care of the world around us.

In the end, the ethics of care probably requires engineering students to be educated in the politics of technology and society—something that the relatively apolitical engineering tradition might find hard to swallow. This is not about indoctrinating students into any particular political view. We find aspects of the Maintainers both in certain forms of conservatism, which argue that we have a moral duty to care for what we have inherited from our ancestors, and in certain forms of progressivism, which assert that healthy capitalism requires active intervention, particularly around issues such as pollution, safety, and the well-being of public works. Care requires holistic, or systems, thinking that goes far beyond the individualist fantasies of innovation-speak with its pantheon of great white men: Gates, Jobs, Bezos, Zuckerberg, Thiel, Musk. It requires us to realize that we are dependent on each other and on the technological systems and infrastructures that many, including those who have come before us, have erected; these systems and infrastructures now require our attention and safekeeping, even when such work bores our pants off and pales in the light of nifty, new, glittering gadgets.

Conclusion: Making Maintainers

In this chapter, we have argued that since the 1960s, American society has increasingly become dominated by innovation-speak, an ideology that glorifies technological change as the answer to society's problems. Countering innovation-speak is important not because it is an annoying way of talking, though that is true enough, but because of two important reasons: First, innovation-centrism offers at best a partial view of human life with technology. Second, reforms made in innovation's name—including changes made to all levels of education—are at best questionably effective and at worst deeply damaging to the traditional roles and practices of institutions.

We have also argued that there are better ways of thinking about ordinary life with technology, which start by focusing on the bulk of human practices with things, including maintenance, repair, and mundane labor. The differences between these two views have important implications for education, and we have tried to articulate how maintenance-centered thinking can be used to reform and improve engineering education. We have tried to show how the ethics of care can provide a holistic vision of engineering education that includes both upkeep and innovation but does not overly privilege the latter.

We have also discussed how engineering education requires more focus on values and ends, and we find innovation-speak particularly lacking on this front. Innovation is not a value in itself, although it is often treated like one in contemporary society. Yet there is one area where innovation-speak currently outpaces the more grounded vision of technology put forward in this chapter, and that is when it comes to positive visions of the future. Certainly, one thing about the current imagery and ideas around innovation that captures young minds is the techno-utopian fantasy of a better future, to which individual innovators can and will contribute. Consider, for instance, the excitement generated by Elon Musk's announced plans to go to Mars.

As yet, the focus on maintenance and maintainers has nothing comparable. In part, this stems from an image problem: maintenance and infrastructure aren't sexy. Comedian John Oliver pointed this out in a segment on infrastructure on his show, *Last Week Tonight*. At the end, he argued, "No one has made a blockbuster movie about the importance of routine maintenance and repair," and he went on to imagine a star-studded nonaction film titled *Infrastructure*. In the real world, we see this difference between innovation and maintenance play out when elected officials have incentives to take part in photo ops and stand in front of ribbon cuttings for new infrastructure but little incentive or opportunity to take credit for existing things working well. Moreover, we are living in a moment of perceived cynicism and pessimism: as the philosopher Slavoj Zizek suggests, we have few utopian, or at least nondystopian, visions of tomorrow in popular culture, which often seems to consist primarily of zombie stories and tales of environmental apocalypse. For a variety of reasons, then, we lack a picture of a positive future that includes a well-ordered and maintained technological society that does not involve radical technological change.

Yet we believe it is incumbent on all of us to put forward such a positive vision. Students in all fields, including engineering students, should be involved in such visionary practices from the start of their educations. But current techno-utopian visions are far too focused on innovation and radical technological change, and basically ignore politics and conservation of the ordinary and mundane. Here are some examples that can be used to kick-start grounded discussions of a positive tomorrow. The American Society of Civil Engineers regularly gives American infrastructure low grades in its infrastructure report card. What would it look like if the country got straight A's? How would we get there? These same questions can be asked of overhauling American drinking water systems. After experts established that the water system in Flint, Michigan, was poisoned with lead, the same situation was found in hundreds of other water systems around the nation. As a massive political and engineering project, how can we transform our current systems and ensure clean drinking water for all? Finally, many, perhaps most, existing homes and buildings throughout the United States are extremely energy inefficient, and yet decreasing energy use is one of the most important ways to manage global climate change. How could we create a program to rehabilitate all existing buildings and bring them in line with energy standards such as LEED? How would such a program work?

In asking these big questions, and putting them in the form of a challenge to students, we are expressing our confidence that we can come up with compelling answers together. In many cases, we suspect that there is a place for innovation and novelty in some of the projects we describe above. Wouldn't it be nice to see innovation put to work in the service of maintaining and caring for our ailing technological society? Indeed, the nation faces a vital moral imperative to make maintainers.

Notes

1. Michael Bierut, "Innovation Is the New Black," *Design Observer*, 20 November 2005, accessed 17 July 2017, http://designobserver.com/feature/innovation-is-the-new-black/3857.

2. Michael O'Bryan, "Innovation: The Most Important and Overused Word in America," *Wired*, accessed 17 July 2017, https://www.wired.com/insights/2013/11/innovation-the-most-important-and-overused-word-in-america/; Leslie Kwoh, "You Call That Innovation?" *Wall Street Journal*, 21 May 2012, accessed 17 July 2017, https://www.wsj.com/articles/SB10001424052702304791704577418250902309914.

3. Lee Vinsel, "How to Give Up the I-Word," parts 1 and 2, *Culture Digitally*, 22 and 23 September, 2014, accessed 17 July 2017, http://culturedigitally.org/2014/09/how-to-give-up-the-i-word-pt-1/ and http://culturedigitally.org/2014/09/how-to-give-up-the-i-word-pt-2/.

4. Deirdre N. McCloskey, *Bourgeois Dignity: Why Economics Can't Explain the Modern World* (Chicago: University of Chicago Press, 2010); Joel Mokyr, *A Culture of Growth: The Origins of the Modern Economy* (Princeton, NJ: Princeton University Press, 2016).

5. David A. Hounshell, "The Evolution of Industrial Research in the United States," in *Engines of Innovation: US Industrial Research at the End of an Era* (Cambridge, MA: Harvard Business Press, 1996), 51–56.

6. See Godin (chapter 9) in this volume and Benoît Godin, *Innovation Contested: The Idea of Innovation over the Centuries* (New York: Routledge, 2015).

7. See Wisnioski (chapter 1) in this volume.

8. Philip Mirowski, *Science-Mart* (Cambridge, MA: Harvard University Press, 2011).

9. See Carlson (chapter 16) in this volume.

10. See Fasihuddin and Britos Cavagnaro (chapter 3) in this volume.

11. "Charles Stark Draper Prize for Engineering," National Academy of Engineering, accessed 17 July 2017, https://www.nae.edu/Projects/Awards/DraperPrize/DraperWinners.aspx.

12. "A Whole New Engineer," *Big Beacon*, accessed 4 September 2018, http://wholenewengineer.org.

13. David E. Goldberg and Mark Somerville, *A Whole New Engineer: The Coming Revolution in Engineering Education* (Douglas, MI: Threejoy, 2014).

14. David Edgerton, *Shock of the Old: Technology and Global History since 1900* (London: Profile Books, 2011); Svante Lindqvist, *Changes in the Technological Landscape: Essays in the History of Science and Technology* (Sagamore Beach, MA: Science History Publications/USA, 2011).

15. Karl August Wittfogel, *Oriental Despotism: A Study of Total Power* (New Haven, CT: Yale University Press, 1957); Albert O. Hirschmann, *The Strategy of Economic Development* (New Haven, CT: Yale University Press, 1958).

16. Ruth Schwartz Cowan, *More Work for Mother: The Ironies of Household Technology from the Open Hearth to the Microwave* (New York: Basic Books, 1983).

17. Edgerton, *Shock of the Old*, 79; "Gross Domestic Spending on R&D," OECD, accessed 10 October 2017, https://data.oecd.org/rd/gross-domestic-spending-on-r-d.htm.

18. Kevin L. Borg, *Auto Mechanics: Technology and Expertise in Twentieth-Century America* (Baltimore: Johns Hopkins University Press, 2010); Christopher Henke, "The Mechanics of Workplace Order: Toward a Sociology of Repair," *Berkeley Journal of Sociology 44* (1999): 55–81; Jérôme Denis and David Pontille, "Material Ordering and the Care of Things," *Science, Technology, and Human Values* 40, no. 3 (2015): 338–367; Steven J. Jackson, "Rethinking Repair" in *Media Technologies*, ed. T. Gillespie, P. J. Boczkowski, and K. A. Foot (Cambridge, MA: MIT Press, 2014), 221–240. See also the recent special issue (vol. 6, no. 1, 2017) on repair edited by Steven Jackson, Daniela Rosner, and Lara Houston of the online journal *Continent*, which includes several essays, including ones by the editors and one by myself, accessed 17 July 2017, http://www.continentcontinent.cc/index.php/continent/issue/view/2.

19. Nathan Ensmenger, "When Good Software Goes Bad: The Unexpected Durability of Digital Technologies," presented at the Maintainers conference, 9 April 2016, accessed 17 July 2017, http://themaintainers.org/s/ensmenger-maintainers-v2.pdf.

20. Carol Gilligan, *In a Different Voice* (Cambridge, MA: Harvard University Press, 1982).

21. John D. Sutter, "Poor Kids of Silicon Valley," CNN, accessed 30 September 2017, http://www.cnn.com/interactive/2015/03/opinion/ctl-child-poverty/.

22. "Code of Ethics," America Society of Mechanical Engineers, accessed 17 July 2017, https://community.asme.org/colorado_section/w/wiki/8080.code-of-ethics.aspx.

23. Samuel P. Hays, *Conservation and the Gospel of Efficiency: The Progressive Conservation Movement, 1890–1920* (Pittsburgh: University of Pittsburgh Press, 1999).

24. Gail Baura, *Engineering Ethics: An Industrial Perspective* (Cambridge, MA: Academic Press, 2006).

25. Angela R. Bielefeldt, "Ethic of Care and Engineering Ethics Instruction," presented at the 2015 meeting of the American Society for Engineering Education Rocky Mountain Section Conference, accessed 17 July 2017, https://www.academia.edu/29646405/Ethic_of_Care_and_Engineering_Ethics_Instruction.

26. Geoff Taggart, "Compassionate Pedagogy: The Ethics of Care in Early Childhood Professionalism," *European Early Childhood Education Research Journal* 24, no. 2 (2016):

173–185; Vicki D. Lachman, "Applying the Ethics of Care to Your Nursing Practice," *Medsurg Nursing* 21, no. 2 (2012): 112.

27. Lachman, "Applying the Ethics of Care," 113.

28. The Maintainers is a global, interdisciplinary research network with an interest in the concepts of maintenance, infrastructure, repair, and the myriad forms of labor and expertise that sustain our human-built world. See http://themaintainers.org.

29. This and other quotes are from an email that Melinda Hodkiewicz sent to the Maintainers listserv on 26 January 2017.

III   Reformers

The imperative to innovate is as much about reform as it is novelty. Champions of innovation assure us that the challenges of the twenty-first century—including hunger, climate change, and inequality—are not unstoppable forces. They assert that with the right expertise and the right models, these global problems are solvable. The message is personal and human-centered: in a risky and uncertain world, we all are empowered to enact change.

However, as the critics in the previous section demonstrate, the progressive ideal of innovation can be naively optimistic and even pernicious. Fear rather than aspiration often drives calls to innovate: we must "change or die!" lest we fall behind as individuals, companies, communities, and nation-states.[1] The innovation economy systematically neglects large segments of the population and ignores necessary infrastructures. As innovators tackle existing problems, they create new ones. Even true believers must admit that replicating and sustaining innovation is difficult; initiatives rise and decline, and models that work in one context may not in another.

Despite conflicting visions, one core idea unites both champions and critics of innovation: for both camps, innovation is about *change*. Reforming the world at large is the central goal driving the innovator imperative. Bringing attention to possible areas of reform—how innovators are recruited, trained, and deployed—is the goal of many of its critics. Is there common ground in these different visions of reform? If so, how might that reform work in practice?

This section assembles reformers who are reflective about innovation's problems but still optimistic about its potential for social change.

Contributors include educators who build online communities for children, historians who teach future entrepreneurs, feminist technologists who remake workplace cultures, and ethnographers who engage with laboratory scientists. These reformers grapple with the innovator imperative by combining the "How might we?" mentality of innovation's champions with the "Why?" questions of its critics. They ask: *How should we* prepare generations of innovators in the classroom, workplace, and laboratory?

A controversial feature of the innovator imperative is its extension into the education of young children. Champions of innovation argue that creativity is a vital life skill (McManus and MacDonald, chapter 4) and that exposure to innovation in childhood has a causal effect on a child's propensity to become an innovator.[2] Many advocates, however, decry that high-stakes testing, one-size-fits-all curricula, and underfunded public schools rob children of their inherent creativity.[3] As a remedy, they create informal STEM initiatives, including hands-on science centers, maker spaces, and after-school robotics clubs.[4] But these programs are not equally distributed either economically or geographically, they can perpetuate the fear that the nation's children are falling behind, and many are tied to for-profit schemes. Educational reformers ask: *How should we* nurture children's creative expression amid the standardization and corporatization of STEM education?

In chapter 15, "Designing Learning Environments That Engage Young People as Creators," MIT Media Lab researcher Natalie Rusk explains how she introduces millions of children to "playful invention." Rusk describes two related projects from the lab's Lifelong Kindergarten Group: the Computer Clubhouse, a global network of after-school sites that serve 20,000 underprivileged youth in twenty countries; and Scratch, a visual programming language and online community with 21 million users that enables kids to create multimedia projects by snapping together colorful blocks of code.[5] Both are intentional spaces where children "learn by designing," "follow their interests," "build a community," and "foster respect and trust."[6]

Rusk and her colleagues prioritize values of care (Russell and Vinsel, chapter 13) in the design and operation of their initiatives. They codesign their programs with the children they serve, they emphasize peer community-building, and they seek to avoid technocentrism. However, these values sometimes stand in tension with their programs' focus on digital tools and coding skills. Moreover, initiatives that began as opportunities for self-discovery now overlap with the workforce goals of the innovator imperative. Indeed, while Computer Clubhouses and Scratch are nonprofit endeavors, they are underwritten by Intel and Best Buy. More fundamentally, how young is too young to encourage a child to be innovative?[7]

Colleges and universities are especially heated sites of debate regarding the training of innovators. Innovation experts propose new pedagogical and institutional models to upend what they view as an ineffective educational system. The University Innovation Fellows program (chapter 3), and the NSF's I-Corps (chapter 5), for example, imply that traditional academic approaches exert a stifling influence on would-be entrepreneurs and innovators. Yet many contributors in this volume demonstrate how scholarly inquiry better captures how innovation happens in practice. Given the tensions between learning through action and critical inquiry, *how should* universities train engineers and entrepreneurs?

In chapter 16, "Using the Past to Make Innovators," historian W. Bernard Carlson describes how engaging with history improves the education of future engineers. In the University of Virginia's Engineering and Society program, Carlson integrates empirical patterns from his research on Thomas Edison, Alexander Graham Bell, and Nikola Tesla into the engineering classroom. Working against misperceptions that invention is a "mysterious, unknowable activity," he helps students recognize the cognitive, technological, and sociological processes of innovation. As Carlson teaches aspiring innovators through a combination of scholarly inquiry and hands-on activities, students both analyze entrepreneurial networks (Feldman, chapter 6) and practice customer discovery (Arkilic, chapter 5).

Critical inquiry into the past may seem incompatible with the pursuit of novelty, but Carlson demonstrates how students can benefit from the practical application of history (Hintz, chapter 10). He argues that hackathons and other forms of experiential learning will only be effective if they rest upon a solid foundation of critical scholarship on the theory and practice of innovation. However, Sebastian Pfotenhauer (chapter 11) would remind Carlson that his theoretical model of innovation will inevitably mutate in practice. Moreover, Andrew L. Russell and Lee Vinsel (chapter 13) would argue that the University of Virginia misleads its engineering students by ignoring maintenance and focusing so intensely on invention, entrepreneurship, and commercialization.

While debates about the nuances of innovator training persist, participation in these initiatives remains unequally distributed. As numerous contributors in part I demonstrate, jobs in the innovation economy can lead to fulfilling careers and financial success. However, as Lisa Cook observes in part II (chapter 12), women and minorities struggle to access STEM education and careers. This disparity is a problem for everyone.[8] But closing the "innovation gap" requires more than just diagnosing and balancing workplace demographics; it requires changing institutional cultures and practices. So *how should we* eliminate prejudices in the workplace and improve innovation's outcomes through diversity?

In chapter 17, "Confronting the Absence of Women in Technology Innovation," Lucinda M. Sanders and Catherine Ashcraft describe their efforts to address women's systematic underrepresentation in innovation. In 2004, they founded the National Center for Women and Information Technology (NCWIT) to "significantly increase women's meaningful participation in computing." From the outset, they recognized that it would be insufficient to simply "add women to the pot and stir." Rather, they employ cognitive psychology and feminist theory to identify unconscious biases and structural barriers that reinforce women's absence in the innovation economy. They then utilize this analysis to help universities and high-tech companies implement actionable practices such as equitable recruiting and mentorship programs.

NCWIT's theoretically informed interventions offer a programmatic response to misogyny, sexual harassment, and assault in the workplace.[9] NCWIT advises major high-tech firms—including Apple, Microsoft, Google, Intel, and Facebook—that have some of the most egregious gender and diversity records in the IT industry.[10] But these firms are also NCWIT's underwriting sponsors; while this corporate patronage may represent an earnest attempt to improve, NCWIT must guard against being co-opted by its sponsors' interests.[11] Another concern is the glacial pace of change in the technical professions. A 2017 study of patentees, for example, found that at the present rate, the slowly declining gender gap in innovation will take 118 years to close.[12]

As reformers promote inclusiveness in innovation, they also question what innovators are actually *doing*. Champions of innovation pursue emerging technologies such as gene editing, artificial intelligence, and the internet of things with the conviction that these advances bring important social benefits. But all innovations come with intended and unintended costs that most scientists and engineers are ill-equipped to address in their daily work.[13] Also, the broader public typically only can engage with new technologies once they have already been developed, making it difficult to alter their trajectory.[14] So *how should we* account for the costs, benefits, and ethical dimensions of new innovations, from idea to implementation?

In chapter 18, "Making Responsible Innovators," Erik Fisher, David Guston, and Brenda Trinidad offer a model for shaping the moral vision of researchers in training. At Arizona State University's School for the Future of Innovation in Society, ethnographers embed themselves in the laboratories of scientists and engineers who work on innovations with high uncertainty and potentially significant social impact. The humanists who conduct this Socio-Technical Integration Research (STIR) encourage self-reflection by observing scientists and asking them a series of practical and philosophical questions across the research and development process.

According to Fisher, Guston, and Trinidad, "responsible innovation" challenges scientists and engineers to consider the social implications of their innovations and to alter their research practices. Science and technology studies (STS) scholars working alongside scientists and engineers impart their disciplinary knowledge, equipping these researchers to reflexively question their work. Yet such "midstream modulations" may be overwhelmed by the scale of the innovator imperative. These ethical interventions may already be too late; adding "responsible" as a modifier signals that innovation has come to mean the opposite (Godin, chapter 9).

Participating critically from inside the innovation enterprise, contributors in this section examine the value and the shortcomings of the innovator imperative. These hybrid experts have academic research backgrounds but often take on the role of practitioners. They use their expertise in disciplines such as child development, history, gender studies, and ethnography to critique, redefine, and reshape the image and practices of innovators.

The hybrid identities of these reformers are a source of inherent tension in their work. For one, the reformers pursue their efforts at elite universities and in consultation with high-tech firms. These reformist projects are only possible given the academic freedom and financial resources their positions provide. Similarly, interdisciplinary STS programs have always had a dependent and precarious relationship with the pro-innovation institutions that support them, requiring constant rejustification for their survival. Because of the reformers' reliance on corporate grants and consulting contracts, there is always the risk of a special interest's influence. As a result, we might ask: Are these interventions making a difference, or are they merely adding an ethical gloss to the imperative they purport to reform?

Overall, these reformers chart a compromise between the champions' optimism and the critics' skepticism about innovation as a source of social change. As national efforts to cultivate innovators grow, the contributors profiled here seek alternative approaches to who those innovators are, what problems they address, and which methods they employ.

Notes

1. Matthew Wisnioski, "'Change or Die!': The History of the Innovator's Aphorism," *Atlantic*, 12 December 2012, https://www.theatlantic.com/technology/archive/2012/12/change-or-die-the-history-of-the-innovators-aphorism/266191/.

2. Alex Bell, Raj Chetty, Xavier Jaravel, Neviana Petkova, and John Van Reenen, "Who Becomes an Inventor in America? The Importance of Exposure to Innovation," NBER working paper no. 24062, December 2017, http://www.nber.org/papers/w24062.pdf, 2–3.

3. A host of books call for an overhaul of school systems to better train innovators. See Margaret Honey and David E. Kanter, *Design, Make, Play: Growing the Next Generation of STEM Innovators* (New York: Routledge, 2013); Ken Robinson and Lou Aronica, *Creative Schools: The Grassroots Revolution That's Transforming Education* (New York: Viking Penguin, 2015); Tony Wagner and Ted Dintersmith, *Most Likely to Succeed: Preparing Our Kids for the Innovation Era* (New York: Scribner, 2015).

4. Marilyn Fenichel and Heidi A. Schweingruber, *Surrounded by Science: Learning Science in Informal Environments* (Washington, DC: National Academies Press, 2010); Jacie Maslyk, *STEAM Makers: Fostering Creativity and Innovation in Your Elementary Classroom* (Thousand Oaks, CA: Corwin, 2016); Colleen Graves, Aaron Graves, and Diana L. Rendina, *Challenge-Based Learning in the School Library Makerspace* (Santa Barbara, CA: Libraries Unlimited, 2017).

5. Mitchel Resnick, *Lifelong Kindergarten: Cultivating Creativity through Projects, Passion, Peers, and Play* (Cambridge, MA: MIT Press, 2017).

6. "Learning Model," Clubhouse Network, accessed 3 May 2018, http://www.computerclubhouse.org/model.

7. Christopher Mims, "How Young Is Too Young to Learn to Code?" *MIT Technology Review*, 26 February 2012, https://www.technologyreview.com/s/427064/how-young-is-too-young-to-learn-to-code/.

8. A growing body of evidence suggests that teams with racial and gender diversity are associated with greater sales revenue, market share, profits, customer satisfaction, and worker productivity. Scott E. Page, *The Difference: How the Power of Diversity Creates Better Groups, Firms, Schools and Societies* (Princeton, NJ: Princeton University Press, 2008); Cedric Herring, "Does Diversity Pay? Race, Gender, and the Business Case for Diversity," *American Sociological Review* 74, no. 2 (2009): 208–224.

9. NCWIT's work is more important than ever. Between 2015 and 2018, as this volume was taking shape, hundreds of courageous women (and a few men) came forward as part of the #MeToo movement to reveal serious cases of sexual harassment and assault, resulting in the firings, resignations, and convictions of several abusers. See, for example, Mary Schmich, "2017 Was the Year of the Reckoning," *Chicago Tribune*, 5 December 2017, http://www.chicagotribune.com/news/columnists/schmich/ct-met-word-of-year-mary-schmich-20171205-story.html.

10. For example, after years of public pressure to release its diversity data, Google disclosed in 2014 that in technical roles, women accounted for just 17 percent of the company's employees. Hispanics made up 2 percent and African-Americans 1 percent of the technical workforce. Sheelah Kolhatkar, "The Tech Industry's Gender Discrimination Problem," *New Yorker*, 20 November 2017, https://www.newyorker.com/magazine/2017/11/20/the-tech-industrys-gender-discrimination-problem.

11. "Capture" is a phenomenon in which experts, critics, or government regulators begin to serve the interests of those they should be objectively studying, critiquing, or overseeing. For an overview, see Daniel Carpenter and David A. Moss, eds., *Preventing Regulatory Capture: Special Interest Influence and How to Limit It* (Cambridge: Cambridge University Press, 2013).

12. Bell et al., "Who Becomes an Inventor in America?" 1.

13. Sheila Jasanoff, *The Ethics of Invention: Technology and the Human Future* (New York: W. W. Norton, 2016).

14. David Guston, "Understanding 'Anticipatory Governance,'" *Social Studies of Science* 44, no. 2 (2014): 218–242.

15   Designing Learning Environments That Engage Young People as Creators

Natalie Rusk

The MIT Media Lab has become known for developing such new technologies as electronic ink, wearable computers, bionic limbs, and social robots. When the list of its top innovations was released for the lab's twenty-fifth anniversary, I was excited to see that the Computer Clubhouse network and the Scratch programming language were included. I was glad that not only technological developments but also learning initiatives were named as innovations. Through my work on the Clubhouse and Scratch, I had seen how children and teens thrive when they are engaged in the process of designing their own projects. I was encouraged because this announcement signaled a broader recognition of the value of empowering young people to become creators and innovators.

Graduate students and other researchers at the MIT Media Lab design projects that span multiple disciplines—including art, architecture, science, music, and engineering—often through a process that involves playful experimentation, collaboration, and iteration. The Clubhouse and Scratch initiatives seek to open up the opportunities for learning through playful invention and creative expression to a broader and more diverse group of people.

The Clubhouse provides a creative learning environment where young people ages ten to eighteen design projects that build on their interests. When we opened the first Clubhouse in Boston in 1993, it was one of the first after-school centers where young people from low-income neighborhoods could use new technologies to develop their own projects.[1] Since its launch, the Clubhouse has grown into a global network with nearly one hundred program sites in twenty countries that provides in-depth learning experiences and mentoring for more than 20,000 youth in underserved communities each year (figure 15.1).[2]

Figure 15.1
Young people in a Computer Clubhouse in Johannesburg, South Africa. Photo: Sci-Bono Clubhouse.

At the same time, the movement to support youth as creators with new technologies has become widespread. Increasingly, educators, business leaders, and policymakers are recognizing the need for young people to develop skills that go beyond memorizing facts from textbooks. The National Science Foundation and other federal agencies have called for every student to learn to make projects and code computers, with a focus on involving young people from groups traditionally underrepresented in science, engineering, and technological fields.[3] A growing number of schools, libraries, museums, and community centers are opening "maker spaces," where young people can make projects using a range of new and traditional tools.[4]

One of the most widely used technologies for children and teens to make their own projects is Scratch, which our research group launched in 2007 with support from the National Science Foundation.[5] Scratch is a visual programming language that allows young people ages eight and up to create interactive projects by snapping together colorful blocks of code on the

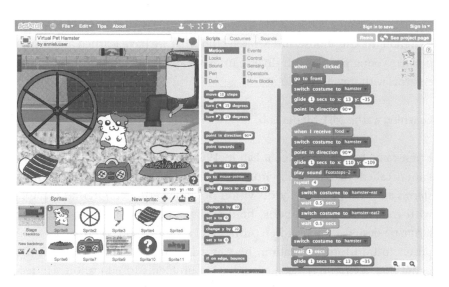

Figure 15.2
A project created using the Scratch programming language.

screen (figure 15.2). Young people around the world use Scratch to program their own interactive stories, animations, and games—and then share their creations with others in the Scratch online community. Participation on the Scratch website has grown rapidly, with more than forty million young people actively engaged in creating, discussing, and sharing projects. While Scratch enables young people to learn to code, it is designed to support the development of broader skills, including the ability to think creatively, reason systematically, and work collaboratively.

The Clubhouse and Scratch initiatives grew out of research in the MIT Media Lab's Lifelong Kindergarten Group, led by Mitchel Resnick. Our research group is inspired by the ways children learn in kindergarten: when they create buildings with wooden blocks, they learn about structures and stability; when they create pictures with finger paint, they learn how colors mix together. We develop new technologies and programs that extend this kindergarten style of learning, so that people of all ages continue to learn through the process of designing, creating, experimenting, and collaborating. The goal of the Lifelong Kindergarten Group is to support the development of a world filled with playfully creative people who can improve lives for themselves and their communities.[6]

In this chapter, I take a closer look at the ideas that have guided the design of the Clubhouse and Scratch, and what we have learned about supporting young people with diverse interests and backgrounds in becoming creators with new technologies.

Learning at the MIT Media Lab

I first came to the MIT Media Lab in 1988 to take a course with Seymour Papert called "Learning Environments." I had read Papert's book *Mindstorms* and was inspired by his vision of empowering children as creators with computers.[7] In the book, he described Logo, the first programming language designed for children. Although Papert saw the potential for computers to transform education, he warned against "technocentrism," the belief that technology itself would solve problems in education. He recognized the importance of learning as part of a community and gave the example of *samba* schools in Brazil as an environment in which people of all ages learn from each other.[8] Unfortunately, the technocentric approach is still prevalent in many discussions of making and coding, which too often focus on which tools to use (e.g., choice of equipment or programming language) and underestimate the importance of developing a supportive learning community.

In Papert's course, I met other researchers and educators with a shared interest in how children think and learn, including Mitchel Resnick, who was then a graduate student. Resnick had collaborated on the design of the first programmable Lego kit, known as Lego/Logo.[9] This construction kit enabled children to build machines using Lego bricks, gears, motors, lights, and sensors, and then program them to move and respond using the Logo programming language. I began helping to facilitate Lego/Logo robotics workshops for children and educators at local museums and schools, and then started to work full-time the following year at the Computer Museum in Boston.

The Origin of the Computer Clubhouse

"*Mira, Mira*! Look at this!" A boy called out excitedly in Spanish and English, eager to show his friends the Lego motor he held in his hands. I had just showed him how he could program it to turn on and off, using a computer. It was school vacation week at the Computer Museum, and I was leading a program for visiting families to make their own interactive machines.

I was motivated to offer this Lego/Logo program to replace the museum's previous "Build a Robot" workshop for families. In that workshop, parents and children followed step-by-step instructions to build model robots by attaching dozens of parts using tiny screws and a screwdriver. Those model robots allowed for no variations in the design and no meaningful learning opportunities beyond how to use a screwdriver and keep track of small parts. Each model robot turned out exactly the same, and there was no way to program them to change how they moved.

That workshop was particularly disappointing to me because it was such a contrast from the creative approach to robotics I had learned when leading workshops with Lego/Logo. When facilitating Lego/Logo workshops, I had seen how excited children became when they designed and programmed their own interactive animals, amusement park rides, cities, and other creations (figure 15.3).

I had observed how, in the process of creating these personally meaningful projects, children could learn powerful math, science, and engineering concepts, such as how to use gear ratios to harness the power of a motor or how to use sensors to respond to changes in the environment. Unlike the prefab model robots, the Lego/Logo construction sets allowed for endless combinations and a wide range of possible projects. These sets included optional instructions, but the instructions were just helpful starting points, suggesting a variety of potential directions that learners could explore and adapt.

The contrast between the model robot workshops and the Lego/Logo robotics workshops made me aware of how educators could use the same words—"a workshop where families learn to make a robot together"—yet be talking about experiences that fundamentally differed in terms of engaging participants and providing learning opportunities.

The boy who had become so excited to program a motor returned each day of the school vacation week with his friends to make interactive machines using Lego motors, gears, and sensors. The following week, with school vacation over, the museum was very quiet. In the mid-afternoon, I saw the museum's large elevator doors open, and inside were four children: the boy and his friends. They recognized me and asked, "Lego/Logo?" They

Figure 15.3
A robotic owl made by an eleven-year-old girl using programmable Lego materials.
Photo: Natalie Rusk.

had come straight from school, hoping they could continue to build and program Lego creations. I had to tell them no, as I had returned the Lego/Logo materials to MIT. So the four friends wandered around the museum trying out the exhibits.

A couple of weeks later, a museum administrator sent an email message to the staff warning them to be on the lookout for a group of kids sneaking into the museum. Employees were urged to alert security if the children were seen. It turned out that these were the same children who had enthusiastically participated in the weeklong robotics program. Now, because they were hanging around the museum without activities to participate in, they were getting into trouble with the museum's security staff.

I asked around to find local after-school centers that could accommodate these children, but there were none in the downtown area. I also visited the first Boston-area community technology centers. These centers had recently opened to address inequities in access to computers. However, at the time, they were primarily offering educational computer games for children and office applications for adults. I found out that there was no place for this group of children to go after school where they could use technology to make their own projects.

To address this need, I began collaborating with Mitchel Resnick to plan a permanent space where young people could use technologies to create projects based on their interests. We envisioned a space that would have the feel of a creative design studio—a combination of an art studio, music studio, video production studio, and robotics lab. Our goal was to provide young people in low-income neighborhoods opportunities to express themselves fluently with new technologies, and in the process, to become motivated and confident learners.

Early on, we identified four guiding principles for the Computer Clubhouse.[10] We applied these principles to set up the first Clubhouse at the Computer Museum. But the principles have continued to play an important role as the Clubhouse has expanded into a global network.[11]

Clubhouse Guiding Principles

Principle One: Learning through Designing

Our work on the Clubhouse and other initiatives is grounded in the educational philosophy known as *constructionism*, a word coined by Seymour Papert, who was building on Jean Piaget's theory that children construct knowledge from their experiences in the world.[12] They do not just passively receive knowledge—they actively build knowledge based on their interactions. Papert took Piaget's theory a step further and argued that people construct knowledge best when they are engaged in *constructing projects*. They might be constructing sand castles, Lego machines, or computer programs, but what matters is that they are actively engaged in creating something, particularly something they find personally meaningful.

Activities at Clubhouses vary widely, from constructing robotic inventions to orchestrating virtual dancers to recording songs. Yet these varied activities are all based on the common framework of learning through design. Young people not only learn how to use the tools but also how to express their ideas.

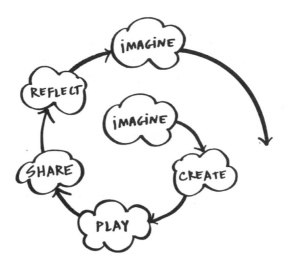

Figure 15.4
Creative learning spiral, illustrating an iterative process for designing projects. Courtesy of Lifelong Kindergarten Group, MIT Media Lab.

As youth work on designing projects, they move through what we call the creative learning spiral (figure 15.4). In this process, they imagine what they want to do, create a project based on their ideas, experiment with alternatives, share their ideas and creations with others, and reflect on their experiences—all of which leads them to imagine new ideas and new projects, and to start a new design cycle. As youth learn to go through this process with a variety of projects, they become increasingly skilled in understanding how to carry out a project from start to finish, solving problems along the way.

Principle Two: Building on Their Interests
To motivate students to learn, many people assume that they need to offer rewards, or turn the subject matter into a competitive game, with prizes for those with the best scores. If you look outside school, however, you can find many examples of people learning—in fact, learning exceptionally well—without explicit rewards. Youth who seem to have short attention spans in school often display great concentration on projects in which they are truly interested. They might spend hours learning to play the guitar or perform tricks on a skateboard. Indeed, many of the most successful

designers, scientists, and other professionals trace their involvement and success in their fields back to a childhood interest.

At first, some youth interests might seem trivial or shallow, but youth can build large networks of knowledge related to their interests. Pursuing any topic in depth can lead to connections to other subjects and disciplines. The educational challenge is to find ways to help youth make those connections and develop them more fully. For example, an interest in riding a bicycle can lead to investigations of gearing, the physics of balancing, the evolution of vehicles over time, or the environmental effects of different transportation modes.

While youth from high-income households generally have many opportunities to build on their interests (for example, music lessons and specialty camps), the youth who typically come to Clubhouses have had few such opportunities. Most do not have the resources or support to help them identify and explore potential interest areas, let alone to build on them.[13]

The approach of helping young people build on their interests works only if the environment supports a great diversity of possible projects and paths. Young people have a wide variety of interests, so creative environments need to provide a wide variety of activities to match those interests.

On the walls and shelves in Clubhouses there are collections of sample projects, designed to spark in participants a sense of what they can make and to provide multiple entry points for getting started. Many youth begin by mimicking a sample project, then working on variations on the theme before developing their own path that stems from their personal interests. At any given time in a Clubhouse, a pair of youth might be using one computer to edit a video about their neighborhood, while another participant at the next computer might be designing a 3-D model airplane.

Principle Three: Developing a Sense of Community

From the beginning, we recognized that Clubhouses needed to provide more than access to technology. Youth also need access to people who know how to use technology in interesting and creative ways. Clubhouses provide a new way for adults in the community to share their skills with local youth. Mentors at Clubhouses do not simply provide support or help; many work on their own projects and encourage Clubhouse youth to join in.

Clubhouse coordinators recruit a team of adult mentors—professionals and college students in art, music, science, engineering, and technology from diverse racial and cultural backgrounds. Mentors act as coaches, catalysts, and consultants, bringing new project ideas to their Clubhouses.[14] For example, an engineering student might be working on robotics projects with Clubhouse participants, a videographer on digital video projects, and a programmer on interactive games. For youth who have never interacted with an adult involved in academic or professional careers, this opportunity to connect with mentors can be pivotal to envisioning themselves following similar career paths.[15]

In today's rapidly changing society, perhaps the most important skill of all is the ability to learn new things. It might seem obvious that in order to become good learners youth should observe adults learning. Yet that is rare in schools, where teachers are typically expected to deliver knowledge to students. At Clubhouses, youth often see adults in the process of learning. For some Clubhouse participants, this is quite a shock. For example, several of them were startled one day when a Clubhouse staff member, after debugging a tricky programming problem, exclaimed, "I just learned something!"

As youth become more fluent with the technologies at Clubhouses, they too start to act as mentors. Over time, youth begin to take on more mentoring roles, helping introduce newcomers to the equipment, projects, and ideas of the Clubhouse.

Principle Four: Fostering Respect and Trust

When visitors walk into a Clubhouse, they are often impressed by the artistic creations and the technical abilities of Clubhouse participants. But just as often, they are struck by the way Clubhouse youth interact with one another. The Clubhouse approach puts a high priority on developing a culture of respect and trust. These values not only make the Clubhouse an inviting place to spend time, but they are also essential for enabling Clubhouse youth to try out new ideas, take risks, follow their interests, and develop fluency with new technologies. Indeed, none of the other guiding principles can be put into practice without staff, mentors, and youth helping to sustain an environment of respect and trust.

The principle of treating everyone with respect is directly stated in the Clubhouse code of conduct, to which each member agrees when joining.

However, even more importantly, new members learn by seeing mutual respect modeled by other members of the Clubhouse community.

Designing Scratch

The Scratch graphical programming language, which is now used by young people around the world, grew out of our work with young people in Clubhouses.[16] Most of these young people had become highly engaged in media creation, whether in the form of editing photos, filming videos, or recording music. Many of them also wanted to make interactive animations and games like those they had viewed online, yet few were developing the programming skills needed to make these types of projects.

Our team recognized a need for a new programming language that would work well for young people in homes, community centers, and other informal learning environments. We identified several ways in which it would need to differ from existing programming languages in order to be more broadly adopted. First, it needed to be "media rich" to allow young people to easily bring in their own photos, music, and other media so that they could make projects that reflected their interests and ideas. Second, it needed to support learning through playful experimentation rather than requiring initial and ongoing instruction.

Our design of Scratch was inspired by Lego bricks: it allows young people to code programs by snapping together colorful blocks on the screen, combining them to program images, text, music, and sounds.[17] Rather than having to write a program and then run it, we designed the blocks to be "tinkerable"—children can experiment and try out each piece as they build up a program. When children build Scratch programs there are no error messages. Instead, they see the effects of their actions and make their own adjustments and corrections until they get the results they want.

In our research, we have found this ability to tinker is one of the key reasons children see Scratch as easy to learn and fun to use; it helps them gain confidence in their ability to solve problems. While the design of Scratch differs in interface from traditional text-based languages, it enables beginners to learn such fundamental computational concepts as sequences, looping, conditionals, and variables, as well as core computational practices such as debugging and modularizing.[18]

Our team designed Scratch to enable young people to create a wide variety of projects to connect with a diversity of interests, from art and music to games and simulations. Although this may sound like an obvious idea, the value of taking a project-based approach to introducing coding is still not widely appreciated. Many of today's introductory coding activities for children adopt the look-and-feel of Scratch blocks yet focus on children solving puzzles correctly. While a puzzle-based approach can provide an easy entry point, it risks giving young people the idea that coding is about getting a single right answer. In contrast, Scratch enables them to explore multiple possibilities and to express their own ideas.

Based on our work in Clubhouses, we recognized the importance of a supportive community for learning, collaborating, and sharing projects. So when we launched Scratch in 2007, we launched the programming language alongside a website for sharing projects. As young people create projects, they can click the Share button to post their project.[19] They can also easily view the code inside any project and remix the project to adapt and customize it.

Young people often say they learn Scratch by experimenting and by looking at each other's code. They also learn by asking questions and exchanging comments on each other's projects and profiles. This dynamic social aspect is missing in most other online coding activities for children.

Research has revealed just how important the Scratch online community is for fostering young people's ongoing participation. Even young people who initially come to Scratch to program their own games or animations often say that their interactions on the site keep them participating. In the words of one young community member, "The very best thing about Scratch to me is the amazing, amazing community of people to work with. I love doing all sorts of collabs with people, and I love seeing what others do as inspiration."[20] Young people who use Scratch identify making friends, sharing projects, receiving feedback, and collaborating with others as important parts of their experience. As one youth said, "I made many friends here who remix my projects, give comments, and have taught me new things."[21]

From our experience in Clubhouses, we understood how important it is to maintain an environment of trust and respect so that young people feel welcome and safe to participate and share their creations. Developing, managing, moderating, and supporting the Scratch online community is an important ongoing aspect of our group's work. The Scratch moderation team—a thoughtful and highly skilled group of online moderators—works hard to promote positive interactions in the community and to ensure that the content remains age appropriate, respectful, and friendly. The Scratch community guidelines—to be respectful, constructive, honest, and friendly—are written in an inviting way and are often referred to by moderators and participants in the community.[22] Young people on the site have initiated a variety of ways to make others feel included, such as managing a Welcoming Committee studio with projects that welcome newcomers to the website.

The process of developing Scratch itself has been guided by the iterative approach represented in the creative learning spiral. Our team continues to develop and revise versions of Scratch for use in after-school programs, schools, and other learning environments (as well as playing and creating with it ourselves). We make improvements based on our observations and feedback from young people in the Scratch community.

Rather than taking a win-or-lose approach in which youth compete against each other to earn a prize for the best project, the Clubhouse and Scratch take a learner-centered approach that supports their longer-term development. We have found that young people are more motivated and willing to persist despite setbacks when they have experiences that enable them to follow their interests, connect with peers, receive encouragement from mentors, and build confidence in their abilities.

The Impact of Clubhouse and Scratch Programs

The Clubhouse has influenced the development of other learning networks, including Digital Youth Network and the YOUMedia network in libraries and museums around the country. Scratch has been adopted worldwide and has inspired the design of many other coding experiences and languages for children and adults.

A growing body of research and evaluation documents the influence of the Clubhouse and Scratch in young people's development of diverse skills and understanding of their own interests and abilities. To better understand the impact of learning environments like the Clubhouse and Scratch, it can help to follow the development of a young person participating in these programs. Jaleesa Trapp began attending the Computer Clubhouse in Tacoma, Washington, when she was thirteen years old. After graduating from college, Trapp became leader of the Clubhouse program in Tacoma and began pursuing graduate studies in human-centered design and engineering. Although Trapp stands out as an exceptionally motivated young leader, she explains that when she first came to the Clubhouse at her aunt's urging, she was bored in school and considered herself lazy. She wasn't yet interested in technology and did not plan to go to college.

Trapp's reflections on her experience highlight some of the key lessons we have learned about creating learning environments that help young people develop their potential.[23] First, Trapp emphasizes that receiving encouragement and mentoring from the director of the Tacoma Clubhouse, Luversa Sullivan, kept her involved in the Clubhouse program. Sullivan recognized Trapp's potential and encouraged her to try new things, including programming her own interactive animations.

Trapp said when she first came to the Clubhouse she spent her time with a simple software program that could distort photos of people's faces. Sullivan believed that Trapp was not living up to her potential. As Trapp says, "She knew I was capable of more than making funny faces and pushed me to do more." Sullivan found tutorials that Trapp could use to learn to program animations, and she encouraged Trapp to learn through experimenting and solving problems, rather than simply giving her answers.

As Trapp developed more skills, she started to program interactive projects for school assignments:

> I really got into making interactive CD-ROMs. The first one that I made was for Black History Month. It was based off of a play called *What If There Were No Black People*. I worked with another member: she helped me draw the characters. And then I put them on the stage and programmed them. It was really exciting, because I could show my mom what I had made.

Trapp says that the ability to share the interactive projects she made with other people in her life (including her family and teachers) motivated her to continue to participate and learn more.

Sullivan also helped Trapp pursue several opportunities beyond the Clubhouse, including applying for scholarships, presenting at technology conferences, and participating in a summer internship program during high school. For the internship, Trapp had to wake up at 4:30 a.m. to catch three different buses to reach Microsoft's offices, where her internship was based. I noted the change from her initial description of herself as "lazy" when she first came to the Clubhouse to her intense motivation in high school to make complex projects and pursue a summer internship. When I remarked on this to Trapp, she said, "I think what changed was that I was able to do something outside of school, and I was able to show people what I was working on. People were like, 'You did that? That was really good!'" Trapp also was motivated to serve as a role model for her younger sister: "My younger sister is two years behind me, and I want her to have someone to look up to—to be able to see, 'If you're doing that, I could do that too.'"

Through her experience as a Clubhouse staff leader, Trapp helped children learn to create projects based on their interests. For example, Trapp encouraged a girl in her Clubhouse to code an animation about her favorite sports team using Scratch. Although reluctant at first, the girl was surprised and proud of what she had made. "I didn't know I could do that!" she said.

Trapp describes the approach she uses for encouraging children to learn to program with Scratch:

> In Scratch workshops, I'll give them a theme and let them run with it. I try not to give them step-by-step instructions. It's just amazing—each group comes up with something totally different. It's their own ideas and something they are proud of. And even though they didn't think they would like programming, they like it because it's something that they did on their own.

Trapp now helps young people exceed their previous expectations, just as her mentor, Sullivan, encouraged her.

Broader Support for Young People as Creators

Providing youth with opportunities to become creative learners and thinkers is not simply an enrichment activity but rather is essential if we are to help young people develop their potential and be prepared to handle challenges that lie ahead. To thrive in a complex and rapidly changing world, young people need to think creatively, reason systematically, and work collaboratively.

The growing maker and coding movements have the potential to make a difference in young people's lives. Yet access to physical spaces, tools, and technologies is not enough. To reach their potential, programs need to go beyond a technocentric approach, focusing on teaching a specific tool or technology (such as how to use a 3-D printer or how to use a text-based language). Young people need encouragement to develop their ideas and interests in a supportive learning community, with peers who became creative collaborators and mentors who provide encouragement and connect youth to further opportunities.

I am hopeful that more people will see the value of learning environments that support children and teens to become creators with technology. We need to ensure that these learning environments, both in-person and online, are welcoming, engaging, and encouraging young people with diverse interests and backgrounds to create and share together.

Notes

1. Natalie Rusk, Mitchel Resnick, and Stina Cooke, "Origins and Guiding Principles of the Computer Clubhouse," in *The Computer Clubhouse: Constructionism and Creativity in Youth Communities*, ed. Yasmin Kafai, Kylie Peppler, and Robbin Chapman (New York: Teachers College Press, 1999), 17–25.

2. Gail Breslow and Mark St. John, "Leveraging STEM Investments for Long-Term Impact: The Clubhouse Network as Case Study," in *STEM Ready America: Inspiring and Preparing Students for Success with Afterschool and Summer Learning*, ed. Ronald Ottinger, March 2017, http://stemreadyamerica.org/article-leveraging-investments/.

3. Obama White House, "Fact Sheet: New Commitments in Support of the President's Nation of Makers Initiative to Kick Off 2016 National Week of Making," 17 June 2016, accessed 13 August 2017, https://obamawhitehouse.archives.gov/sites /whitehouse.gov/files/images/Blog/2016 National Week of Making Fact Sheet.pdf; "CS for All," National Science Foundation, accessed 14 August 2017, https://www .nsf.gov/news/special_reports/csed/csforall.jsp.

4. For an overview of maker spaces for school-age children, see Kylie Peppler and Sophia Bender, "Maker Movement Spreads Innovation One Project at a Time," *Phi Delta Kappan* 95, no. 3 (2013): 22–27. For a description of maker spaces in university education, see chapter 3 in this volume.

5. Mitchel Resnick et al., "Scratch: Programming for All," *Communications of the ACM* 52 (2009): 60–67.

6. Mitchel Resnick, *Lifelong Kindergarten: Cultivating Creativity through Projects, Passion, Peers, and Play* (Cambridge, MA: MIT Press, 2017).

7. Seymour Papert, *Mindstorms: Children, Computers, and Powerful Ideas* (New York: Basic Books, 1980).

8. Papert, *Mindstorms*, 178.

9. Mitchel Resnick and Steve Ocko, "LEGO/Logo: Learning through and about Design," in *Constructionism*, ed. Idit Harel and Seymour Papert (Norwood, NJ: Ablex Publishing, 1991).

10. Mitchel Resnick and Natalie Rusk, "The Computer Clubhouse: Preparing for Life in a Digital World," *IBM Systems Journal* 35 (1996): 431–440; Mitchel Resnick, Natalie Rusk, and Stina Cooke, "The Computer Clubhouse: Technological Fluency in the Inner City," in *High Technology and Low-Income Communities*, ed. D. Schön, B. Sanyal, and W. Mitchell (Cambridge, MA: MIT Press, 1999), 263–285.

11. Gail Breslow, "Participation, Engagement, and Youth Impact in the Clubhouse Network," in Kafai et al., *Computer Clubhouse*, 111–124.

12. Seymour Papert, "Situating Constructionism," in *Constructionism*, ed. Idit Harel and Seymour Papert (Norwood, NJ: Ablex Publishing, 1991).

13. For more on the growing disparity in investments that poor and wealthy families make on enrichment activities, see Greg J. Duncan and Richard J. Murnane, *Whither Opportunity? Rising Inequality, Schools, and Children's Life Chances* (New York: Russell Sage Foundation, 2011).

14. See Yasmin Kafai et al., "The Multiple Roles of Mentors," in Katai et al., *Computer Clubhouse*, 90–99.

15. For an in-depth analysis of childhood experiences that influence who becomes inventors in adulthood, see Alexander M. Bell et al., "Who Becomes an Inventor in America? The Importance of Exposure to Innovation," December 2017, NBER working paper no. 24062, accessed 30 August 2018, http://www.nber.org/papers/w24062.pdf.

16. Resnick et al., "Scratch: Programming for All." Many people have contributed to the ongoing development of Scratch. Mitchel Resnick initiated and directs the Scratch project. John Maloney was lead programmer of Scratch for its first eight years. Andrés Monroy-Hernandez led the development of the original Scratch website. For further credits, see https://scratch.mit.edu/info/credits/.

17. The first research on block-based programming languages in the Lifelong Kindergarten research group is described in Andy Begel, "LogoBlocks: A Graphical Programming Language for Interacting with the World," master's thesis, MIT Media Lab, 1996.

18. John Maloney et al., "The Scratch Programming Language and Environment," *ACM Transactions on Computing Education* 10, no. 4 (2010): article 16, 1–15. See also Karen Brennan and Mitchel Resnick, "New Frameworks for Studying and Assessing the Development of Computational Thinking," paper presented at the Annual Meeting of the American Educational Research Association, Vancouver, Canada, 2012.

19. Andres Monroy-Hernandez and Mitchel Resnick, "Empowering Kids to Create and Share Programmable Media," *Interactions* 15, no. 2 (2008): 50–53.

20. "Collabs" is a term used by youth on Scratch to refer to projects they create with others on the site, such as collaborative animations, where multiple young people create different sections, or collaborative games, where different people take on different roles. See Ricarose Roque, Natalie Rusk, and Mitchel Resnick, "Supporting Diverse and Creative Collaboration in the Scratch Online Community," in *Mass Collaboration and Education*, ed. U. Cress et al. (Cham, Switzerland: Springer, 2016), 241–256.

21. Natalie Rusk, "Motivation for Making," in *Makeology: Makers as Learners*, ed. K. Peppler, E. Halverson, Y. Kafai (New York: Routledge, 2016), 85–108.

22. "Scratch Community Guidelines," accessed 14 August 2017, http://scratch.mit.edu/community_guidelines/.

23. Jaleesa Trapp, panel presentation at the Digital Media and Learning Conference, Boston, 7 March 2014, with a follow-up interview on the same day.

16 Using the Past to Make Innovators

W. Bernard Carlson

It was a quiet afternoon at the workshop that brought most of the authors of this volume together, and the participants, who came from a variety of backgrounds, had been thoughtfully discussing various aspects of the innovation process. The organizers probably figured that the conference would come off without any problems.

But then came the fireworks. Humera Fasihuddin presented an overview of the University Innovation Fellows program, which trains students to act as advocates for innovation and entrepreneurship on college campuses across the United States.[1] Humera's talk was followed by comments from several students who reported firsthand on their experiences as Fellows. Student after student testified that administrators were hostile to introducing entrepreneurship programs and that faculty really did not have anything to offer students in terms of teaching them to be innovators. For students to succeed, they simply needed the opportunity to unleash their creative powers.

After about the fifth student, Maryann Feldman and I finally interrupted. "Wait a minute," we said. "We're both professors and have been working for years at our universities to create programs that foster innovation. You can't accuse all of us of neglecting student entrepreneurship. Some of us have devoted our careers to understanding the innovation process and teaching students to be innovators." Fortunately, both Humera and the students heard us, and the afternoon concluded with a fruitful exchange of views about what the students felt they needed and what we, as scholars, could teach them.

Nevertheless, I came away with a renewed awareness that many people do not appreciate what history can teach us about invention and innovation. Great inventions, it is commonly assumed, come only to incredibly smart people like James Watt, Nikola Tesla, or Steve Jobs; these individuals possessed exceptional mental abilities that allowed them to do the extraordinary. Many people tend to think of invention as a mysterious, unknowable activity. As much as we would like to characterize the brain as a computer, we cannot fathom how humans are able to create beautiful paintings, amazing scientific theories, or inventions that revolutionize daily life.

If you regard invention and innovation as the products of genius, luck, or mystery, it is easy to conclude that innovation cannot be taught: you are either born to be the next Thomas Edison, or you are not. Invention—like all creative acts—is not something that you can teach; you can nurture and inspire the next Elon Musk, but you cannot reduce invention to ten easy lessons. To teach invention, then, is to teach the unteachable.

Despite these popular notions, my colleagues and I at the School of Engineering and Applied Science at the University of Virginia (UVa) have been working to teach the unteachable to students for thirty years. We have approached invention as a process that can be analyzed, using historical cases to create a robust notion of how inventors work. Moreover, we have distilled principles and techniques to teach students how to be innovators. This is why I had such a visceral reaction to what the University Innovation Fellows were saying at the workshop.

Our experience with invention and innovation at UVa Engineering has intertwined research and teaching. In this chapter, I first recount how our research on inventors evolved. I then explain how we used our research findings to shape courses and programs to make innovators. And I close by reporting on some of the outcomes we have achieved and comment on whether what we have learned at UVa can be transferred to other programs.

Building a Cognitive Framework of Invention

Much of my professional life has been devoted to studying such major inventors as Thomas Edison, Alexander Graham Bell, and Nikola Tesla. Inspired by historian Thomas P. Hughes, I decided to study inventors because they often left substantial source material (notebooks, letters, testimony, and artifacts). And because nonacademic audiences are interested in inventors, you can use inventors to talk about a variety of social and ethical issues concerning technology with students, engineers, business leaders, and the public.[2]

I have stayed with inventors for so long because they raise hard questions about the nature of technological change:

- How do we make sense of both individual actions and social forces in history? Do such individuals as Napoleon or Edison "make" history, or are they merely the representatives of various interests?
- Do ideas just exist "out there" in some platonic realm waiting to be discovered or invented by individuals? Or are ideas generally constructed by individuals and groups out of the cultural raw materials available at a given time?
- What kinds of knowledge and skills are involved in creating new technological artifacts? Can we characterize the nonscientific knowledge and skills involved in this creative work?

I began studying inventors in the 1980s by examining Elihu Thomson, a contemporary of Edison. Along with inventing a successful arc-lighting system and doing pioneering work with alternating current, Thomson was significant because, unlike other late nineteenth-century inventors who generally worked alone, he spent his career in a large company, General Electric. This gave me the opportunity to look at how the organizational environment affects the innovation process. Thomson's career showed that innovation is a social process not only in the sense that it involves the interplay of individuals and groups, but also because effective innovation requires the coproduction of technological artifacts, corporate structure, and markets.[3]

But as I read through Thomson's letter books, filled with the memos he wrote to vice presidents and plant engineers, I realized that while I was learning a lot about how he moved his inventions through the company, I wasn't learning as much about how he conceived of his inventions. Thomson viewed dictating letters as being one step removed from the creative work on the benchtop, and I came to agree with him. The research question then became how to get closer to the point of knowledge and artifact production.

To investigate more closely what inventors did at the benchtop, I started a new research project with my colleague Michael Gorman. Gorman is a cognitive psychologist who had been conducting simulations of how people solve scientific problems, and he welcomed the idea of using historical materials to investigate how people developed new technology.[4] Together we looked at the history of the telephone because of the availability of substantial archival materials on Alexander Graham Bell, Elisha Gray, and Thomas Edison.[5]

We undertook a fine-grained examination of work at the inventor's benchtop. We found we could borrow some ideas from the laboratory ethnographies produced by sociologists of science Bruno Latour, Steve Woolgar, and others, and we were encouraged by Peter Galison's study of experimental methods in physics.[6] However, with the exception of a brief study of Leonardo's sketches by Bert S. Hall and Hughes's ideas about the style and methods of inventors, we could not find much from the history of technology that could help us with this investigation.[7] Hence, we turned to a field with which Gorman was familiar, namely, cognitive science.

A dominant issue in cognitive science at the time was the tension between mental models and heuristics. Some major figures in the field—such as Philip Johnson-Laird—believed the key to understanding how people think was to comprehend the meta-ideas, or categories, by which they processed information, or what cognitive scientists call mental models. Other researchers—such as John Anderson and Herbert Simon—argued that cognition is much more about the strategies or procedures that an individual employs in thinking. These strategies and procedures are called heuristics.[8]

We decided to explore how inventors might use both mental models and heuristics in their work. Why privilege one concept over the other? We also sought to find a way to pay attention to the specific objects an inventor manipulates on the benchtop.[9] After all, it seemed possible that inventors might supplement their organizing ideas (mental models) with specific mechanisms, circuits, or materials with which they were familiar.[10] To pay attention to these "building blocks," we introduced mechanical representations as a third category for analysis. Inventors often begin an investigation of a mental model by borrowing components or devices from other projects since they are familiar with how those components perform. Edison, for example, took the drum cylinder he had used on the phonograph and covered it with a photographic emulsion to transform it into a key component of the kinetoscope, his motion-picture device.

With these three categories in mind—mental models, heuristics, and mechanical representations—we set out to study the notebooks, sketches, models, patents, testimony, and correspondence of our three inventors. We hoped that by tracing how Bell, Gray, and Edison each worked on the telephone, we could gain insight into the interplay of ideas and objects (mental models and mechanical representations) as well as the thoughts and actions (mental models and heuristics). By comparing three inventors, we hoped to produce generalizations about how inventors worked, generalizations that could be tested further through case studies of other inventors and technologists.

Gorman and I always saw our investigation as alternating between the theoretical categories and the historical evidence. As we learned more about how our inventors worked, we refined our notions of what constituted a mental model, a heuristic, or a mechanical representation. We were not interested in merely taking an established theory off the shelf and testing it with new cases; rather, we wanted to shape our categories as we worked with the sources.

Creating a Mapping Technique

As we started to study the primary sources, Gorman and I quickly realized that verbal accounts of the invention process (such as patent testimony or recollections) did not always square with the visual and physical sources (notebooks, sketches, and artifacts) produced in working on an invention. Verbal accounts of the invention process typically were created years later when an inventor was seeking to prove that he or she was the first to invent something. Adhering to the popular notion of a eureka moment, inventors often collapse the long process of trial and error into a single clairvoyant moment of insight.

If we were going to get close to what our inventors thought and did at the benchtop, we needed to create techniques for analyzing the visual as well as the written materials. Given the volume of sketches (dozens for Bell and hundreds for Edison) and the diversity of sources (especially with Gray), we decided to develop a computer-based system for organizing and storing the visual sources.

At this point, Gorman and I took advantage of the fact that we teach undergraduate engineering students at UVa. We turned to our best students for help. Over a seven-year period (1989–1996), they worked with us to devise techniques for analyzing the visual materials used by our inventors. Together we created a mapping technique for the invention process.[11]

One way to understand our mapping technique is to think about the workspaces of inventors and artists. If we look around an artist's studio or an inventor's laboratory, what do we find? Both creative spaces are crammed with models, sketches, and notes as well as machines or paintings in various stages of development.[12] Using all these resources, an artist or inventor is trying to merge abstract ideas in the mind with objects in the real world. On the one hand, an inventor may have an idea for an invention and struggle with how to build a device to realize that idea. On the other hand, an inventor may experiment with a device to see what new ideas it produces. An inventor may move from idea to object and vice versa. Invention, in short, is about merging mental models and mechanical representations.

While it is easy for us to picture an inventor tinkering with a device on the benchtop, this is not the only way that he or she can mingle ideas and objects. He or she can do the same with a variety of substitutes; rather than make a full-scale version of a device, an inventor might use a prototype, sketch, a series of calculations, computer simulations, or even a written description. The advantage of these substitutes is that they often can be generated more quickly and manipulated differently than the full-size device. Edison, for example, filled notebooks with hundreds of sketches because drawing frequently allowed him to determine whether a device would work, thereby allowing him to avoid the cost and hassle of building each version that he envisioned. Substitutes that stand in place of the fully developed invention— such as prototypes, sketches, calculations, and written descriptions—are *representations* of an invention.

Invention, then, can be considered an activity in which inventors use a variety of representations to merge an abstract idea in the mind with material objects that exist out in the world. One way to sort through these representations is to create a map. Suppose you allow each permutation of an invention—each prototype, model, sketch, or experiment—to be a box on a map. This "box" can be a piece of paper, a square drawn on a computer, or a Post-it on the wall; it doesn't matter. In our case, we scanned all three inventors' sketches into the computer and then cut-and-pasted them into the boxes on the map. It was critical for us also to stay close to the visual

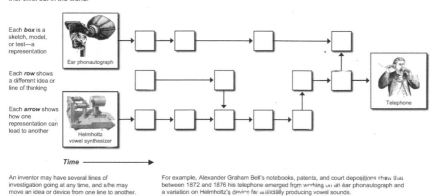

Like artists, inventors engage in a series of activities that facilitate creativity. Both artists and inventors use a variety of representations—sketches, models, written descriptions—to merge an abstract idea in the mind with material objects that exist out in the world.

Each **box** is a sketch, model, or test—a representation

Ear phonautograph

Each **row** shows a different idea or line of thinking

Each **arrow** shows how one representation can lead to another

Helmholtz vowel synthesizer

Telephone

Time

An inventor may have several lines of investigation going at any time, and s/he may move an idea or device from one line to another.

For example, Alexander Graham Bell's notebooks, patents, and court depositions show that between 1872 and 1876 his telephone emerged from working on an ear phonautograph and a variation on Helmholtz's device for artificially producing vowel sounds.

Figure 16.1
Mapping the invention process.

materials and not reduce the sketches to verbal descriptions. We frequently gained insights by placing several sketches side by side on a map to look for similarities and differences.

Once you have the boxes, you can then place them in chronological sequence. In figure 16.1, for example, time moves horizontally from left to right. Arraying the representations in chronological order allows you to look for cause and effect. In some cases, one sketch prompts an inventor to produce another sketch or model, and you can capture these connections by drawing arrows between the boxes. You can also arrange the boxes in different rows, with each row representing a line of investigation or particular invention. Inventors frequently pursue several lines of investigation simultaneously and may move an idea or device from one line to another. For instance, when Edison was working on the telephone in 1877 and 1878, he was simultaneously investigating his quadruplex telegraph as well as the phonograph.[13] In fact, a key part of creativity is the ability or willingness to move ideas and objects from one line to another—to mix things up in unexpected ways.

Gorman and I worked on these maps with our students for seven years. It took such a long time because we developed multiple maps for each individual inventor. To cope with the more than five hundred telephone sketches produced by Edison and his team at Menlo Park, we ultimately created eighteen large maps, each with dozens of boxes and arrows.

What Did We Find Out?

Using the maps, we looked for patterns in the ways that Bell, Edison, and Gray thought about and worked on the telephone. The maps suggested several aspects of the invention process:

1. Different inventors have different kinds of mental models. While Bell was guided by analogies (e.g., make the transmitter like the human ear), Edison tended to work in terms of a functional principle (use variable resistance to convert sound waves into electric current waves).

2. Different inventors use different heuristics. Bell, for instance, worked in a very incremental and methodical fashion; he often had to test each possible variation of his basic design on the benchtop. In contrast, Edison often varied several parameters at once and did not conduct as many benchtop experiments, because sketching allowed him to determine if a particular design would work.

3. We found it fascinating that while Edison generally worked "bottom up," manipulating mechanical representations on the benchtop in order to formulate his mental model, Bell worked "top down," starting with a mental model that he then tested on the benchtop.

Teaching Innovation

As we discerned patterns in our maps, we were determined to draw on our research to teach our engineering students how to invent. Both Gorman and I began moving ideas from our research into our teaching.

Gorman, for instance, teamed up with a colleague in mechanical engineering, Larry G. Richards, to offer a course on invention and design. In early versions of this course, students constructed their own telegraph and telephone systems.[14] Their course had a profound impact on one student, Evan Edwards, who, together with his brother Eric, invented an auto-injector for medicines. After graduating, the two of them launched a company to develop this product.[15]

I transferred ideas from our research to the communications course I was teaching to first-year engineering students. Since the 1930s, UVa Engineering has had a tradition of teaching writing and speaking as part of the school's curriculum. Rather than have engineering students learn writing in classes offered by the College of Arts and Sciences, the Engineering School instead has its own faculty who teach communications, ethics, history, and the social sciences as they relate to engineering.

For me, the challenge in teaching first-year communications was to show my students how writing and speaking were integral to engineering practice. Although I initially had the students study Edison's career for inspiration, I quickly realized the potential of having the students imitate Edison—to try their hand at invention. I took seriously the fundamental results of our research—that invention is the interplay of ideas and objects and that inventors mix up ideas and objects by using a variety of representations. To get objects into the classroom, I began having students build kits, first a pendulum clock and then a robot car.[16] For the representations, I had them sketch ideas in notebooks, write technical descriptions, and ultimately draft a patent application. Whenever possible, I encouraged them to borrow from their mathematics and computer science classes to consider how to represent inventions in terms of equations or computer simulations. Through this teaching, I came to agree with the great cognitive scientist Herbert Simon, who insisted, "Solving a problem simply means representing it so as to make the solution transparent." I went on to articulate a philosophy of engineering-as-representation.[17] This philosophical approach now forms the basis of how we teach communications in the first-year engineering program.

Adding Context to the Cognitive Framework

During the early 2000s, I continued to mull over what Gorman and I had learned about the invention process. In particular, I was not happy with the fact that our cognitive framework did not take into account the social, economic, or cultural context in which inventors work. How might larger, external forces be included in this cognitive framework?

Attempts to understand the process of invention and innovation frequently follow a so-called linear model from ideas to manufacturing to adoption. Engineering textbooks reproduce this ideal to students in block diagram models (figure 16.2).[18] They suggest that inventors move through a specific sequence with no false starts or backtracking. The linear model embodies several dichotomies:

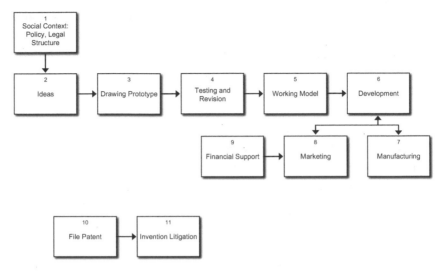

Figure 16.2
Invention as a linear process.

- It assumes that *ideas* come before *devices*, yet some inventors work closely with devices first and in so doing generate new ideas.
- *Social* factors appear only at the *beginning* and *end* of the model, yet inventors may be influenced by all sorts of external factors throughout their work.
- The model draws a distinction between the *technical* work that occurs in the middle steps and *marketing*, which comes at the end. But inventors are often doing both technical and marketing work at the same time.

While a linear model may be useful for introducing students to design, it is problematic for understanding how inventors actually work. For scholars intent on understanding the details of the invention process, using this sort of one-size-fits-all prescriptive model can seem like jamming a square peg into a round hole.[19]

I wondered if it would be possible to build on the work that Gorman and I had already done to create a more robust model that would soften these dichotomies; recognize how inventors are influenced by a variety of ideas, resources, and people; and highlight their day-to-day activities. Would it be possible to generate a nonlinear model of the invention process that would capture the ways that invention is social and cultural?

Rethinking Invention as a Process

I began by reconsidering what it means to say that invention is a process (figure 16.3). Like other processes, invention has both inputs and outputs. In terms of inputs, what does an inventor use in the course of his or her work? Besides tangible resources (money, tools, and materials), he or she also relies on other individuals to provide help with such essential tasks as model building, patenting, manufacturing, and marketing. And an inventor draws on a range of intangible resources, including preexisting knowledge (of both science and other fields), needs seen in the marketplace, legal expertise (patents, regulations, and contracts), and emotional support and encouragement. Inventors may also be influenced by public events (such as wars or business conditions) or their personal situation (love, marriage, health, emotional depression), and so these should be included as inputs. To highlight that invention is an iterative process, there could just as easily be a loop back from the outputs to society, back to the original inputs.

If invention is a process that we can study, then at the very least it consists of inputs, a process, and outputs …

… and we should keep in mind that inventors interact with various people who provide inputs and help produce outputs

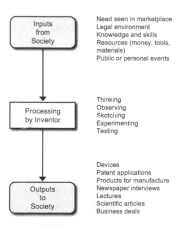

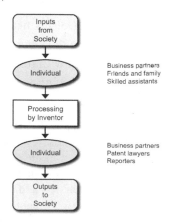

Figure 16.3
Invention as a social process.

In terms of outputs, we expect inventors to produce devices and patent applications, but they also need to promote their creations. Hence, the outputs from the invention process should include interviews, lectures, and publications. And since inventors may sell or license their creations, we should also list business deals.

While inventors can bring some resources into the creative process on their own (for instance, they can read the technical literature or study needs in the marketplace), these inputs and outputs often come into the invention process through interactions with other people. These individuals can include business partners, skilled assistants, family and friends, patent lawyers, and newspaper reporters. All of these intermediaries affect how ideas, resources, and results flow in and out of the invention process.

By considering the inputs and outputs to the invention process and by highlighting the individuals with whom inventors interact, I was placing the cognitive work of invention into a social context. Including an inventor's relationships in the diagram allowed me to show how inventors draw on their social environment and, at the same time, how they seek to shape it.

An Inventor's Flow

As I articulated invention as inputs-process-outputs, I soon realized that what Gorman and I had worked out in our research on the invention of the telephone was all the stuff that belonged in the middle, in the process box on figure 16.3. We had been mapping the activities by which an inventor gets an idea out of his head, refines it through multiple representations, and realizes it on the laboratory bench.

But as I looked with fresh eyes at our invention maps, I was reminded of how Tesla described his feelings in 1882 after he had envisioned his new AC motor using a rotating magnetic field. "For a while," Tesla recalled fondly,

> I gave myself up entirely to the intense enjoyment of picturing machines and devising new forms. It was a mental state of happiness about as complete as I have ever known in life. Ideas came in an uninterrupted stream and the only difficulty I had was to hold them fast. The pieces of apparatus I conceived were to me absolutely real and tangible in every detail, even to the minutest marks and signs of wear. I delighted in imagining the motors constantly running, for in this way they presented to the mind's eye a more fascinating sight. When natural inclination develops into passionate desire, one advances toward his goal in seven-league boots.[20]

What I had not yet appreciated was that invention is not just a process but a *state of mind*.

Like other creative people, inventors intentionally strive to generate a steady stream of ideas and representations, to study them, and to shape these ideas into meaningful inventions. Psychologist Mihaly Csikszentmihalyi calls this effort to generate a steady stream of ideas the "flow" of creativity.[21] Taken as a whole, mapping the various representations used by inventors gives us a visual picture of the flow of their work.

In looking at an inventor's flow, it is important to consider short-term versus long-term goals. Immersed in the creative process, an inventor may not be worried about the ultimate version of an invention but may simply be thinking about how she moves from one representation to the next. Being in the flow means being focused on the immediate opportunities and open to new possibilities as they present themselves. Along the way, inventors may change their mind about the ultimate goal, or the goal may only become clear by doing, by being in the flow. Hence, while it would be tempting to create a map of the invention process in which everything funnels into what we know as the final version, we should instead strive to include the ideas and devices that went nowhere, the false starts, and the wrong turns.

Combining Process and Flow

I now brought these two perspectives—input/output and flow—together on a single page (figure 16.4). I let the process move vertically down the page, with the inputs at the top and the outputs at the bottom. Meanwhile, I had the inventor's flow move horizontally across the page.

This diagram allows us to follow how ideas and resources affect an inventor's flow as well as how prototypes and other products then move out from the inventor's workshop into society. For instance, an inventor may learn about an idea on his own. In this case, we would draw a line and arrow from the idea box down to the sketch or experiment box that it affects. In other cases, an inventor learns about an idea from an individual, such as a friend, business partner, or assistant. In these situations, we add an oval showing the person's name.

In organizing the inputs and outputs depicted in figure 16.4, it is important to arrange them by how they are experienced by the inventor, rather

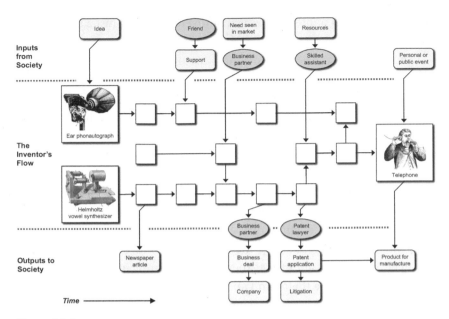

Figure 16.4
Invention as a flow process.

than by some sequence embedded in the model. This is the major advantage of the mapping technique over the old linear model. For example, the inputs—ideas, encouragement, needs, resources, and events—should be placed in the order by which historical sources indicate they affected the invention flow. In situations where a version of an invention is reported in a newspaper or written up in a patent application, we can connect a box in the flow that indicates outputs. Rather than squeezing the historical materials into the model, the model and diagrams are modified to reflect what the inventor is actually doing.

Yet we should be clear that the story portrayed in the historian's flow process model can be very different than the one an inventor might tell about how she came to create a particular device. It is important to recognize that inventors may edit the flow process down to a simple narrative to secure strong patents and to suggest that there was a logical progression in their thought processes. Moreover, an inventor may simply not be able to reconstruct all the twists and turns he took on the way to a particular outcome. Edison, for instance, introduced hundreds of sketches of the telephone as evidence in patent litigation in the late 1870s, but in testifying he was hard-pressed to explain how he moved through this entire corpus of work.[22]

What Did This New Model Capture?

The flow model helps us capture three aspects of the invention process:

- The people, ideas, and resources that shape an inventor's work
- The activities that inventors use to develop new inventions (thinking, sketching, experimenting, and testing)
- The results of the invention process (patents, prototypes, processes, products, and publicity)

It is important to note that this model is not intended to be restrictive. No "one size" fits all inventors. Indeed, the point of the model is to draw out the unique and special qualities of each inventor. The model serves to help us ask consistent questions across different episodes of invention. It should help us discern both the social and the cognitive dimensions of creativity.

This model permits us to capture more of the details of the invention process and allows us to construct richer narratives of how inventors combine ideas and objects and merge the technical with the social. It also helps us to understand not only successful inventions but also the dead ends and ideas that fail. But, above all, this new model should enlarge our vision of what invention is and challenge us to better understand how people can create remarkable technology that changes the world.

The Entrepreneurial Turn in Research and Teaching

I found this new model of innovation as a nonlinear flow process to be enormously helpful in understanding how Nikola Tesla worked, but perhaps even more so in my approach to teaching innovation.[23] One of the most powerful features of the model is that it reveals the intellectual and social work required to convert a benchtop device into a commercial product. Left on their own, inventors often prefer to concentrate on the flow, on moving from sketch, to model, to new experiment; recall the sublime pleasure that Tesla felt as he imagined a stream of new electric motors. In the diagram for the model, inventors want to keep moving in a horizontal direction. When the time comes to convert a benchtop device into a commercial product, however, the inventor must change direction; rather than put his or her energy into generating new variations, he or she has to work on connecting an invention with the wider world via publicity, business deals, or manufacturing. Often these connections cannot be made by the inventor alone but require the assistance of partners and patent lawyers. On the diagram, the movement of an invention from the benchtop to the marketplace means making a turn from the horizontal to the vertical. Hence, the diagram shows major decisions—where the technical gets connected to the social—as a *turn*.

This entrepreneurial turn brings into crisp focus the importance of partnerships in technological creativity. Careful observers will notice that partnerships combining technical skills with business acumen are scattered across the history of technology. In the case of Tesla and his AC motor, he needed the help of his business partners, Charles Peck and Alfred Brown, to know which motor ideas to patent, demonstrate in public lectures, and ultimately sell to George Westinghouse.[24]

The notion of an entrepreneurial turn also had powerful implications for teaching innovation. Several key ingredients are required to bring innovations into the world:

- There must be creative people—inventors, designers, engineers, and scientists—who are allowed to generate ideas and prototypes. The British lovingly call these creative types "boffins."
- The boffins must be protected from external distractions so they can be in the flow. This may mean giving them their own resources and space in an organization.
- At the same time, there must be individuals—the entrepreneurs—who can select the most promising variations from the flow and move them out into the world.

These three ingredients became, first, the kernel of a new course, and then a new technology entrepreneurship program at UVa.

In a new class, "Engineers as Entrepreneurs," I guided students as they explored the habits and practices required to place themselves in a creative flow: how to observe the world; represent it in words, sketches, and simulations; and generate innovations. My students had to understand that at times engineers—like artists or musicians—must fully immerse themselves in a technically rich stream of activities. This is the existential pleasure of working in a lab or design studio.[25] At the same time, I challenged them to pay attention to what customers and other stakeholders might need or want by conducting interviews and market research. Altogether, they would have to use their technical expertise to understand the flow of ideas in the laboratory and draw out those ideas that might be used to satisfy the needs of customers.

"Engineers as Entrepreneurs" was well received by UVa students and caught the attention of both the dean of engineering and engineering alumni who wanted to promote innovation within the school. Working as venture capitalists in Silicon Valley and elsewhere, several alumni had learned that the most important people to meet from a potential start-up were not the "suits" (the MBAs), but rather the boffins, the engineers. Indeed, these alumni had become convinced that the most promising future start-ups would not be led by MBAs who employed engineers, but rather by engineers who knew enough about entrepreneurship to run their own enterprises.

With guidance and financial support from these alumni, we launched UVa's Technology Entrepreneurship Program (TEP) in 2010. TEP provides courses and activities that allow faculty, staff, and students to develop as entrepreneurs and innovators.[26] The program offers at least six undergraduate courses every year that cumulatively attract three hundred students. Students take these classes as part of their engineering major or in pursuit of a minor in entrepreneurship. In these courses, we teach customer discovery and employ Steve Blank's Business Model Canvas.[27] In most classes, students work in teams and practice pitching their ideas to investors and entrepreneurs from the local region. TEP also provides a variety of cocurricular activities to help aspiring entrepreneurs. Through a program called Works in Progress, students meet outside of class to brainstorm, form teams, and network with local entrepreneurs. To support our entrepreneurial students, we have two dedicated workspaces in the engineering school complex, and alumni have endowed several funds that support student projects.

Three core ideas underlie these courses and activities. First, we believe that technology entrepreneurship is different from the entrepreneurship taught in business schools. Entrepreneurship programs in business schools tend to emphasize demand-pull innovations: find an untapped market and then create a product or service that satisfies that market. In contrast, we believe the breakthrough technologies that alter everyday life are often supply-push innovations that start at the laboratory benchtop and then move to the marketplace. Using the flow model of innovation, we are training engineers how to spot the most promising ideas percolating in labs—what one UVa alumni entrepreneur, Robert S. Capon, calls the "noble asset." We then guide the students as they write patent applications, develop business plans, and work to launch ventures around these assets.

Second, it is critical to teach students about customer discovery. All too frequently, engineering students will come up with an innovation in the laboratory that is really cool to them but totally unrelated to a customer need. We show students how to talk to real customers, insisting that they not mention their idea but rather focus on the problems and wishes the customers might have. In fact, we tell our most promising student entrepreneurs that they really don't know what the market will be until they have spoken to at least one hundred potential customers. It is only through this process that students are able to construct a market that fits their innovation. Not surprisingly, customer discovery often prompts our student entrepreneurs to redesign their inventions or to move their laboratory research in new directions. Customer discovery is at the heart of the National Science Foundation's I-Corps program, and in 2017, we secured funding for an I-Corps program to train UVa faculty and graduate student to be entrepreneurs.[28]

The third big idea guiding TEP is that entrepreneurs need a community around them.[29] Popular mythology often depicts entrepreneurs, like inventors, as solitary individuals who work alone. However, as the flow model reminds us, inventors rely on a variety of people—friends, family, assistants, investors, and entrepreneurs—to be creative. In the same way, we have found that entrepreneurs do not work alone but rather need to interact with other entrepreneurs who provide ideas, guidance, and motivation. Across history, at the heart of the most innovative economies, you will find communities of entrepreneurs—whether it be Amsterdam's coffeehouses in the seventeenth century, Philadelphia textile mills in the nineteenth century, Detroit machine shops in the early twentieth century, or Silicon Valley in the late twentieth century. In each of these places, entrepreneurs gathered to share and support one another.[30]

By drawing on history and listening to our students, we are nurturing a community of entrepreneurs. We help students who are just learning about entrepreneurship, but we also go out of our way to mentor those young entrepreneurs who are ready to launch an enterprise. We create opportunities for these founders to meet regularly so they can learn from each other, and we provide our founders with dedicated work space.[31] Whenever possible, we encourage cross-disciplinary teams to form by working with students from the business, policy, and humanities arenas. We are especially proud that our founders frequently place highly in entrepreneurship competitions; for the past several years, our student teams have finished first in UVa events, including the E-Cup and the Darden Business Plan Competition. In 2016, Contraline, a student team whose members had invented a new contraceptive for men, placed second in the Atlantic Coast Conference's InVenture Prize. Another student team, Agrospheres, won the 2016 Collegiate Inventors Competition and took top place in the 2017 InVenture Competition. Agrospheres has since raised $750,000 to support development for using nanospheres to deliver pesticides precisely and safely to crops. And in 2018, Ashwinraj Karthikeyan not only won the ACC Inventure Competition again for UVa but also received the $50,000 Pike Prize for Engineering Entrepreneurship. Karthikeyan has developed a new bandage for treating diabetic foot ulcers and he will be deploying it in India.[32]

Conclusion

How should innovators be trained? Can you use the past to teach the supposedly unteachable? What we have learned is that invention and innovation are not simply about genius, luck, or confidence. By studying Edison, Tesla, and Bell, you can find patterns in the ways they thought and worked. Using these patterns, you can formulate a framework that suggests insights to teach students. If you want students to be innovative, they need to know that invention is the interplay of ideas and objects. Students can learn and practice using different representations—notes, sketches, prototypes, and simulations—to move ideas from the mind's eye into the material world. We can help them understand that inventors seek to be in a creative flow, striving to learn from each representation and to keep generating new variations. And student innovators can master the entrepreneurial turn—what it takes to recognize the most promising idea in a creative flow and to guide it from the laboratory bench to the end user. The past can be used to create curricula for innovation and entrepreneurship.

Our experience in researching and teaching innovation and entrepreneurship at UVa offers several lessons for faculty and administrators who are creating similar programs at their institutions. First and foremost, there is a body of knowledge and a set of skills that can be taught and that improve the chances of students successfully converting an idea into a product. This knowledge and these skills can be introduced in formal courses, but more importantly, you need to create opportunities and spaces outside the classroom where students can practice, practice, practice. As we often remind our students, Roland H. Macy failed four times before he came up with a winning formula for his New York department store. Is it not better for entrepreneurs to fumble a few times during their student years, before they face the vicissitudes of the real world?

To ensure that practice leads to results, students need guidance, mentoring, and resources. Students cannot teach students to be entrepreneurs and innovators; they need to learn from both scholars of entrepreneurship as well as experienced entrepreneurs. An effective entrepreneurship program does not assume that students will automatically "get it" and keep coming back for more. Administrators and faculty must be committed to helping individual students connect what they learn in the entrepreneurship classroom with cocurricular activities. Because these connections are built one student at a time, a successful entrepreneurship program is labor-intensive and ultimately expensive. Our experience has taught us that you can have all sorts of competitions, guest speakers, and get-togethers promoting entrepreneurship, but students only become innovators by working closely with dedicated staff and faculty. It is about the relationships created, not the events.

But do programs emphasizing entrepreneurial knowledge, skill, and practice demonstrably increase innovation? To be sure, several UVa students have become full-fledged innovators; we are proud of Evan Edwards and his auto-injector, and we expect great things from student teams like Contraline and Agrospheres. I would suggest that it's not simple to measure direct cause and effect, that "concept X, taught to Y number of students, resulted in an increase in innovation by Z percent." What one sees when working with student innovators is much more subtle—a better notebook sketch here, a more persuasive rocket pitch there, or a student team coming together and gaining confidence in themselves. Ultimately, what we teach is a mindset, an attitude that change is possible. Our students know that they can develop the skills needed to innovate. Innovation, at its heart, is about nurturing the human spirit and fostering a faith in progress. None of our students have gone on to become the next Nikola Tesla—yet—but we see signs that we are on the right track.

Notes

1. See Fasihuddin and Britos Cavagnaro (chapter 3) in this volume.

2. W. Bernard Carlson, "From Order to Messy Complexity: Thoughts on the Intellectual Journey of Thomas Parke Hughes," *Technology and Culture* 55, no. 4 (October 2014): 945–952.

3. W. Bernard Carlson, *Innovation as a Social Process: Elihu Thomson and the Rise of General Electric, 1870–1900* (New York: Cambridge University Press, 1991).

4. Michael E. Gorman, *Simulating Science: Heuristics, Mental Models, and Technoscientific Thinking* (Bloomington: Indiana University Press, 1992).

5. In addition, there was controversy associated with the case, as historians had already suggested that Gray, not Bell, might be the true inventor of the telephone. See David A. Hounshell, "Bell and Gray: Contrasts in Style, Politics, and Etiquette," *Proceedings of the IEEE* 64 (1976): 1305–1314; Seth Shulman, *The Telephone Gambit* (New York: W. W. Norton, 2008); and Bernard S. Finn, "Bell and Gray: Just a Coincidence?" *Technology and Culture* 50 (January 2009): 193–201.

6. Bruno Latour and Steven Woolgar, *Laboratory Life: The Construction of Scientific Facts* (Princeton, NJ: Princeton University Press, 1986); Peter Galison, *How Experiments End* (Chicago: University of Chicago Press, 1992).

7. See Bert S. Hall and Ian Bates, "Leonardo, the Chiarvelle Clock, and Epicyclic Gearing: A Reply to Antonio Simoni," *Antiquitarian Horology* 9 (1976): 910–917. On the concepts of an inventor's style and method, see the following studies by Thomas P. Hughes: *Elmer Sperry: Inventor and Engineer* (Baltimore: Johns Hopkins University Press, 1971); "Edison's Method," in *Technology at the Turning Point*, ed. W. B. Pickett (San Francisco: San Francisco Press, 1977); and *Networks of Power: Electrification in Western Society, 1880–1930* (Baltimore: Johns Hopkins University Press, 1983).

8. See Philip N. Johnson-Laird, *Mental Models* (Cambridge, MA: Harvard University Press, 1983); John R. Anderson, *The Architecture of Cognition* (Cambridge, MA: Harvard University Press, 1983). For a review of the cognitive science literature around the time that we launched this project, see Gorman, *Simulating Science*.

9. W. Bernard Carlson and Michael E. Gorman, "Understanding Invention as a Cognitive Process: The Case of Thomas Edison and Early Motion Pictures, 1888–1891," *Social Studies of Science* 20, no. 3 (August 1990): 387–430.

10. As Reese V. Jenkins observed, "Any creative technologist possesses a mental set of stock solutions from which he draws in addressing problems." See Jenkins, "Elements of Style: Continuities in Edison's Thinking," *Annals of the New York Academy of Sciences* 424, no. 1 (May 1984): 153.

11. Critical to this effort was Matthew M. Mehalik, who worked closely with Gorman and me all through his undergraduate years and subsequently earned his PhD in systems engineering at UVa with Gorman as his dissertation advisor.

12. See McManus and MacDonald (chapter 4) in this volume.

13. In the cognitive science literature on creativity and discovery, the collection of multiple lines of investigation is called a network of enterprises. See Howard E. Gruber and Katja Bodeker, *Creativity, Psychology, and the History of Science*, Boston Studies in the Philosophy of Science, vol. 245 (Dordrecht, Netherlands: Springer, 2005), 22–24, 52–57, and 89–104.

14. Michael E. Gorman and J. Kirby Robinson, "Using History to Teach Invention and Design: The Case of the Telephone," *Science and Education* 7, no. 2 (1998): 173–201.

15. "Taking Their Best Shot: Kaléo Makes Big Strides in Auto-Injectors," Lemelson Foundation, 24 June 2015, accessed 1 September 2017, http://www.lemelson.org /resources/success-stories/Kaleo.

16. W. Bernard Carlson and Karin Peterson, "Making Clocks: A First-Year Course Integrating Professional Communications with an Introduction to Engineering," *Proceedings of the American Society for Engineering Education*, CD-ROM, 1996.

17. See Herbert A. Simon, *The Sciences of the Artificial*, 3rd ed. (Cambridge, MA: MIT Press, 1996), 132, and W. Bernard Carlson, "Toward a Philosophy of Engineering: The Fundamental Role of Representation," *Proceedings of the American Society for Engineering Education*, CD-ROM, 2003.

18. When the Smithsonian Institution's Lemelson Center for the Study of Invention and Innovation retained me as a consultant in 2006, I had an opportunity to revisit our cognitive framework. Shortly after its founding in 1995, the center had developed a linear model of the invention process that it used to generate questions and ideas about what sorts of materials it should collect to document the creative work of inventors. The Lemelson Center brought me in as a consultant not only because of my research on inventors but also because of my experience consulting with the R&D group at Corning. There I drew on the approach I had developed with Gorman to prepare a number of proprietary case studies of the innovation process. I captured my initial findings in a Lemelson Center white paper, "Understanding Invention and Innovation: A Documentary Study," last updated 25 August 1998. As the center matured, so did its model of the invention process, especially as presented in Spark!Lab, its hands-on invention center for children aged 6–12 and their

families. Spark!Lab breaks the invention process into seven nonlinear steps: identify a problem or need (Think It); conduct research (Explore It); make sketches (Sketch It); build prototypes (Create It); test the invention (Try It); refine the invention (Tweak It); market the invention (Sell It). See "About Spark!Lab," Lemelson Center for the Study of Invention and Innovation, accessed 1 May 2018, https://invention.si.edu/about-sparklab.

19. As Eugene S. Ferguson observed in his study of design and technological creativity, "Block diagrams imply division of design into discrete segments, each of which can be 'processed' before one turns to the next." Eugene S. Ferguson, *Engineering and the Mind's Eye* (Cambridge, MA: MIT Press, 1992), 37.

20. Nikola Tesla, *My Inventions: The Autobiography of Nikola Tesla*, ed. Ben Johnston (Williston, VT: Hart Brothers, 1982), 65.

21. Mihaly Csikszentmihalyi, *Flow: The Psychology of Optimal Experience* (New York: Harper & Row, 1990).

22. Several inventors made this point during a Lemelson Center workshop on prototypes I helped organize in the fall of 2006. See W. Bernard Carlson, "Documenting Invention: Prototypes and Invention—An Inquiry into How Inventors Think and Communicate," 1 September 2007, accessed 1 September 2017, https://invention.si.edu/documenting-invention-prototypes-and-invention-inquiry-how-inventors-think-and-communicate.

23. While the flow model does not appear explicitly in my biography of Tesla, it particularly guided my thinking on how he evolved his plan in the 1890s for wirelessly broadcasting power through the earth. See W. Bernard Carlson, *Tesla: Inventor of the Electrical Age* (Princeton, NJ: Princeton University Press, 2013), especially chapters 10 through 15.

24. Carlson, *Tesla*, 76–105.

25. Samuel Florman, *The Existential Pleasures of Engineering*, 2nd ed. (New York: St. Martin's Griffin, 1994).

26. "Entrepreneurship and Business," accessed 1 September 2017, https://engineering.virginia.edu/departments/engineering-and-society/entrepreneurship-and-business.

27. Steve Blank and Bob Dorf, *The Startup Owner's Manual: The Step-by-Step Guide for Building a Great Company* (Pescadero, CA: K&S Ranch Press, 2012); Alexander Osterwalder, *Business Model Generation: A Handbook for Visionaries, Game Changers, and Challengers* (Hoboken, NJ: John Wiley, 2010); Alexander Osterwalder et al., *Value Proposition Design* (Hoboken, NJ: John Wiley, 2014).

28. See Arkilic (chapter 5) in this volume.

29. See Hintz (chapter 10) in this volume.

30. See Feldman (chapter 6) in this volume.

31. Elizabeth P. Pyle and Alexander J. Zorychta, "The Social Mechanisms of Supporting Entrepreneurial Projects beyond the Classroom," *Proceedings of American Society for Engineering Education*, Columbus Meeting, 2017, paper no. 19983.

32. "Agrospheres Announces Close of Oversubscribed $750,000 Seed Round," 21 August 2017, accessed 1 September 2017, http://www.agrospheres.com/news/2017/8/21/agrospheres-announces-close-of-oversubscribed-750000-seed-round; Elizabeth Thiel Mather, "Major Wins Propel UVA Engineering Entrepreneurship Team to Future Business Success," 8 May 2018, accessed 9 September 2018, https://engineering.virginia.edu/news/2018/05/major-wins-propel-uva-engineering-entrepreneurship-team-future-business-success.

17 Confronting the Absence of Women in Technology Innovation

Lucinda M. Sanders and Catherine Ashcraft

Only 19 percent of all software developers are female, and very few are in technology leadership roles that would enable them to make truly innovative contributions. Consider that 88 percent of all information technology patents (1980–2010) have male-only invention teams while only 2 percent have female-only invention teams.[1] These statistics and others imply that a largely homogeneous group is creating the technology the world uses today—US white males (and increasingly Asian males). Women, and especially women of color, are essentially absent from technology innovation—absent because of low participation, absent because the world does not experience their potential contributions, and absent because when women do make technical contributions, they are often ignored, not recognized, or not given credit for their ideas. This is especially troubling given ample evidence of the critical benefits diversity brings to innovation, problem-solving, and creativity. Indeed, innovation springs from diversity—diversity of ideas, perspectives, voices, and in short, a diversity of people.[2]

However, numerous social and cultural influences are increasingly impeding women's contributions to technical creation in today's tech workforce. Recognizing women as technical contributors requires explicit, conscious effort. Simply adding women to the pot and stirring is not going to make their ideas recognized or used. Technical design teams need to employ democratic principles and techniques for making sure ideas are heard and discussed. Managers and supervisors (both men and women) need to perform as champions for their female technologists. And they need to be informed and equipped to do so effectively, with a clear understanding of both the values and unique challenges to gender inclusion embedded in our current systems and operations.

A pressing need to address these factors led to the formation of the National Center for Women and Information Technology (NCWIT) in 2004, a National Science Foundation (NSF) funded effort to "*significantly* increase women's *meaningful* participation in computing." We emphasize "significantly" because clearly the numbers were quite low then and "meaningful" because this effort was not intended to be merely a numbers game. We must have women in the roles critical to the invention of future technology (architect, lead designer, etc.) and not only in the jobs that support those who create it (such as project management and system verification).

In this chapter, we look briefly at the latest research documenting the benefits that diversity brings to innovation, identify what "meaningful participation" really means, and explore the underlying psychological and cultural mechanisms behind women's continued absence in technical innovation. We describe actionable practices that not only can mitigate these impacts, but also help organizations adopt pro-innovation strategies through effective diversity and inclusion efforts involving the important work and practical resources of NCWIT.

Why Does Addressing Women's Absence Matter?

Women's current underrepresentation spells trouble for the tech industry and for the future of technical innovation, especially in light of an increasing body of research documenting the significant benefits that diversity brings to innovation. Some of these key benefits are summarized below. Let's consider a few examples:

- A 2009 study of five hundred US-based companies found that higher levels of racial and gender diversity were associated with increased sales revenue, more customers, greater market share, and greater relative profits. Racial diversity, in particular, was one of the most important predictors of a company's competitive standing within its industry.[3]

- A 2007 study by the London Business School of one hundred teams at twenty-one different companies found that work teams with equal numbers of women and men were more innovative and more productive than teams of any other composition. They attributed this finding to the fact that members on these teams were less likely to feel like "tokens" and better able to meaningfully contribute their ideas and efforts.[4]

- Using computer and mathematical modeling, Scott E. Page of the University of Michigan has demonstrated that diverse teams consistently outperform even teams comprising only the "highest-ability" agents.[5]
- Another recent study of work teams revealed that the intelligence level of individual team members was not a predictor of the collective intelligence of the team. However, one of the key predictors of the collective intelligence of a team was a larger number of female team members.[6]

Additional studies have found benefits of gender diversity for start-up companies as well. For example, an analysis of more than 20,000 venture-backed companies showed that successful start-ups have twice as many women in senior positions as unsuccessful companies.[7] In addition, in a study of all investments made in US-based companies between 2000 and 2010 by US-based venture capital (VC) firms, those that invested in women-led businesses saw an improvement in their VC firm's performance.[8]

In light of research studies such as these, technical organizations cannot afford to continue losing out on the benefits that gender diversity and other kinds of diversity can bring to technical advancements. Technology innovation is a creative process that involves teams; often many teams, both large and small, work on a single product or service from front-end requirements generation, through design and development, to product rollout and support. Clearly, it matters who sits at the table on these teams, and all of them would benefit from the increased participation of women. But as we noted earlier, by and large, women are *not* participating in technology innovation roles, but rather in the roles that support others who are engaging in leading-edge technology design and development. This point is not intended to be an evaluative statement about the importance of these support roles—indeed, they are critical. However, they are often not the roles leading to new technology breakthroughs. NCWIT partners with many of the world's largest companies, and it is rare to find a woman, let alone a racial minority, leading any significant technical development effort. We need a diverse array of people contributing to both support and creation roles.

Toward this end, we need to understand *why* women are absent in the first place. Here, it is important first to clarify that women (and other underrepresented groups) are not simply *absent* from technical environments, but that even the few women who *are* present are leaving the field. A 2008 study found that 74 percent of technical women report "loving their work," yet 56 percent of these women leave midcareer (after ten to twenty years). For the most part, they are not leaving, as many often suspect, for familial or other kinds of obligations. Of the 56 percent of women who leave, 75 percent remain in the workforce full time. Fifty percent remain in technical jobs, but not in the private sector; they either take positions in government, nonprofits, or start-ups. Only 25 percent leave the workforce. These patterns suggest that women leave not because they do not like tech or because they want time out of the workforce, but rather because of factors in the private sector's corporate technical environment itself. Imagine what would happen if we could change these factors and stem the tide of female attrition. The same study estimates that reducing female attrition by 25 percent would return hundreds of thousands of women to science and technology workplaces.[9]

The good news is that we have extensive knowledge about the factors causing women to leave technology or preventing them from entering in the first place. An understanding of these barriers is the first step in addressing them. In the next section, we examine some of the principal reasons for women's absences, most notably the pervasive and subtle effects of unconscious bias.

Why Does Women's Absence from Technology Persist?

We contend that for lasting change to occur, companies need to understand and focus serious attention on the role that unconscious bias plays in women's persistent absence from technology contexts. Doing so is crucial for at least three reasons. First, unconscious bias is arguably the most pervasive, but least understood, factor affecting women's participation in technology. As we shall see, these subtle biases are encoded in both institutional barriers and in everyday instances in technology workplaces, creating myriad stealth barriers to women's meaningful contributions.[10] In an encouraging turn of events, recent public conversation around these biases has increased, but we have a long way to go to help company leaders, managers, and employees understand the impact of these biases on the technical workforce.[11] Second, an understanding of how unconscious bias works helps to quell the "blame game" and other kinds of unproductive conversation so often present in discussions around women in tech.[12] Addressing these issues is not about pitting groups against each other; it is not about "good" people who are enlightened and "bad" people who are biased. We *all* share these biases, women and men alike, and while at first glance this may seem rather discouraging, the good news is that we can all work together to address these biases in meaningful ways. Finally, it is worth pointing out that unconscious biases can be relatively inexpensive to address. While larger company programs are also needed to fully address these biases, individual managers and leaders can often make a significant difference by altering their behaviors in small but significant ways.

When NCWIT presents this body of knowledge in technical organizations, the reaction is largely one of relief and confidence in a new way of talking about these issues that avoids pitting groups against each other. We stress, however, that it is important to move beyond awareness and training and toward action. As a result, we often see significant action to "operationalize" this newfound awareness and address biases in everyday business processes.

We also wish to note that while the focus of this chapter is on the experiences of women, it is important to remember that women vary in terms of race, class, gender, ability, sexual orientation, and other points of difference. As a result, considering how these unconscious biases play out for a *diverse range* of women is crucial. Likewise, increasing the participation of all underrepresented groups is also important. Although most of our examples focus on women (and will, therefore, sometimes gloss over important differences among women), it is vital that we keep in mind that these unconscious biases affect members of all underrepresented groups in various ways. Importantly, many of the strategies that we recommend will also improve the work environment for members from a variety of underrepresented groups.

What Is Unconscious Bias, and How Does It Work?

Unconscious biases are the direct result of schemas—or maps that we all have in our head—that help us quickly filter new information and categorize it in meaningful ways. These schemas are vital and necessary; without them we would be paralyzed by all the information we receive in a given day. For example, we have schemas for simple concepts such as a tree, a car, or a mall. We may drive by a strange group of buildings we have never seen before, but because we have developed a schema for "mall," subtle cues let us know that this set of buildings is another example of a mall. In other words, we do not have to start from scratch and examine new objects closely each time we see them. We also have schemas for more complex things like "leader" or "technical person." These schemas shape our definitions of what makes a good leader or what a talented technical person looks like. We also have schemas related to gender, race, class, and other intersecting categories of identity. These schemas subtly influence our perceptions about what is appropriate behavior for women, men, and so on.

Of course, because these schemas help us filter information, they can also lead us to filter in ways that result in misrecognition, misinterpretations, or misunderstandings. When it comes to more complex interactions in the workplace, these schemas can cause us to misinterpret people's behaviors or to miss certain strengths, talents, or characteristics that do not fit our schemas for, in this case, a good leader or a good technical person. Likewise, such schemas also can cause us to characterize women as "too aggressive" and advise them to "tone it down" when the same behavior from men is often deemed more acceptable.[13]

A great deal of research has shown that society has significant biases about gender, science, and technology.[14] Likewise, these biases pervade popular culture—for example, from overt displays in advertisements and children's books to more subtle messages of omission, such as the lack of media representations of women in key technical roles.[15] Keep in mind that we do not necessarily need to buy into or consciously believe these messages to be affected by them. A lifetime of exposure to these messages affects our schemas even if we *consciously* believe that anyone can do science or technology. To test this out yourself, try taking the Implicit Association Test created in 1998 by researchers Tony Greenwald, Mahzarin Banaji, and Brian Nosek.[16] The test related to gender and science/technology measures the speed of associations test takers make between male and female terms and science, technology, or liberal arts terms. Over the past two decades, thousands have taken the test, with nearly 90 percent having some kind of masculine-associated bias when it comes to gender and technology. We too have taken the test, and though we work at NCWIT, we both still test moderately or mildly biased.

Similar patterns of bias also have been demonstrated repeatedly in myriad studies.[17] For example, several studies have shown that evaluators consistently score résumés lower when they are assigned a female name instead of a male name, even though the résumés are exactly the same.[18] Similar patterns have also been documented when it comes to race, where résumés with stereotypically "white-sounding" names received 50 percent more callbacks than résumés with stereotypically "black-sounding" names.[19] In some of these studies, the participants making the biased evaluations were psychologists themselves—individuals who study and are especially well versed in this kind of phenomenon![20] In addition, both female and male evaluators tend to make lower assessments of women candidates. The important takeaway here is that this is not about "enlightened" versus "prejudiced" people; we all (both women and men) share these subtle biases and need to work together to address them.

How Do These Biases Play Out Every Day in Workplaces—and in Particular, Technology Workplaces?

Biases are already circulating in society at large, and individuals encounter them in a variety of contexts even before they enter organizations. We then bring these biases into our organizational cultures. They shape organizational cultures in two ways: (1) *subtle dynamics* or everyday interactions that may seem small in the moment but that add up over time, creating an exhausting or unwelcome environment; and (2) *institutional barriers*, formed where these biases become unconsciously embedded in the organization's policies and programs, making these systems appear as simply the "natural order of things."

Subtle Dynamics

Let's consider some examples of both of these kinds of biases, beginning with subtle dynamics. Micro-inequities are one powerful example of these kinds of subtle instances (see box 17.1).[21]

All of these examples are subtle, tiny jabs that can slowly erode a woman's sense of belonging, confidence, and her sense that she "fits in." For those making the comments, it is also subtle; remember, this differs from overt discrimination. These interactions stem from unconscious biases that cause us to misjudge, misread, or be unaware of the effects of these kinds of comments. While any one instance can seem small, like a dripping faucet, the effect adds up over time, making women feel as if they do not really belong and undermining what is often thought of as a meritocratic organization.

Another pervasive example of these subtle biases surfaces in a phenomenon called stereotype threat—that is, the fear that our performance or actions will confirm a negative stereotype about an identity group to

Box 17.1

Micro-Inequities in the Workplace

Slight: "Actually [surprised tone], Susan has a good idea."

Exclusion: "Oops, I forgot to cc her on the email about the architecture review."

Recognition: "No, I'm pretty sure Jane would not have had the idea to use a link algorithm."

Isolation: "Dude, let's talk about it over a beer."

---

which we belong. In nearly two hundred studies with a variety of different populations, this phenomenon has been shown to reduce confidence, performance, and risk-taking.[22] For example, elderly people perform worse on memory tests when told the test is designed to increase understanding about connections between age and senility. Likewise, students of color perform worse on tests when racial stereotypes about intelligence are invoked ahead of time, and girls or women perform worse on math tests when gender/math stereotypes are called to their attention. In fact, simply moving the race or gender question to the end of a standardized test has been shown to dramatically increase the scores for women and students of color.[23]

In a particularly interesting study, the researchers wanted to see if a single instance of stereotype threat could induce similar effects. To do so, they conducted an experiment with white male engineering students at Stanford University, all of whom had high math scores. Half of the students were simply told that they were taking a math ability test while the other half were told that they were part of a study to understand why Asian students scored better on math tests than white students, thereby invoking the one context where white men might experience stereotype threat around math. As you might imagine, the students in the latter group did significantly worse than those who were not exposed to the stereotype threat. This study demonstrated that even one (or relatively few) instance of exposure to stereotype threat can have powerful effects.[24] Imagine the effect that a lifetime of exposure has on individuals.

So how does stereotype threat show up in the technical workplace? It is important to recognize that being a minority in a majority environment can be a significant trigger for stereotype threat, reducing confidence and risk-taking and resulting in not speaking up in meetings, a reluctance to take on leadership roles, and overly harsh evaluations of one's work products and personal performance.

Often other colleagues attribute these types of actions as evidence of "innate personality" traits of these individuals. You have perhaps heard this expressed in comments such as "so-and-so just isn't very confident" or "isn't much of a risk taker." But it is important to remember that these incidents are often about the *environment* and not the *individual*.[25] Reducing stereotype threat and making the environment more welcoming goes a long way toward increasing confidence and performance and eliminating these kinds of survival strategies. This points to the importance of building structures and strategies that create productive team environments, such as soliciting the opinions of quieter employees and creating environments where everyone can be heard, not just the loudest speakers. It also involves intervening when someone gets credit for an idea someone else offered earlier and ensuring that quieter employees get credit for their work.

NCWIT spends considerable time explaining stereotype threat research and its practical applications within organizations, and we find the gained understanding can quickly shift the institutional focus from that of "fixing" individuals to one of changing the environment. Of course, individual employees need professional development, and individual people must take action to change the environment, but this is distinctly different from interventions that assume women are not risk-takers or that they just need to be more confident.[26] While these kinds of "fix the individual" approaches pervade popular discourse, they often ignore important environmental reasons why women may seem or choose to be "less assertive" or to not take risks.[27] Without addressing these fundamental environmental or systemic issues, real change will not occur.[28]

Likewise, encouragement goes a long way toward reducing the effects of stereotype threat. Countless women have told us how encouragement from a colleague was the critical factor in their decision to apply for a promotion or an award; without this encouragement, they would have felt it was too risky. If they receive the promotion or award, this also further reinforces their sense of confidence and belonging. Research also shows that encouragement is one of the most effective strategies for increasing the retention of women in computing.[29]

It is important to point out here that strategies for confronting stereotype threat will benefit all employees. For example, quieter men will also benefit from more inclusive team meetings. These strategies make the environment better for everyone, but they are especially important for minorities working in a majority environment.

Institutional Barriers

A great deal of research also highlights how biases become encoded in larger policies and programs in the workplace, thereby forming institutional barriers that supersede individual behaviors or interactions. Earlier, we discussed how these biases subtly influence hiring with the evaluation of female and male candidates' résumés. But other institutionalized hiring practices also can reflect hidden biases. For example, relying primarily on personal contacts, referrals, or recommendations when hiring tends to perpetuate the status quo. When it comes to referrals and recommendations, people tend to recommend people much like themselves, a phenomenon known as "assortative matching."[30] According to a study for the Federal Reserve Bank of New York, 64 percent of employees recommended candidates of the same gender, while 71 percent referred candidates of the same race or ethnicity.[31] At least one study has found that women referred for entry-level tech jobs are significantly more likely to be hired than women without referrals; the same study found that for executive high-tech jobs, referred candidates are much more likely to be men than women.[32]

These biases also shape performance evaluations and promotion processes. For example, evaluations for men tend to be longer and to contain more comments about skills and individual achievement, while letters for women are shorter and contain more comments about "softer," stereotypically feminine skills such as communicating and collaborating. Evaluations for women tend to attribute achievement to hard work or luck rather than talent or intelligence, and to contain more "doubt raisers."[33] Similarly, men tend to receive more "constructive criticism" related to skills, whereas women tend to receive more of this kind of criticism related to personality issues, with comments about sometimes coming across as "abrasive" or needing to be less "judgmental" in tone. One investigation found that these kinds of comments appeared in seventy-one of the ninety-four critical reviews received by women but only twice in eighty-three critical reviews received by men.[34]

Finally, and especially important for our concerns here about women's participation in innovation, biases also pervade the kinds of jobs or tasks women take on or are asked to do. As a result, it is important to look for unintended biases in the assignment of particular tasks—who gets assigned the "high visibility" tasks and who gets assigned the more mundane tasks or the "higher risk, scapegoat" kinds of projects. These kinds of subtleties are more difficult to measure, and we are only beginning to understand how they play out in technical workplaces. One study found that nearly half of all women in technology workplaces felt that women were more often pushed into "execution" roles and had less access to "creative" roles.[35] There is anecdotal evidence to suggest that technical women are told they are "better communicators" and "better team players" than men—in essence, that they are more valuable in those roles than in the lead technical creative roles. More research to understand these kinds of barriers is vital for fostering women's involvement in technical invention.

At this point, we have mostly discussed the bad news. Thankfully, there is also good news. We can employ strategies that help expand our schemas and that reduce the effects of these biases. Consider one powerful example from a different industry. In the 1970s and 1980s, many orchestras began implementing a "blind audition" screening process. Previously, hopeful musicians would perform in person, allowing the evaluators to see the individual giving the performance. With blind auditions, performers were required to audition behind a curtain, and great care was taken to avoid unintentional gender cues (e.g., removing shoes so that one couldn't hear the click of high heels). This simple change increased the chances that female performers would make it out of preliminary screening rounds by 50 percent and resulted in a 25 percent increase in the number of females ultimately hired.[36] In the next section, we explore how we might work toward change when it comes to increasing women's participation in technical innovation.

Call to Action: Addressing Biases and Creating Inclusive Organizational Cultures

Over the past ten years, NCWIT has learned quite a bit about ways to address these existing biases and to ultimately develop more inclusive technical cultures where a diverse range of employees can thrive. In what follows, we first focus on promising practices and systemic reform in corporate technical culture. We call particular attention to this arena because, to date, it is one of the most often overlooked areas for change. Typically, change efforts tend to focus on the educational pipeline, from engaging girls earlier in school to recruiting and retaining them in postsecondary education. These efforts will do little good, however, if we do nothing to change the conditions currently causing women and other underrepresented groups to leave tech. Changing these cultures to improve retention and advancement is equally if not more important and needs to be given more attention than it currently receives. At the same time, of course, the pipeline is also important. We conclude then with a look at some of NCWIT's national and local efforts to support organizations across the pipeline in creating change.

Corporate Organizational Strategies for Action

Drawing from the available research, Catherine Ashcraft developed an industry change model for how companies can take an "ecosystem" or multipronged approach that addresses the primary biases and factors that affect women's participation in information technology (figure 17.1).[37] Such an approach is necessary for lasting change to occur. Instituting piecemeal practices may be helpful in some cases but will not result in sustained, systemic change.

The two elements in the center of the model are vital for the sustained success of all other efforts: (1) establishing top leadership support and institutional accountability, and (2) improving the managerial relationship. What gets measured is what gets done. Accordingly, leaders must give more than lip service to these efforts and must institute "accountability metrics" that track progress. In addition, educating and resourcing managers in order to create inclusive environments is critical because managers can make daily life difficult for employees, even if many inclusive policies are formally on the books. Without these two foundational efforts, other change efforts are less likely to have the desired impact. Over years of working with companies on these efforts, we have developed a set of resources to equip managers and senior leaders with strategies for reducing biases in recruitment and selection, employee development, team management, and performance evaluation and promotion.[38]

The six areas in the outer part of the circle indicate the key areas where change is typically needed to create an inclusive ecosystem. We encourage companies to engage in data collection and strategic planning to identify

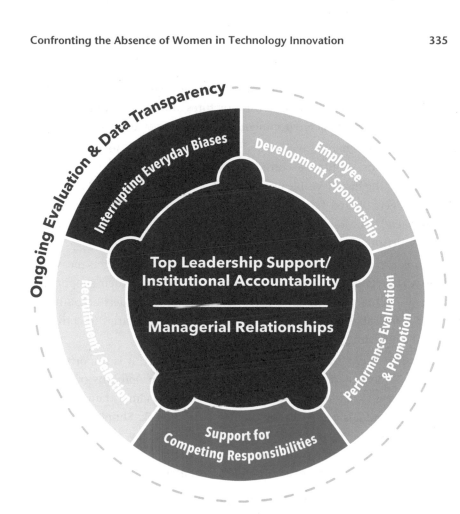

Figure 17.1
NCWIT's Industry Change Model. Courtesy of NCWIT.

the areas in which they are weaker and stronger.[39] Below are a few central
practices that relate to different areas of the Industry Change Model.

- *Analyzing job descriptions and interview practices.* Resources exist to help
  ensure that job ads are free of subtle biases in language, that they are
  engagingly written, and that they contain clear criteria about what skills
  are required versus "preferred."[40] Also, interview teams should include
  diverse representation and ensure that interview questions really probe
  the skills needed for the job, as opposed to assessing vague qualities such
  as "fit" for the organization. The latter is a red flag for the presence of
  biases.

- *Auditing physical environment and recruitment materials/practices.* Ensure that recruitment materials, website pictures and text, and pictures and décor in the physical environment represent a diverse range of people.
- *Analyzing performance reviews and assigning tasks.* The following types of questions should be considered: Are particular employees always assigned the highly visible tasks instead of the high-risk tasks? Who routinely gets offered "stretch" assignments? Do performance reviews for men tend to be longer than for women? Are there differences in the kinds of criticism offered or the kinds of skills highlighted?
- *Setting accountability metrics.* As the old adage goes, "What gets measured is what gets done." Top leaders need to demonstrate their support for inclusive practices, holding themselves and other leaders accountable for measurable change. This involves establishing metrics for progress and connecting these to the business.[41]
- *Creating healthy team environments.* As noted earlier, it is important for team leaders (and everyone, really) to solicit the opinions of quieter employees either during meetings or after. It is also important to ensure that individuals get credit for their ideas and work. Simple comments such as "I'd like to hear more about what she has to say" or "I believe she mentioned that same point earlier, and I'd like to hear more about what she was thinking" can make an enormous difference in team climate.

It is also important to point out that in many ways these types of practices are also just good leadership practices. Conveniently, this means that they are good for men; however, they are even more important for women and other underrepresented groups—for anyone who is a minority in a majority group environment.

NCWIT Strategies and Platforms for Action

We have focused up to this point mostly on private-sector technology innovation, even though such progress also emanates from universities and public-sector institutions. We also acknowledge the importance of inclusion in technical education; hence, we will discuss NCWIT's work across the full computing pipeline.

Founded in 2004, NCWIT is a change-leader network whose mission is to significantly increase the meaningful participation of women and girls in technology. We employ a three-pronged approach to effecting this change:

(1) convene, (2) equip, and (3) unite. First, as a capacity-building organization, we *convene* more than 650 universities, companies, nonprofits, and government organizations nationwide working across K–12 education, higher education, industry, and policy contexts to increase women's participation in computing and technology. Member organizations participate in one or more of five NCWIT alliances spanning the educational "pipeline" (K–12, academic, workforce, entrepreneurial, and affinity group alliances). As part of their membership in these alliances, they are entered into a change-leader community and connected to myriad other organizations working on similar efforts to make computing curriculum, as well as educational and workplace environments, more inclusive.

Members meet annually at the NCWIT National Summit to learn about and share promising practices and the latest research related to gender, diversity, and technology. Throughout the year, we employ a "personal trainer model," helping these organizations assess their needs and plan strategies for change in their own organizations. In so doing, we emphasize changing the overall culture around computing and technology, as opposed to promoting professional development efforts that tend to focus on helping women fit into existing cultures (e.g., creating a personal brand, developing an executive presence, speaking more assertively). While professional development efforts can help women survive their current organizational climate, we believe that, given the extensive research regarding the pervasiveness of bias, lasting change will not occur unless we work to change existing technology cultures.

The second prong in our approach revolves around *equipping* our members with research-based resources to help them implement practices that we know work for women and other underrepresented groups in computing. The NCWIT Social Science Team ensures that these resources and recommendations are grounded in the latest and best available research. According to our last member survey in 2013, our resources have been successful in raising awareness, increasing knowledge, and prompting action. Evaluation shows that eight out of ten members reported learning something new from NCWIT resources, and six out of ten used a new strategy because of a resource. Ninety percent of our members have shared a NCWIT resource with someone else.

Finally, the third prong of our approach involves *uniting* our members. We combine the efforts of programs such as Counselors for Computing, sitwithme.org, and NCWIT Pacesetters to accelerate progress through national platforms and campaigns. Each combines the collective strength of our membership; NCWIT provides the infrastructure, tool kits, and project management to support these efforts, and members participate locally, tailoring these efforts to their own cultures and environments.

Our Counselors for Computing campaign aims to raise awareness among K–12 counselors, providing workshops and resources that build awareness about educational pursuits and careers in computing. Our Aspirations in Computing program is a pipeline program reaching thousands of girls each year from high school through college. Seed Funds and Extension Services support postsecondary practice implementation. The NCWIT Pacesetters program brings corporations and universities together to focus on quantifiable progress. Sit With Me is a platform to raise awareness about the important contributions technical women make. In providing a platform to showcase the stories of successful women in technology, this program also aims to challenge cultural images that associate technology with men or masculinity. A number of other NCWIT efforts are working to rewrite these cultural images, including our *Entrepreneurial Heroes* podcast series and Technolochicas, a new website that highlights the experiences of successful Latinas in computing.[42]

Throughout NCWIT's history we have focused on a wide variety of both aspirational and remedial efforts in each one of our strategic prongs, working with our membership and the general public at large to source ideas. For example, at the NCWIT summit, we acknowledge, inspire, and celebrate change leadership while discussing research and promising practices for change. The Aspirations in Computing program is inspirational and instructional, while Extension Services is remedial. Although it is difficult to prioritize and implement such a wide range of efforts, we have found that this range is critical to national progress. In other words, there is no "silver bullet"—the stakeholder base is broad and requires a wide spectrum of support.

Conclusion

Recognizing women as technical innovators requires explicit, conscious effort. To reiterate, simply adding women to the pot and stirring is not going to make their ideas recognized or used. The NCWIT approach (convene, equip, unite) has resulted in practices and programs that have combined to achieve notable educational pipeline growth. We see this growth accelerating in the coming years. We also see public awareness of these important issues increasing rapidly. Note the recent attention to these matters in Silicon Valley and the subsequent release of diversity data by a number of companies.[43] We are excited to see this momentum and believe that it bodes well for a turning point in the conversation. While the numbers are often dismal, being transparent about the problem and providing a starting point from which to measure progress is vital. We cannot afford to let this momentum subside if we truly want to change existing conditions for women and other underrepresented groups in computing. We encourage organizations to take advantage of this moment and to work together across education, industry, government, and nonprofit sectors, ensuring that women no longer remain absent—that they instead are able to make vital and meaningful contributions to enhance the technology of future generations.

Notes

1. We examined all US information technology (IT) patents during this time period. We also used a multistep process to identify inventor gender. For more details on methodology, see Catherine Ashcraft and Anthony Breitzman, "Who Invents IT? Women's Participation in Information Technology Patenting," 2012 update, 19 July 2012, accessed 29 January 2016, http://www.ncwit.org/patentreport.

2. A solid body of research in computing and in other fields documents the enhanced performance outcomes when a diverse range of women are present in meaningful and creative roles. Lecia Barker, Cynthia Mancha, and Catherine Ashcraft, "What Is the Impact of Gender Diversity on Technology-Business Performance? Research Summary," 29 May 2014, accessed 29 January 2016, http://www.ncwit.org/businesscase/.

3. Cedric Herring, "Does Diversity Pay?" *American Sociological Review* 74, no. 2 (2009): 213.

4. Lynda Gratton, Elisabeth Kelan, Andreas Voigt, Lamia Walker, and Hans-Joachim Wolfram, "Innovative Potential: Men and Women in Teams," Lehman Brothers Centre for Women in Business, London Business School, 2007, accessed 29 January 2016, https://www.lnds.net/blog/images/2013/09/grattonreportinnovative_potential _nov_2007.pdf.

5. Scott E. Page, *The Difference: How the Power of Diversity Creates Better Groups, Firms, Schools, and Societies* (Princeton, NJ: Princeton University Press, 2008).

6. Anita Williams Woolley, Christopher F. Chabris, Alex Pentland, Nada Hashmi, and Thomas W. Malone, "Evidence for a Collective Intelligence Factor in the Performance of Human Groups," *Science* 330, no. 6004 (2010): 686–688.

7. Jessica Canning, Maryam Haque, and Yimeng Wang, "Women at the Wheel: Do Female Executives Drive Start-up Success?" Dow Jones and Company (2012).

8. JMG Consulting LLC and Wyckoff Consulting LLC for the US Small Business Administration, "Venture Capital, Social Capital and the Funding of Women-led Businesses," April 2013, accessed 29 January 2016, https://www.sba.gov/content /venture-capital-social-capital-and-funding-women-led-businesses.

9. Sylvia Ann Hewlett, Carolyn Buck Luce, Lisa J. Servon, Laura Sherbin, Peggy Shiller, Eytan Sosnovich, and Karen Sumberg, "The Athena Factor: Reversing the Brain Drain in Science, Engineering, and Technology," Center for Work-Life Policy, Harvard Business Review Research Report no. 10,094, 1 June 2008.

10. Catherine Ashcraft, Brad McLain, and Elizabeth Eger, "Women in Tech: The Facts," National Center for Women and Information Technology, 2016, http://www .ncwit.org/thefacts.

11. For example, see Sheryl Sandberg and Adam Grant, "Speaking While Female: Sheryl Sandberg and Adam Grant on Why Women Stay Quiet at Work," *New York Times*, 12 January 2015; Farhad Manjoo, "Exposing Hidden Bias at Google," *New York Times*, 24 September 2014.

12. Catherine Hill, Christianne Corbett, and Andresse St. Rose, *Why So Few? Women in Science, Technology, Engineering, and Mathematics* (Washington, DC: American Association of University Women, 2010); Howard Ross, "Exploring Unconscious Bias," *CDO Insights—Diversity Best Practices* 2, no. 5 (August 2008): 1–16, accessed 29 January 2016, http://www.cookross.com/docs/UnconsciousBias.pdf.

13. For more in-depth examples of how schemas work and how they affect the participation of women and other underrepresented groups in technology workplaces, see NCWIT's video "Unconscious Bias and Why It Matters for Women and Tech," 19 May 2014, accessed 29 January 2016, http://www.ncwit.org/resources/unconscious -bias-and-why-it-matters-women-and-tech.

14. Jane Margolis et al., *Stuck in the Shallow End: Education, Race, and Computing* (Cambridge, MA: MIT Press, 2008); Jane Margolis and Allan Fisher, *Unlocking the Clubhouse: Women in Computing* (Cambridge, MA: MIT Press, 2001); Corinne A. Moss-Racusin, John F. Dovidio, Victoria L. Brescoll, Mark J. Graham, and Jo Handelsman, "Science Faculty's Subtle Gender Biases Favor Male Students," *Proceedings of the National Academy of Sciences* 109, no. 41 (2012): 16474–16479; Christine Wenneras and Agnes Wold, "Nepotism and Sexism in Peer Review," *Nature* 387, no. 6631 (1997): 341–343.

15. "11 Sexist Tech Ads," BuzzFeed News, 13 June 2012, accessed 29 January 2016, http://www.buzzfeed.com/copyranter/11-incredibly-sexist-tech-ads#.ubqGKoKPk; Stephanie Mlot, "Mattel Apologizes, Pulls Sexist Computer-Engineer Barbie Book," *PC Magazine*, 21 November 2014, accessed 29 January 2016, http://www.pcmag.com /article2/0,2817,2472529,00.asp.

16. "Demonstration Site," Project Implicit, accessed 29 January 2016, https:// implicit.harvard.edu/implicit/. For more on the methodology underlying the IAT, see Mahzarin Banaji and Anthony Greenwald, *Blind Spot: Hidden Biases of Good People* (New York: Delacorte Press, 2013).

17. Shelley J. Correll, Stephen Benard, and In Paik, "Getting a Job: Is There a Motherhood Penalty?" *American Journal of Sociology* 112, no. 5 (2007): 1297–1338; Joshua Correll, Bernadette Park, Charles M. Judd, Bernd Wittenbrink, Melody S. Sadler, and Tracie L. Keesee, "Across the Thin Blue Line: Police Officers and Racial Bias in the Decision to Shoot," *Journal of Personality and Social Psychology* 92, no. 6 (2007): 1006–1023; Frances Trix and Carolyn Psenka, "Exploring the Color of Glass: Letters of Recommendation for Female and Male Medical Faculty," *Discourse and Society* 14, no. 2 (2003): 191–220; Wenneras and Wold, "Nepotism and Sexism in Peer Review."

18. Moss-Racusin et al., "Science Faculty's Subtle Gender Biases Favor Male Students"; Rhea E. Steinpreis, Katie A. Anders, and Dawn Ritzke, "The Impact of Gender on the Review of the Curricula Vitae of Job Applicants and Tenure Candidates: A National Empirical Study," *Sex Roles* 41, no. 7 (October 1999): 509–528.

19. Marianne Bertrand and Sendhil Mullainathan, "Are Emily and Greg More Employable Than Lakisha and Jamal? A Field Experiment on Labor Market Discrimination," *American Economic Review* 94, no. 4 (September 2004): 991–1013.

20. See, e.g., Steinpreis, Anders, and Ritzke, "Impact of Gender on the Review of the Curricula Vitae."

21. Mary P. Rowe, "Barriers to Equality: The Power of Subtle Discrimination to Maintain Unequal Opportunity," *Employee Responsibilities and Rights Journal* 3, no. 2 (June 1990): 153–163.

22. See Steven Stroessner and Catherine Good, "Bibliography," accessed 29 January 2016, http://www.reducingstereotypethreat.org/articles/.

23. Claude M. Steele and Joshua Aronson, "Stereotype Threat and the Intellectual Test Performance of African Americans," *Journal of Personality and Social Psychology* 69, no. 5 (1995): 797–811; Lawrence J. Stricker and William C. Ward, "Stereotype Threat, Inquiring about Test Takers' Ethnicity and Gender, and Standardized Test Performance," *Journal of Applied Social Psychology* 34, no. 4 (April 2004): 665–693.

24. Joshua Aronson, Michael J. Lustina, Catherine Good, Kelli Keough, Claude M. Steele, and Joseph Brown, "When White Men Can't Do Math: Necessary and

Sufficient Factors in Stereotype Threat," *Journal of Experimental Social Psychology* 35, no. 1 (January 1999): 29–46.

25. Steele and Aronson, "Stereotype Threat"; Stricker and Ward, "Stereotype Threat."

26. Jessica Valenti, "The Female 'Confidence Gap' Is a Sham," *Guardian*, 23 April 2014, accessed 29 January 2016, https://www.theguardian.com/commentisfree /2014/apr/23/female-confidence-gap-katty-kay-claire-shipman.

27. NCWIT, "The Problem: Fix the Woman Approaches to Diversity," *Critical Listening Guide*, accessed 29 January 2016, https://www.ncwit.org/resources/draft-critical -listening-guide#Fix.

28. Karen Lee Ashcraft, "The Glass Slipper: 'Incorporating' Occupational Identity in Management Studies," *Academy of Management Review* 38, no. 1 (January 2013): 6–31.

29. Mark Guzdial, Barbara J. Ericson, Tom McKlin, and Shelly Engelman, "A State-wide Survey on Computing Education Pathways and Influences: Factors in Broadening Participation in Computing," in *Proceedings of the Ninth Annual International Conference on International Computing Education Research (ICER '12)* (New York: ACM, 2012), 143–150.

30. Nelson D. Schwartz, "In Hiring, a Friend in Need Is a Prospect, Indeed," 27 January 2013, accessed 29 January 2016, http://www.nytimes.com/2013/01/28/busi ness/employers-increasingly-rely-on-internal-referrals-in-hiring.html?src=me&ref =general.

31. Meta Brown, Elizabeth Setren, and Giorgio Topa, "Do Informal Referrals Lead to Better Matches? Evidence from a Firm's Employee Referral System," *Journal of Labor Economics* 34, no. 1 (January 2016): 161–209.

32. Roberto M. Fernandez and Santiago Campero, "Gender Sorting and the Glass Ceiling in High Tech," MIT Sloan research paper no. 4989–12, January 2012, accessed 29 January 2016, https://ssrn.com/abstract=2067102.

33. Trix and Psenka, "Exploring the Color of Glass"; Virginia Valian, *Why So Slow? The Advancement of Women* (Cambridge, MA: MIT Press, 1999).

34. Trix and Psenka, "Exploring the Color of Glass"; Kieran Snyder, "The Abrasiveness Trap: High-Achieving Men and Women Are Described Differently in Reviews," *Fortune*, 26 August 2014, accessed 29 January 2016, http://fortune.com/2014/08/26 /performance-review-gender-bias/.

35. Hewlett et al., "Athena Factor."

36. Claudia Goldin and Cecilia Rouse, "Orchestrating Impartiality: The Impact of 'Blind' Auditions on Female Musicians," National Bureau of Economic Research,

working paper no. 5903, January 1997, accessed 29 January 2016, http://www.nber .org/papers/w5903.

37. Catherine Ashcraft, "Strategic Planning for Increasing Women's Participation in the Computing Industry," NCWIT, 1 April 2010, accessed 29 January 2016, https:// www.ncwit.org/resources/strategic-planning-increasing-women%E2%80%99s -participation-computing-industry.

38. "Supervising-in-a-Box Series: Full Series," NCWIT, 3 May 2012, accessed 29 January 2016, https://www.ncwit.org/supervising.

39. "Recruiting, Retaining, and Advancing a Diverse Technical Workforce: Data Collection and Strategic Planning Guidelines," NCWIT, 21 January 2015, accessed 29 January 2016, http://www.ncwit.org/datacollectionguide.

40. "NCWIT Tips for Writing Better Job Ads," NCWIT, accessed 29 January 2016, https://www.ncwit.org/resources/ncwit-tips-writing-better-job-ads.

41. "Recruiting, Retaining, and Advancing a Diverse Technical Workforce," NCWIT, accessed 27 September 2017, http://www.ncwit.org/datacollectionguide.

42. "Entrepreneurial Heroes," NCWIT, accessed 29 January 2016, https://www .ncwit.org/heroes; NCWIT and Televisa Foundation, "Technolochicas," accessed 29 January 2016, http://technolochicas.org/.

43. Julia Love, "Jesse Jackson Gathers Tech Leaders for Frank Conversation about Diversity," *San Jose Mercury News*, 10 December 2014, accessed 29 January 2016, http://www.mercurynews.com/business/ci_27111004/jackson-gathers-tech-leaders -frank-conversation-about-diversity.

18 Making Responsible Innovators

Erik Fisher, David Guston, and Brenda Trinidad[1]

Although the premise behind the innovator imperative is that the world needs more innovators, we argue here that it is important to make *responsible* innovators. Simply supporting research and development in the name of societal benefits and then waiting for the market to sort it out is not enough. Rather, we have a responsibility to shepherd science and technology through the innovation process to ensure that innovation outcomes do not adversely impact the societies we intend to advance. Accordingly, we need to look beyond traditional educational models in order to instill the capacities that will help innovators be more socially responsible in their daily activities.

Responsible innovation is a contemporary response to an old question of the production of novelty: Is it good, or just new? There is no doubt that the world is awash in challenges that require innovation in response. The Millennium Development Goals and their successor Sustainable Development Goals from the United Nations are good examples of sociotechnical challenges for which the right knowledge-based innovations might be incredibly helpful.

Responsible innovation also recognizes, however, that knowledge-based innovation is one of the forces that got us into our current mess. The legacies of nuclear waste, industrial chemistry and pesticides, e-waste, automobiles, and so forth reinforce the idea that novelty is not an unalloyed good. We should give significant forethought to both the kind of innovations we want to introduce and take up into use, as well as the kind of world we want to be building by choosing one kind of innovation over another kind of innovation, or even another category of response.

Responsible innovation approaches are particularly important in addressing the innovations propelled by emerging science and technologies such as nanotechnology, synthetic biology, artificial intelligence and robotics, and neurotechnologies. These emerging technologies—characterized by their high stakes, high uncertainty, and contested novelty—come with few data and controversial experience for guidance.[2] Responsible innovation is a way to think about guidance or, more directly, governance of emerging technologies even as they have yet to emerge fully.

As a term of art, responsible innovation builds upon decades of policy experience, educational aspirations, and scholarly insight around the complex interactions of science, technology, and society. The precise meaning of responsible innovation and that of its various cognates is still emerging and contested, but we outline what we see as three distinctive features of responsible innovation as a practical and concerted attempt by science policy actors and others to address societal dimensions of science and innovation. We then briefly outline some of the implications for (re)training scientists, engineers, and others involved in various aspects of innovation. Finally, we reflect on some of the integrative conditions and outcomes of the Center for Nanotechnology in Society at Arizona State University (CNS-ASU) as a basis for informing experimental institutional designs aimed at building capacities for responsible innovation.

What Is Responsible Innovation?

Although policies, practices, definitions, and frameworks for responsible innovation continue to emerge, several formulations have been particularly influential. In one such definition by policy scholar René von Schomberg, it is

> a transparent, interactive process by which societal actors and innovators become mutually responsive to each other with a view to the (ethical) acceptability, sustainability and societal desirability of the innovation process and its marketable products (in order to allow a proper embedding of scientific and technological advances in our society).[3]

In another definition that we helped craft, it is

> a commitment of care for the future through collective stewardship of science and innovation in the present.[4]

Cognate terms include "responsible development," used by the US National Nanotechnology Initiative in its strategic plans;[5] "responsible research and innovation (RRI)," often used in the European context;[6] and "prudent vigilance," used by the US Presidential Commission for the Study of Bioethical Issues in conjunction with its inquiry into synthetic biology.[7]

Diverse research, policy, and educational efforts have recently been devoted to developing, implementing, and observing responsible innovation. These include formal policies and programs, transnational meetings, and large-scale research projects and consortia.[8] Erik Fisher and Arie Rip review a number of related developments around responsible innovation, primarily across Europe and North America, from the standpoint of multilevel governance and its dynamics. They find that responsible innovation has been an explicit focus of policy discourse and directives in numerous nations and cross-national activities.[9] In short, responsible innovation is having effects on concrete decisions at the levels of both science policymaking and scientific practice.

As mentioned, much of what is called responsible innovation derives from decades of policy experience around science-society interactions. Therefore, it is worth reflecting on how it compares with previous and existing policy programs and mechanisms that have proliferated internationally. Widely adopted "policy for science" programs and mechanisms have sought to raise the ethical awareness of researchers, mitigate undesirable impacts of scientific research, enlist societal research in a collaborative manner, and enable public participation. These include institutional review boards (IRBs), which oversee research on human subjects at institutions that apply for federal funds; the responsible conduct of research (RCR), which articulates standards for laboratory conduct around fraud, falsification, plagiarism, and other inappropriate practices; technology assessment, which has been institutionalized (and deinstitutionalized) in some legislative settings in the United States and Northern and Western Europe; ethical, legal, and societal implications research, which has accompanied research on the Human Genome Initiative, the National Nanotechnology Initiative, and similar national research initiatives in the United States and Europe; and environmental impact assessments, which provide a process for public inquiry into the environmental ramifications of large-scale projects. To the extent that these functions are also associated with responsible innovation, it is fair to question whether that moniker brings anything of its own to the family of explicitly normative "policies for science." We think that it does.

What Is Different about Responsible Innovation?

The definitions, frameworks, and practices associated with responsible innovation exhibit three relatively unique features in comparison to previous internationally adopted policy approaches to addressing the societal aspects of science and innovation. In our view, responsible innovation is unique in (1) its comparatively broad stance toward science and innovation, (2) the societal context within which it embeds science and innovation, and (3) the active role assigned to experts who participate in the innovation process, including scientific researchers, technology developers, and industrial designers.

To begin with, as the term itself implies, in responsible innovation the object of attention (or in policy terms, the object of governance) appears to be shifting from the societal "impacts," "implications," or "dimensions" of a *specific* research program—such as genomics or nanotechnology—to those of science and innovation *writ large*. Consider the responsible innovation framework recently adopted by the UK Engineering and Physical Sciences Research Council (EPSRC), which applies at the generic level of science and innovation funding, rather than being confined to one particular program. In other words, responsible innovation is less about *whether* to conduct societal research, technology assessment, or an environmental impact assessment than about *how* innovation should take place in light of its social, environmental, and political contexts.

Closely related to this first point is a second one that concerns the societal context of innovation and its governance in society. The idea that the societal dimensions of innovation need to be taken into account throughout the innovation process and as a general proposition resonates with theoretical claims about the centrality of science and technology to modern societies. As more and more aspects of modern life become dependent upon science and technology for their routine operation, so too is the success of science and technology increasingly dependent upon complex social dynamics and political phenomena. This is perhaps most clear in the case of technological risk and its management. Due to complex interactions between social and technological systems, risk is no longer something we can think of as dispensed with through expert reasoning and public assurances. Rather, risk is a condition of innovation and must be dealt with accordingly. This means treating risk and other societal dimensions not merely as a matter of expert calculations and public communication, but also as matters of societal engagement and technological adjustments.

This brings us to our third point, which has to do with the wide variety of participants in the innovation process who share in the collective responsibility for the outcomes of innovation. While academic commentators have long argued for such a perspective, the appearance of more socially engaged modes of scientific research—and recent science policy measures that insist on these—suggests that what counts as good scientific research is slowly starting to change as a result of the central role that scientists, engineers, and innovators play in contributing to the complex developments of social life.

As a result, on one level of sociotechnological governance, responsible innovation suggests an increased vigilance toward asking questions of choice and impact on the part of the scientific community, rather than waiting for triggering events and subsequent political pressure to place such questions on the policy agenda.[10] Such vigilance can be seen in the subtle yet significant difference between the National Science Foundation's broader impacts criterion, which asks researchers to explain how their research might "benefit society and contribute to the achievement of specific, desired societal outcomes" and the EPSRC framework, which acknowledges that scientific research may have "often ambiguous…impacts, beneficial or otherwise."[11] In other words, responsible innovation assumes that individual scientists and innovators are participants in the social processes by which innovation outcomes are produced, and it accordingly asks them to be more aware of this fact.

A responsible innovation framework locates research and innovation within a broader process of social shaping that helps to construct the very pathways that technological development takes. This framework contrasts with traditional compliance-based approaches such as RCR, which seeks to avoid egregious ethical and professional deviations, or IRBs, which are meant to protect human research subject rights. For example, US federal legislation calling for the "responsible development of nanotechnology" breaks rank with traditional science policy models that rely on a division of moral labor between scientific research and societal assessment.[12] Not only does the 2003 law imply that nanotechnology researchers and developers can influence the direction of technological development, but it explicitly prescribes the integration of societal research into nanotechnology research, development, and commercialization precisely for this purpose.[13]

One of the most remarkable ways in which responsible innovation differs from previous science policy approaches is clear from the intense focus it places on scientific and engineering experts. Quite simply, modern societies have increasingly come to depend upon research and innovation practices and institutions as resources with which to govern themselves. Of course, addressing the societal dimensions of innovation also requires the concerted efforts of numerous actors operating within numerous sites and at multiple scales, many of whom contribute to complex sociotechnological outcomes without being solely responsible for them. This distributed approach to innovation implicates multiple floors of action and centers of production that interact across government, public, and private sectors, in which the overall capacity to respond to numerous and diverse inputs, including normative questions of trust, accountability, and responsibility, is more important than is any one particular input to the system.

How Can Responsible Innovators Be Educated?

From these fundamental observations, we derive three corresponding objectives for responsible innovation training and, more broadly, education. Regardless of the specific content, methods, and goals that a given pedagogical program may associate with responsible innovation, the technical experts and practitioners that these programs aim to produce will, ideally, recognize (1) that science and innovation entail societal dimensions as a matter of course; (2) that these dimensions are coproduced by science, technology, and their societal contexts; and (3) that these dimensions can and should be attended to at every mundane stage and level of innovation, including bench-level research, development, and design practices.

While these guiding objectives will be familiar to scholars in the fields of applied ethics and science and technology studies (STS), they are less evident within comparable policy programs and are often completely absent from traditional science and engineering training programs. More to the point, the skills that these guidelines call for are often underdeveloped and sometimes missing altogether in the skill sets of technical experts.[14] This is unsurprising, for most science and innovation policy and education programs still subscribe to what has been called the "modernist" two-track regime of promotion and control, in which the scientists and science policies that are responsible for promoting innovation have little to no interaction with those who are tasked with controlling its undesirable effects.[15] It is also to be expected that operationalizing these guidelines in a productive and meaningful manner poses significant challenges when they run up against traditional cultural and institutional practices of science and engineering.

Without underestimating these challenges, and based on our experience working with research groups, design teams, and laboratory cultures, we suggest that educating innovators in light of these three objectives can build responsible innovation capacities while at the same time adding practical value to existing and emerging science and innovation programs. In our experience, by building and exercising their capacities of anticipation, reflexivity, inclusiveness, and responsiveness through innovative pedagogy and training programs, innovators become better able to attend to the complex and uncertain societal dimensions of their technical work.[16] Much like the research process itself, the capacity of responsiveness in particular is an ongoing and dynamic process that requires both conceptual and hands-on practical learning.

Big Social Science—It Takes a Center

Among other things, our view of responsible innovation calls for an experiential approach to training. In other words, the practical experience gained in the laboratory and in other sites of science and innovation is a vehicle for building and exercising the fundamental capacities of responsible innovation. Such experience allows for relevant connections to be forged between science and innovation practices and their broader societal dimensions. We have found that, while these dimensions are often at first invisible and easily overlooked in practice, interdisciplinary sociotechnical collaborations can elucidate them rather effectively. Such collaborations create something of a charged atmosphere where, much more so than in classroom environments, sustained interactions between innovators-in-the-making and science and society scholars lead to dialogue, inquiry, and firsthand learning about the societal context of technical practices.

Sociotechnical collaborations and the pedagogical atmosphere they create need to make connections between contemporary science and society not only intellectually interesting to technical students and practitioners, but also practically relevant. Although sociotechnical collaborations can spring up spontaneously, they often face barriers and can be significantly aided by supportive organizational contexts and institutional environments. For instance, on university and college campuses, dedicated STS or liberal arts programs that are embedded within engineering schools allow students to become familiar with the study of sociotechnological interactions not only in the classroom but, ideally, also through capstone coursework and committees. Similarly, centers and institutes can build bridges across multiple science and engineering departments and schools through crosscutting humanistic and social-scientific research activities. In what follows, we report on our experiences with one such center, the CNS-ASU, which not only allowed for the participation of science and engineering students and faculty in our research and outreach activities, but in many cases actually depended upon it.

In 2004, NSF held a competition for a nanoscale science and engineering center dedicated to studying the societal aspects of nanotechnology and contributing to the National Nanotechnology Initiative's (NNI) strategic goal of responsible development of nanotechnology.[17] In 2005, NSF made awards to Arizona State University and the University of California, Santa Barbara, to create centers for nanotechnology in society. CNS-ASU received approximately $6.2 million in its initial award and $6.5 million in a subsequent renewal, as well as additional funds in supplements over its eleven years of operation (2005–2016).[18]

In an effort to interact closely with nanoscale science and engineering (NSE) practice, CNS-ASU included not only science and engineering faculty but postdocs, graduate students, and undergraduates in numerous research and engagement activities. Center activities included workshops, research collaborations, public events, and various other extracurricular activities—in addition to more standard research and pedagogical activities involving interviews, ethnographies, bibliometric analysis, surveys, seminars, and courses—that generally fall under the anticipatory governance areas of *foresight, engagement,* and *integration.*[19]

CNS-ASU's ensemble of engaged research activities had the effect of complementing classroom pedagogy with practical insights and experience that not only deepened student understanding of the societal contexts that structure and inform their own bench-level research, but that could provide them with social-scientific and humanistic tools and techniques to open up their own practices for more deliberate participation within these contexts such as called for by responsible innovation. More specifically, CNS-ASU consciously created an ensemble of engaged social science research activities that were current with, adept in, and proximate enough to the very science and engineering projects in which students were already participating so as to shed critically interpretive and practically informative light on their practical choices and future societal outcomes.

As reflected in the center's publications, some of its interactions were with undergraduates somewhat in the mode described above of a liberal arts program embedded in an engineering school, in which cross-functional teams of engineers, business students, and graphic and industrial designers use an "integrated innovation model," akin to responsible innovation, to design prospective new products and create marketing plans and engineering models.[20] Other center activities involved immersive experiences in which graduate students were introduced to the workings of science policy in order to gain an informed perspective of "science outside the lab."[21] In the remainder of the chapter, we will focus on the role of integrative collaborations in pursuing responsible innovation within the laboratory.

Socio-Technical Integration Research (STIR)

Socio-Technical Integration Research, or STIR, served as the center's flagship integration activity. After a pilot study at the University of Colorado, Boulder, and thanks to a separate grant from the NSF, the STIR program expanded in 2009 into dozens of laboratories at Arizona State University and in universities and research organizations around the world. The STIR process helps to elucidate the societal dimensions of science and engineering research and, thereby, to create opportunities for laboratory practitioners to participate more deliberately in the governance of science, technology, and innovation. STIR involves collaborative inquiry between "embedded humanists" or social scientists and the scientists, engineers, and others who host them. Fisher coined the term "embedded humanist" in order to emphasize that, while always somewhat out of place in the lab, the humanities scholar nevertheless is meant to become an accepted and functioning member of the scientific team. In STIR, laboratories and other research institutes are deemed to function "midstream" between authorization and dissemination of science-based activities and their eventual outcomes.[22] The collaborative inquiry takes place during routine research and innovation activities, generating feedback learning that, in turn, leads scientists and engineers to "modulate" these activities over time. Published findings from STIR studies demonstrate that "midstream modulation" productively disrupts and creatively enhances the conditions under which research and innovation practitioners engage in critical reflections on the social contexts of their work.[23]

Following Fisher's pilot study, the initial STIR project coordinated thirty studies in laboratories working in emerging areas such as nanotechnology, synthetic biology, and neuroscience.[24] The collaborations usually lasted twelve weeks and the interactions took place between doctoral students in the humanities and social sciences and doctoral students in science and engineering. The heart of the STIR process involves a decision protocol that the embedded humanist uses intensively to engage with day-to-day technical practices in order to open these practices to collaborative description, inquiry, deliberation, and subsequent adjustment (figure 18.1).

STIR "laboratory engagement studies" differ in several ways from more traditional laboratory ethnographies, which are modeled after the study

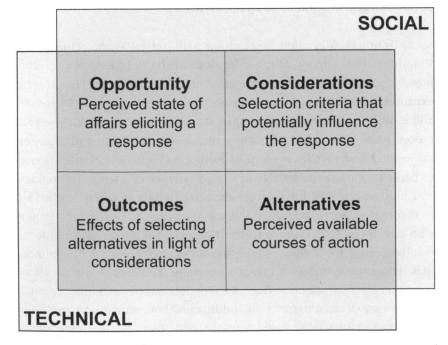

Figure 18.1
The STIR decision protocol. Courtesy of Erik Fisher.

of foreign cultures and in which a social researcher observes scientists in their "native" laboratory environment. The primary purpose of such ethnographies is for the social scientist to generate knowledge. STIR studies, however, also generate high-impact reflection and practical adjustments that inform both social science and the application of science and engineering. They do this by embedding their collaborative techniques and feeding back their findings and perspectives directly into the field of study, in real time. In short, embedded humanists in the STIR program "give their stories away" rather than keeping their insights to themselves until the time of publication. This is not to say that their stories don't continue to develop and evolve by the time they are published, but that the interactions between the collaborators involve a transparent process of inquiry into technical problems and their social contexts.

Two short vignettes illustrate how STIR works. In one study, embedded humanist Fisher concluded that the mechanical engineering researchers participating in his study, who stated on multiple occasions that they do not make decisions and hence have no responsibility for the lab's research projects or their outcomes, actually generate both data and ideas that help inform the lab's research directions. Soon afterward, he shared his observations—for instance, examples of graduate students deciding how, when, and even why to conduct research as well as crucial contributions that they made to the direction of several research projects—with the entire laboratory group. Some in the group readily agreed with Fisher's examples, while others continued to maintain their original position. This led to continued discussion and debate, not only between Fisher and the members of the lab but also separately among members of the lab, about what constitutes a choice in research. Not only did sharing his observations and insights with the group help Fisher to refine his own understanding of the nature of the laboratory research he was studying and to acquire a considerable degree of interactional expertise, but by the end of the study he was told by even the most skeptical members that his observations were both accurate and useful.[25]

In another case, Daan Schuurbiers described how he shared his observations of and questions about the safety practices of a biotechnology lab in which he was embedded with the larger group (figure 18.2). As in the previous example, his observations and perspectives at first triggered a lively debate among the lab group, in this case about when and whether to wear

Figure 18.2
Daan Schuurbiers uses the STIR decision protocol to converse with a researcher working on a synthetic biology project in 2009. Courtesy of Daan Schuurbiers.

lab coats. A short time later, the entire team of laboratory researchers had
collectively altered their safety practices. What is particularly of interest in
this example from Schuurbiers's study is that the entire lab group reorga-
nized itself as a result of what originally began as a philosophical discussion
about the role of science in society.[26]

As both of these short vignettes suggest, both individuals and small
groups of laboratory researchers who participate in STIR activities tend to
become more conscious of their role in shaping sociotechnical outcomes as
a result of their interactions with the embedded humanist. As a result of the

ongoing use of the STIR protocol, these innovators-in-the-making engage in productive reflection that often leads them to change their laboratory practice accordingly. Although in both of these vignettes the feedback was provided in a group setting, STIR also occurs in the form of one-on-one exercises that allow collaborators to engage in detailed analysis and reflection that in turn helps build mutual trust and understanding.

The STIR project asked to what extent sociotechnical integration is possible, as well as to what extent it is useful for multiple actors; in this way, the project sought to investigate and empirically test controversial propositions about the social construction and shaping of science and technology and, by implication, about the effects of critically engaging this construction and shaping on the laboratory floor. In addition to finding that such engagement is indeed possible within laboratory cultures, STIR studies have also documented the practical benefit of embedded and sustained sociotechnical interactions.[27] Significantly, STIR interactions fostered both first-order reflective learning, which consists of reflection on the instrumental means to accomplish prior (often unquestioned) research values and goals, and also second-order reflective learning, which consists of reflection on which of the values and goals ought to be considered in the first place.[28]

In addition to finding strong indications of both the possibility and the utility of collaborative sociotechnical integration as a route toward responsible innovation, STIR studies have identified the following kinds of integration capacity-building modulations:

- *Reflexive awareness:* Laboratory researchers have become more aware of the choices they are making and of their connection to broader societal values and contexts.

- *Value deliberations:* Laboratory researchers have expanded the social values and stakeholder perspectives that relate to their experimental and design choices, as evidenced in examples of both first- and second-order reflective learning.

- *Practical adjustments:* Interactions with embedded humanists and social scientists have sparked new research ideas, catalyzed laboratories to engage in outreach activities, and inspired changes in bench-level and strategic laboratory practices—from synthetic biology safety practices and nanomaterial waste disposal practices to enhancements in productivity and project management.

While STIR project goals were originally focused on probing laboratory capacities for reflexivity and responsiveness in order to inform institutional and policy design for responsible innovation, the STIR process has since been used in a number of different environments and for a number of different purposes. Based on documented instances of reflexive learning, value deliberations, and practical adjustments, we can identify two general purposes to which such sociotechnical collaborations can contribute. First, these collaborations can alter the material shape and direction of emerging research and technological pathways. Second, such collaborations can build the longer-term capacities of anticipation, reflexivity, and responsiveness for innovators-in-the-making.[29] Clearly, while the first purpose resonates closely with the core scholarly insights that helped to eventually give rise to responsible innovation in the first place, the second (pedagogical) purpose speaks to the need to establish a foundation for responsible innovation with science and innovation practices, cultures, and—to the extent that they are aggregate functions of individual behaviors—institutional structures.[30]

Center-Level Impact

In order to document and assess the influence of center activities—including STIR as well as the others listed above—on the nanotechnology researchers with whom we collaborated, we implemented an interview protocol on an annual basis for six years. This protocol focused on the knowledge, identity, and practices of our collaborating scientists, particularly around their understanding of the societal aspects of their work. We conducted baseline research in 2006 and subsequent rounds annually from 2007 to 2011, when we switched to a survey format in order to reach nearly eight hundred center participants. The following sample responses taken across this period suggest that, over time, center activities became more embedded within nanotechnology research culture and infrastructure and the CNS-ASU was thus increasingly in a position to contribute to capacity-building for responsible innovation.

During the first year of the center's operation (2005–2006), interview respondents reported being aware of social issues in general but did not see a connection with their own research. Knowledge of social aspects had minimal, if any, impact on their scientific or professional practices. This pattern began to show signs of change starting in the CNS-ASU's second year (2006–2007), when some tenured professors and all graduate students reported high familiarity and moderate to high involvement. By contrast, all untenured faculty and postdocs reported low familiarity and no involvement. More importantly, respondents suggested that the CNS-ASU was beginning to have an impact, insofar as they associated high familiarity and high involvement with significant changes in knowledge and initial changes in practice.

In year three of CNS-ASU operations (2007–2008), there was a critical change, with the center starting to become embedded and serving as both a resource and a catalyst for changes in knowledge, identity, and practice. As one nanoscale science faculty member stated:

> I do notice the students taking time [to participate with CNS-ASU], and apparently enjoying it in the sense that they certainly don't come complaining to me about 'oh dear all the time I spent on CNS things,' [emphatically] they don't, I mean they really don't, despite the fact that I am putting pressure on them to do lots of other things in the lab. So I would say that it is clear that there is an impact at the level of students. And in some ways, actually, that is the most important thing to go for.... The students are the most valuable output of this place [i.e., the university] anyway, right?

More specifically, impacts on students appeared to interrelate changes in knowledge with changes in identity. As one faculty member stated:

> [CNS-ASU] adds color to the environment, and I think because the students are learning science, the impact on them is greater. So to give you a very positive example, [one student who would not have been satisfied as a natural or a social scientist has shifted her career plans and] has been so happy at this interface of the public and science, which CNS sort of moved her into.

In the words of a graduate engineering student:

> Normally, the PhD research in environmental engineering, the stuff I would be doing, is just the straight research on tracking nanomaterials in the environment. But now I'm focusing on this policy stuff. I'm taking trips to DC, I'm interacting with people involved in the policy process—it's a whole section of education on the policy implications of research that there's no way I would get unless I came and talked with [CNS-ASU faculty members] because the CNS is here. So it's just a whole section of education that I wouldn't get in the engineering department.

This statement from a faculty member suggests that the CNS-ASU was also facilitating interrelated changes in knowledge and practice:

We're about to start to decide what technologies to use to do [medical] detections.... If it's focused on the user and not the clinical community it's just a different thing.... That [CNS-ASU workshop] was a mind opening experience!

Respondents from subsequent years indicated similar interrelated changes, not only in language ("I've worked with students who haven't been involved with these things at CNS...and you *never* hear them talking about the issues that you guys bring up") but also in thinking about research practices. For instance, one researcher mused that he "might tweak some of the experiments that I do...to make them more policy relevant," and another speculated that "if we're going to evolve a new enzyme to perform a certain reaction to replace an existing [one], I wouldn't just think of that—I would think of the entire supply chain."

In the seventh year of the center (2011–2012), annual interviews included two comparable subgroups: ASU-based NSE respondents (six faculty/administrators and three doctoral students in life sciences and engineering) and an international group of STIR project participants (five faculty/administrators and three doctoral students in natural science and social science/humanities). The annual interviews ask several questions that center around interdisciplinarity. Although interdisciplinarity is not a sufficient condition for responsible innovation, it is most likely a necessary one, especially in the case of responsiveness. An interesting comparison emerged between the two subgroups in relation to this topic: in almost all cases, ASU and STIR subgroup participants agreed that working in an interdisciplinary-oriented environment was valuable and beneficial. Moreover, both subgroups reported that interdisciplinary collaboration advanced in stages, and that its value was realized only after initial language and/or cultural barriers were somehow transformed or transcended.

However, the results and perceived value of interdisciplinary collaboration varied distinctly between the two subgroups. STIR respondents reported increased reflective capacities and richer dialogues as language or other barriers were negotiated over their twelve-week laboratory engagements. Respondents felt that the project design itself "required" transdisciplinary barriers to be overcome and enabled integrative effects, such as changes in practice, to occur. They also reported experiencing a built-in motivation for all parties to "hang in there" and stay committed to the STIR process in the form of a common goal that was larger than, yet also encompassed, their research. In the ASU group, the disciplinary boundaries respondents discussed were those among the natural scientists themselves, and there were fewer indications that language and cultural barriers were effectively negotiated. In fact, some ASU respondents reported that the interdisciplinary structure of some research units actually served to fragment members of the unit, while one suggested that cultural and language tensions between biologists and engineers persisted after approximately two years, presenting obstacles to research progress.[31]

While the learning outcomes and changes in practices reported by the participants in the STIR program are encouraging from the standpoint of responsible innovation, the fact that participants in other programs report very different experiences suggests that interdisciplinary collaboration does not simply improve over time. Rather, it suggests that we need to pay more attention to other conditions that enable interdisciplinary collaboration, such as the nature of daily interactions and how they may or may not help advance a shared sense of purpose. It also refocuses us on the larger challenge of institutionalizing the conditions for education in responsible innovation. If we take interdisciplinarity as an example, it is clear that even established institutional structures—such as the interdisciplinary makeup of the membership and physical layout of research labs in a university institute for biology-based research and innovation—may sometimes have difficulty achieving their intended effects. It is therefore pleasantly surprising to find that transdisciplinary collaborations across the "two cultures," which in some cases took place within the same university and even the same institute, can lead to both purposeful and productive interdisciplinary outcomes.[32]

As a final reflection, we suggest that the bulk of the learning and changes in practice that we were able to document over nearly a ten-year period were the result of scientists and engineers choosing to do things differently because they perceived value in making such changes. In other words, as in the STIR process, a "soft" and voluntary approach to educating for responsible innovation is likely to find more traction in science and engineering practices than a "hard" or compliance-based approach. Perhaps counterintuitively, it is precisely because the STIR collaborations are *voluntary* and of *temporary duration*—even though they are also intensive and sustained—that they are able to productively disrupt and add value to laboratory research practices. By implication, efforts aimed at responsible innovation capacity-building may well have to exist in their own nether realm, populated by liminal agents who are embedded in but not beholden to the projects and organizations in which they operate. We suggest that both the STIR program and, more generally, the Center for Nanotechnology in Society at ASU fostered such an environment, and we hope they serve as inspirations if not models for future efforts.

Conclusion

This chapter offers a correction to the innovator imperative by encouraging educators and others to embrace the challenge of making *responsible* innovators. While no single individual can take responsibility for the collective effects of innovations in the future, those who research and develop novel ideas and applications; those who seek out, fund, and work to disseminate them; and those who provide the resources and training that help make these efforts possible are all in unique positions to make a difference. Creating the conditions that enable socially responsible innovation to become the norm in business and university settings is one of our greatest contemporary challenges, and one that organizations like the Center for Nanotechnology in Society demonstrate are indeed possible.

The NSF funding for the CNS-ASU expired in 2016, leaving us in the position of having piloted what could become a prototypical boundary organization for university-wide capacity-building in responsible innovation even as we recognize that such an organization and its component activities must, to be effective, exist and operate at numerous spatial, temporal, social, and organizational boundaries. Although dynamic and uncertain, such a position is perhaps entirely befitting of the larger enterprise of collective socio-techno-scientific responsibility it is meant to serve.

Meanwhile, many of the ASU-based personnel and many of its programs are preserved intact in the recently created School for the Future of Innovation in Society (SFIS). SFIS houses a preexisting doctoral program in the human and social dimensions of science and technology, offers a master of science and technology policy program that CNS-ASU helped create, and offers a graduate certificate in responsible innovation in science, engineering, and society that CNS-ASU also nurtured. The school has also created an undergraduate program that seeks to inculcate the skills and concepts of responsible innovation for a variety of students who are in many cases just beginning their university training. Additionally, ASU has created the Institute for the Future of Innovation in Society, a companion to SFIS that will house the research capacity spun off from CNS-ASU, including the Virtual Institute for Responsible Innovation, the Center for Engagement and Training in Science and Society, the Center for the Study of Futures, and other established centers that continue and extend its mission. These various institutional developments are informed by our experiences with the CNS-ASU, experiences that point to the rewarding benefits of learning to take collective responsibility for the novelties that we, as individuals and as a society, originate, make, and disseminate.

Notes

1. The authors would like to acknowledge the contributions of our departed colleague David Conz, whose expertise and insights helped shape the initial foundations of the annual interviews conducted by the CNS-ASU. We would also like to thank the editors for their patience and assistance. This material is based upon work supported by the National Science Foundation awards #0849101, #0937591 and #1257246.

2. David Guston, "Understanding 'Anticipatory Governance,'" *Social Studies of Science* 44, no. 2 (2013; 2014): 218–242.

3. René von Schomberg, "Towards Responsible Research and Innovation in the Information and Communication Technologies and Security Technologies Fields," in *Towards Responsible Research and Innovation in the Information and Communication Technologies and Security Technologies Fields*, ed. René von Schomberg (Luxembourg: Publications Office of the European Union, 2011), 9.

4. Richard Owen, Jack Stilgoe, Phil Macnaghten, Michael Gorman, Erik Fisher, and David Guston, "A Framework for Responsible Innovation," in *Responsible Innovation: Managing the Responsible Emergence of Science and Innovation in Society*, ed. Richard Owen et al. (Chichester, UK: Wiley, 2013), 27–50.

5. National Science and Technology Council, Committee on Technology: Nanoscale Science, Engineering, and Technology Subcommittee, *The National Nanotechnology Initiative Strategic Plan* (Washington, DC: National Nanotechnology Coordination Office, December 2004).

6. Hilary Sutcliffe, *A Report on Responsible Research and Innovation*, MATTER and the European Commission, 2011; Richard Owen, Phil Macnaghten, and Jack Stilgoe, "Responsible Research and Innovation: From Science in Society to Science for Society, with Society," *Science and Public Policy* 39, no. 6 (2012): 751–760.

7. Presidential Commission for the Study of Bioethical Issues, "New Directions: The Ethics of Synthetic Biology and Emerging Technologies," 2010, https://bioethics archive.georgetown.edu/pcsbi/synthetic-biology-report.html; Arnim Wiek, David Guston, Emma Frow, and Jane Calvert, "Sustainability and Anticipatory Governance in Synthetic Biology," *International Journal of Social Ecology and Sustainable Development* 3, no. 2 (2012): 25–38.

8. The European project ResAgora, for example, provides a useful resource for much of these efforts (http://res-agora.eu/rri-resources/). See also the website of the NSF-funded Virtual Institute for Responsible Innovation (VIRI; https://cns.asu.edu/viri /blogs).

9. Erik Fisher and Arie Rip, "Responsible Innovation: Multi-Level Dynamics and Soft Intervention Practices," in Owen et al., *Responsible Innovation*, 165–183.

10. For instance, in the case of nanotechnology, see Bill Joy, "Why the Future Doesn't Need Us," *Wired* 8, no. 4 (2000): 238; Ira Bennett and Daniel Sarewitz, "Too Little, Too Late? Research Policies on the Societal Implications of Nanotechnology in the United States," *Science as Culture* 15, no. 4 (2006): 309–325.

11. Richard Owen, "The UK Engineering and Physical Sciences Research Council's Commitment to a Framework for Responsible Innovation," *Journal of Responsible Innovation* 1, no. 1 (2014): 113–117.

12. Erik Fisher and Roop L. Mahajan, "Contradictory Intent? U.S. Federal Legislation on Integrating Societal Concerns into Nanotechnology Research and Development," *Science and Public Policy* 33, no. 1 (2006): 5–16.

13. Erik Fisher, "Lessons Learned from the Ethical, Legal, and Social Implications Program (ELSI): Planning Societal Implications Research for the National Nanotechnology Program," *Technology in Society* 27, no. 3 (2005): 321–328; *21st Century Nanotechnology Research and Development Act*, Pub. L. 2003, https://www.govtrack.us /congress/bills/108/s189/text/enr, 108–153.

14. Louis L. Bucciarelli, *Designing Engineers* (Cambridge, MA: MIT Press, 2014); Erin Cech, "Culture of Disengagement in Engineering Education?" *Science, Technology and Human Values* 39, no. 1 (2014): 42–72; Erik Fisher and Clark Miller, "Contextualizing the Engineering Laboratory," *Engineering in Context*, ed. Steen Hyldgaard Christensen, Bernard Delahousse, and Martin Maganck (Palo Alto, CA: Academic Press, 2009), 369–381.

15. Johan Schot, "The Contested Rise of a Modernist Technology Politics," in *Modernity and Technology*, ed. Thomas J. Misa, Philip Brey, and Andrew Feenberg (Cambridge, MA: MIT Press, 2003), 257–278.

16. Jack Stilgoe, Richard Owen, and Phil Macnaghten, "Developing a Framework for Responsible Innovation," *Research Policy* 42, no. 9 (2013): 1568–1580.

17. For more about the NSF funding of the CNS and its relation to NNI strategic goals, see Guston, "Understanding 'Anticipatory Governance.'"

18. For more information and for publications from the center, see https://cns.asu.edu/.

19. Daniel Barben, Erik Fisher, Cynthia Selin, and David Guston, "Anticipatory Governance of Nanotechnology: Foresight, Engagement, and Integration," in *Handbook of Science and Technology Studies*, ed. Edward J. Hackett, Olga Amsterdamska, Michael E. Lynch, and Judy Wajcman (Cambridge, MA: MIT Press, 2008), 979–1000.

20. Catalogues of CNS-ASU courses, degree programs, and so forth may be found in the center's extensive online library, in its comprehensive annual reports, and in other publications dedicated to educational activities (http://cns.asu.edu/library); Cynthia Selin and Prasad Boradkar, "Prototyping Nanotechnology: A Transdisciplinary Approach to Responsible Innovation," *Journal of Nano Education* 2, no. 1–2 (2010): 1–12.

21. Michael J. Bernstein, Kiera Reifschneider, Ira Bennett, and Jameson M. Wetmore, "Science Outside the Lab: Helping Graduate Students in Science and Engineering Understand the Complexities of Science Policy," *Science and Engineering Ethics* 23, no. 3 (2016): 1–22.

22. Erik Fisher, Roop L. Mahajan, and Carl Mitcham, "Midstream Modulation of Technology: Governance from Within," *Bulletin of Science, Technology and Society* 26, no. 6 (2006): 485–496.

23. Erik Fisher and Daan Schuurbiers, "Socio-technical Integration Research: Collaborative Inquiry at the Midstream of Research and Development," in *Early Engagement and New Technologies: Opening Up the Laboratory*, ed. N. Doorn et al. (Dordrecht, Netherlands: Springer, 2013), 97–110.

24. Erik Fisher, "Ethnographic Invention: Probing the Capacity of Laboratory Decisions," *Nanoethics* 1, no. 2 (2007): 155–165; Erik Fisher and Roop L. Mahajan, "Midstream Modulation of Nanotechnology in an Academic Research Laboratory," *ASME 2006 International Mechanical Engineering Congress and Exposition*, 2006, 189–195.

25. Fisher and Mahajan, "Midstream Modulation of Nanotechnology in an Academic Research Laboratory."

26. Fisher and Schuurbiers, "Socio-technical Integration." Cf. Daan Schuurbiers, "What Happens in the Lab: Applying Midstream Modulation to Enhance Critical Reflection in the Laboratory," *Science and Engineering Ethics* 17, no. 4 (2011): 769–788.

27. Erik Fisher, Simon Biggs, Stuart Lindsay, and Jie Zhao, "Research Thrives on Integration of Natural and Social Sciences," *Nature* 463, no. 7284 (2010): 1018; Fisher and Schuurbiers, "Socio-technical Integration"; Steven M. Flipse, Maarten C. A. van der Sanden, and Patricia Osseweijer, "Midstream Modulation in Biotechnology

Industry: Redefining What Is 'Part of the Job' of Researchers in Industry," *Science and Engineering Ethics* 19, no. 3 (2013): 1141–1164; Steven M. Flipse, Maarten C. A. van der Sanden, and Patricia Osseweijer, "Improving Industrial R&D Practices with Social and Ethical Aspects: Aligning Key Performance Indicators with Social and Ethical Aspects in Food Technology R&D," *Technological Forecasting and Social Change* 85 (2014): 185–197; Kaylie McTiernan, Brian Polagye, Erik Fisher, and Kiki Jenkins, "Integrating Socio-Technical Research with Future Visions for Tidal Energy," 2016 Council of Engineering Systems Universities (CESUN) Symposium, George Washington University, 27–29 June 2016; Schuurbiers, "What Happens in the Lab."

28. Schuurbiers, "What Happens in the Lab"; cf. Brian Wynne, "Lab Work Goes Social, and Vice Versa: Strategising Public Engagement Processes," *Science and Engineering Ethics* 17, no. 4 (2011): 791–800.

29. Wiebe Bijker, *Of Bicycles, Bakelites, and Bulbs: Toward a Theory of Sociotechnical Change* (Cambridge, MA: MIT Press, 1995).

30. Ibid. See also Robin Williams and David Edge, "The Social Shaping of Technology," *Research Policy* 25, no. 6 (1996): 865–899.

31. On the difficulties of interdisciplinary teamwork, see McManus and MacDonald (chapter 4) in this volume.

32. C. P. Snow lamented the polarization between scientists and humanists; see *The Two Cultures and the Scientific Revolution* (Cambridge: Cambridge University Press, 1959).

19   Remaking the Innovator Imperative

Matthew Wisnioski, Eric S. Hintz, and Marie Stettler Kleine

Does America need more innovators? We posed the question to highlight how innovation has become a national imperative pursued through the transformation of people. Societal goals such as regional development and international competitiveness take shape through initiatives to make innovators. Contributors have shown that such programs are ubiquitous and pervasive. Innovator initiatives target all age groups and career stages, from kindergarteners to senior scientists. Innovators train in formal and informal educational settings, supported by public and private funding. The innovator imperative operates at all scales, from individual garage inventors to interdisciplinary teams, regional innovation districts, and global federations.

But asking the question implies doubt. It calls attention to the fact that the demand for innovation is at a crossroads. The contributors to this volume join journalists, policy advocates, and scholars in challenging the assumptions and impact of innovator initiatives. They have demonstrated that innovation training programs are historically prone to failure, they have questioned the efficacy of supposedly universal models, they have documented gender and racial disparities across the enterprise, and they have argued that innovation—once a means for solving societal problems—has become an end unto itself.

Finally, we inquired about the need for innovators to open a dialogue about the purpose of innovator initiatives and whom they serve. We assembled champions, critics, and reformers to explore innovation's contradictory dimensions; to engage practitioners directly; and to do so via a reflective approach that treated participants symmetrically. Contributors collectively contextualized the assumptions, goals, practices, and consequences of the demand for innovators. This dialogue fosters opportunities for seeing how the imperative can be remade.

What Drives the Imperative?

The volume's contributors reveal several reasons why the call for innovators enjoys widespread support. Innovator initiatives thrive because they promise to cultivate the skills, mindsets, and human capital needed to address broad societal challenges. How, for example, should future generations of children learn to live and work in a digital age? How can companies, universities, and governments successfully develop new technologies in a global market? How can local communities, regions, and nations achieve cultural growth and economic prosperity?

Innovator initiatives offer reproducible methods to solve these societal challenges across interconnected scales. Programs featured in this book teach *individuals* to acquire change-management skills, to bring an idea to market, or to cultivate a mindset for lifelong creativity. These personal objectives support *organizations* as they seek advantage over competitors or as they enhance opportunities for once-excluded populations. These institutional interventions support *the nation's* reform efforts: they produce millions of young coders, incubate thousands of start-ups, and generate technological breakthroughs that will maintain international competitiveness. These methods provide a sense of empowerment and control. Across all levels, innovation experts contend that with the right people, the rights models, and the right technology, society's thorniest problems can be solved.

Stakeholders with very different motivations pursue innovator initiatives united by a broad vision of innovation as progress achieved through social and technological means. These programs, in turn, generate different ideals of the innovators they seek to produce. The image of innovators as young, cosmopolitan risk-takers first codified in the 1950s is still dominant among many champions of innovation. However, the programs featured in this volume demonstrate a dramatic expansion of who counts as an innovator.[1] Indeed, a key tenet of the imperative is that anyone can innovate.

A small set of powerful institutions, however, underlies the imperative's democratic ethic. A "triple helix" of government agencies, large corporations, and elite research universities provide funding, expertise, and direction to the innovator imperative.[2] As the creator of policy and through granting agencies such as the NSF, the federal government is the major sponsor for most innovator initiatives.[3] High-tech companies such as Google and Microsoft and research universities such as MIT and Stanford also drive the imperative in a symbiotic relationship with the government. Philanthropies such as the Kauffman, Lemelson, and Sloan Foundations are yet another important contributor. These institutions even underwrite the research of innovation's critics and reformers; indeed, nearly all of the contributors to this book have received such support.[4] Pro-innovation organizations do not speak with one voice; however, the hundreds of institutional signatories on the AAAS's 2015 report, "Innovation: An American Imperative," reveal an increasingly shared vision.[5]

Innovator initiatives operate with the urgency and mindset of a social movement. Champions of innovation define their cause in opposition to some unmet social need or untenable situation with the status quo. Initiates recruit others to the cause through personal contact and media campaigns that highlight how individuals can make a difference. They develop a distinct vocabulary and worldview through formal training and rituals.[6] Lastly, they are guided by the faith that they will change the world. This righteous optimism is the innovator imperative's driving strength.

Consequences of the Imperative

The demand for innovation produces energetic students, new technologies, and regional economic growth, but it also generates a series of undesirable consequences.[7] Contributors to this volume demonstrate that the growing critique of innovation coalesces around three overarching concerns: hegemony, inequality, and hubris.

Champions of innovation portray themselves as insurgent outsiders, but innovation is a widely supported ideology with significant cultural and institutional impacts. As described above, innovator initiatives succeed with the support of major corporations, research universities, the federal government, and philanthropic foundations. These interests have helped to disseminate the language, methods, and models of innovation. One consequence of this proliferation is that when innovation-speak describes everything, it can mean nothing. More insidiously, innovation's aura of progress and empowerment frequently is deployed to obscure the free-market ideology of various institutional reforms. For example, research universities now boast of engineering entrepreneurship courses, technology transfer offices, and student-led pitch contests. Similarly, the NSF increasingly has turned from basic science toward technology commercialization. Innovation can exist in all political and economic systems, but the growth of innovator initiatives often reinforces a neoliberal vision of progress that exacerbates inequality.

The innovation economy is hampered by profound gender, racial, and economic disparities.[8] These disparities are rooted in centuries of racism and discrimination but are reflected and structurally reinforced at every step in the personal formation of innovators, from STEM education to unconscious bias and harassment in the workplace. Moreover, job losses due to automation and other innovations are disproportionately borne by women and people of color with lower incomes and education levels. Finally, the allure of economic growth leads localities to invest in innovation and to defer maintenance, a strategy that sometimes creates middle class jobs but can leave poorer communities saddled with failing infrastructures.

Champions of innovation display several varieties of hubris. First, they engage in technological solutionism, a naive optimism in the power of innovation to solve any problem.[9] Second, innovation experts assume the efficacy and replicability of their toolkits and recipes, from the "MIT model" to the LUMA Institute's human-centered design methods; however, experience shows that these "best practices" are not universally applicable. Third, in their zeal for disruption, innovators' can become overconfident and forget that innovation is an inherently risky enterprise that often ends in bankruptcy, layoffs, and failure.

As critics of innovation level these judgments, they sometimes adopt a negative, polemical stance that matches the optimism and fervor of innovation's most evangelical champions. Critics risk their own hubris in dismissing the motives of innovator initiatives and the efficacy of methods such as human-centered design. Similarly, by characterizing all pro-innovation initiatives as fraudulent or as tools of neoliberalism, they ignore the complex motivations that draw people to innovator programs and the outcomes they produce. Nevertheless, oppositional critique is valuable and necessary because it has the potential to reshape the innovation enterprise.

Remaking the Imperative

Can the tensions underlying the innovator imperative be reconciled? Society benefits from citizens who have the confidence, expertise, and acumen to generate beneficial technologies, challenge outdated dogmas, and contribute to economic growth. But innovation is not a panacea. At its worst, the innovator imperative perpetuates racial and gender inequality, misallocates resources, and produces arrogant, irresponsible innovators.

Reformist contributors to this volume present strategies for engaging the trade-offs inherent in the innovator imperative, which we have characterized as critical participation.[10] These reformers integrate scholarly critique with reflective practice to intervene in the training of innovators. Critical participants value complexity and promote questioning in order to combat issues of power, inequality, and hubris. Their goal is to help would-be innovators avoid blindly pursuing innovation as an end in itself; rather, they ask for whom, and to what ends, innovation is deployed. This process of questioning, critique, and iterative reform encourages humility through an appreciation of others' values, because no one person remains the expert throughout the multiple stages of reform.

Critical participation thus begins with a recognition of the complex and even contradictory motivations that attract people to innovation programs. These include awareness of the dialectical relationship between innovation and maintenance, and the complexities in balancing personal empowerment with community support and care. Natalie Rusk's work with the Computer Clubhouse and Scratch is a particularly striking case. Even as it provides individual children with coding skills that MIT, Microsoft, and the NSF demand, it creates child-centered peer communities motivated by self-expression, empathy, and collaboration.

Critical participation involves sustained personal engagement with stakeholders to address the shortcomings of the innovator imperative. For example, to reduce gender and racial inequality in the IT industry, NCWIT collaborates with firms such as Intel and Facebook to implement inclusive practices, such as unconscious bias training, equitable recruiting practices, and mentoring programs. Similarly, social scientists from Arizona State University work side by side with scientists and engineers to encourage the ethical implementation of emerging technologies. These reformers recognize the importance of personal engagement, because sustained social change only occurs when they can be held accountable by their collaborators.

Critical participation involves discomfort and risk. It requires honest reflection about the motivations and moral commitments of one's work. It requires working alongside people with different backgrounds, motives, and values. Those who take on the challenge face internal crises of identity and external hostility. Practitioners may dismiss ethical reflection as a waste of time, while scholars may label critical participants as sell-outs. Critical participants supported by pro-innovation institutions must also constantly guard against conflicts of interest, capture, and self-censorship. Finally, reformers must be vigilant against regarding "reflection in action" as its own form of solutionism.

We believe that the engagement is worth the risk. This volume has been a critical intervention in the innovator imperative. The outcome is not a cookbook but a set of insights into how the imperative operates, its beneficial and problematic attributes, and how it might be reformed.

Possible Futures

The competing perspectives captured in this volume suggest multiple possible directions for the innovator imperative. In one scenario, a national movement of innovators vanquishes society's "wicked problems" and reduces critics to mere naysayers. Another outcome finds the imperative on its last legs, discredited and irredeemable as governments, universities, and corporations reorganize according to alternative social values. Both of these scenarios seem equally implausible. Innovation-speak likely will wane in the face of emerging critiques. However, the systematic pursuit of innovation is unlikely to subside anytime soon. Educators, legislators, and advocates will continue to call for more innovators, and many of the initiatives presented in this book will grow domestically and abroad.

The contributors to this volume also demonstrate how the innovator imperative already is being remade. As the imperative evolves, it is incumbent on those who cultivate innovators to do so with critical reflection. Through willing exposure to criticism, the leaders of innovator programs and the individuals they mentor can see the limitations of pro-innovation rhetoric and practices. Similarly, we hope that innovation's critics can acknowledge the social needs, progressive desires, and daily challenges of those who educate would-be innovators.

Does America need more innovators? Only if pursued in the service of a different kind of imperative—one that reveals to would-be innovators the assumptions and powers that shape their futures; one that demands the equal valuing of those who care for existing cultures and infrastructures with those who build new things; one that trains scientists, engineers, and entrepreneurs to anticipate the implications of their innovations; and one that cultivates technologists who approach their work with humility in addition to optimism. We hope those who train and deploy the next generation of innovators heed this imperative. The people they seek to transform depend on it.

Notes

1. There are many different "imaginaries" among innovation's champions, and even starker differences when we look across the spectrum of critics and reformers. In this volume's introduction, we described the work of Everett Rogers, who in the 1950s first characterized innovators as young, cosmopolitan risk-takers. That image remains dominant among many of innovation's champions, including, for example, MAYA's expert design consultants (chapter 4). However, the White House's Jenn Gustetic imagines sixth-graders who search for "debris disks" in space telescope images (chapter 7). Benoît Godin's innovator (chapter 9) is a sixteenth-century religious heretic, while Andrew L. Russell and Lee Vinsel (chapter 13) envision a neoliberal huckster. Natalie Rusk's innovator (chapter 15) is an underprivileged ten-year old, while NCWIT (chapter 17) envisions a woman denied access to a career in IT. On "sociotechnical imaginaries," see Sheila Jasanoff and Sebastian Pfotenhauer, "Panacea or Diagnosis? Imaginaries of Innovation and the 'MIT Model' in Three Political Cultures," *Social Studies of Science* 47, no. 6 (2017): 783–810.

2. Henry Etzkowitz and Chunyan Zhou, *The Triple Helix: University-Industry-Government Innovation and Entrepreneurship* (New York: Routledge, 2008).

3. Mariana Mazzucato, *The Entrepreneurial State: Debunking Public vs. Private Sector Myths* (London: Anthem Press, 2013).

4. For example, all three editors and fourteen of the nineteen contributors have administered or benefited from NSF grants: Fasihuddin, Britos Cavagnaro, Arkilic, Feldman, Pfotenhauer, Cook, Vinsel, Rusk, Carlson, Ashcraft, Sanders, Fisher, Guston, and Trinidad. The volume itself is partially funded by the NSF and the Smithsonian's Lemelson Center, which in turn, is supported by the Lemelson Foundation. We remind the NSF that the project is linked to the broader impacts of NSF award no. 1354121. Any opinions, findings, conclusions, or recommendations belong to the editors and individual contributors, and do not necessarily reflect the views of the NSF.

5. "Innovation: An American Imperative," American Academy of Arts and Sciences, 23 June 2015, http://www.amacad.org/content/innovationimperative/.

6. The NSF I-Corps program, for example, formally recruits previous NSF grantees to apply for the program (chapter 5), while faculty advisors initiate new University

Innovation Fellows during an official and individual induction "pinning ceremony" (chapter 3).

7. The classic example of the exploration of these consequences is Everett M. Rogers, *Diffusion of Innovations*, 5th ed. (New York: Free Press, 2003), 436–472.

8. Several initiatives profess explicit diversity goals, including OSTP's aspiration that the "nation's STEM graduates reflect the full diversity of America" and UIF's emphasis on "strength in human diversity." However, the dismal diversity statistics in the STEM sectors suggest that these and other efforts have been slow to make change.

9. For a critique, see Evgeny Morozov, *To Save Everything, Click Here: The Folly of Technological Solutionism* (New York: PublicAffairs, 2013).

10. For a discussion of the meaning and origins of critical participation see Wisnioski (chapter 1) in this volume.

Contributors

Errol Arkilic leads M34 Capital, a private investment firm that focuses on seed-stage investments in companies being spun out of academic and corporate research labs. Previously, he was the founding lead program director of the National Science Foundation I-Corps. Prior to his government service, he was founder and CEO of Strata-Gent Life Sciences. He has a PhD in Aero/Astronautics from MIT.

Catherine Ashcraft is director of research and a senior research scientist with the National Center for Women and Information Technology (NCWIT) at the University of Colorado, Boulder. Her research focuses on gender, diversity, and technology; organizational change and curriculum reform; and media representations and youth identity. She has an MA in organizational communication and a PhD in education from the University of Colorado.

Leticia Britos Cavagnaro is cofounder and codirector of the University Innovation Fellows program, an initiative of the Hasso Plattner Institute of Design (d.school) at Stanford University. She is also an adjunct professor at the d.school and was the deputy director of the NSF-funded National Center for Engineering Pathways to Innovation (Epicenter). She received a PhD developmental biology from Stanford's school of medicine.

W. Bernard Carlson is the Vaughan Professor of Humanities and chair of the Department of Engineering and Society, with a joint appointment in the Corcoran Department of History at the University of Virginia. He is the author of several books on the history of technology and entrepreneurship, including *Tesla: Inventor of the Electrical Age* (Princeton University Press, 2013). Carlson holds an AB from College of the Holy Cross and an MA and PhD from the University of Pennsylvania.

Lisa D. Cook is an associate professor of economics and international relations at Michigan State University. Her research interests include economic growth and development, financial institutions and markets, innovation, and economic history. She served as president of the National Economic Association from 2015 to 2016 and currently serves as codirector of the American Economic Association Summer Training Program. She holds a PhD from the University of California, Berkeley.

Humera Fasihuddin is cofounder and codirector of the University Innovation Fellows program, an initiative of the Hasso Plattner Institute of Design (d.school) at Stanford University. She founded the program after a decade of experience at a nonprofit where she helped expand entrepreneurship and venture creation in academia. Her skills in building innovation movements come from seven years in industry and four years in economic development. She began her career with a liberal arts education at Smith College (mathematics, minor in economics) and expanded her interests in innovation at UMass Amherst (MBA).

Maryann Feldman is the Heninger Distinguished Professor in the Department of Public Policy at the University of North Carolina. Her research and teaching interests focus on innovation, technological change, and the commercialization of academic research. The author of *The Geography of Innovation* (Springer, 1994), her recent work explores emerging industries, entrepreneurship, and the process of regional transformation.

Erik Fisher is an associate professor in the School for the Future of Innovation in Society and the Consortium for Science, Policy, and Outcomes at Arizona State University. He also serves as the chair of the Human and Social Dimensions of Science and Technology PhD program. He is the editor-in-chief of the *Journal of Responsible Innovation*. Fisher studies the governance of emerging technologies from "lab to legislature." He holds a PhD in environmental studies and an MA in classics from the University of Colorado, and a BA in liberal arts from St. John's College.

Benoît Godin is a professor of science studies at the Institut national de la recherche scientifique in Montreal. His projects on the history of science and technology statistics and the intellectual history of innovation have produced many publications, including his books *Innovation Contested: The Idea of Innovation over the Centuries* (Routledge, 2015) and *Models of Innovation: The History of an Idea* (MIT Press, 2017). Godin holds a PhD in science studies from the University of Sussex.

Jenn Gustetic is the Small Business Innovation Research (SBIR/STTR) program executive at NASA and former assistant director of open innovation at the White House Office of Science and Technology Policy. She has worked with several governmental agencies to scale open innovation and entrepreneurial engagement as means of enrolling America's innovators in technically innovative projects. Gustetic holds a BS in aerospace engineering from the University of Florida and an MS in technology policy from MIT.

David Guston is Foundation Professor and founding director of the School for the Future of Innovation in Society at Arizona State University, where he is also codirector of the Consortium for Science, Policy, and Outcomes. He is a fellow of the American Association for the Advancement of Science and holds a BA from Yale and a PhD from MIT.

Eric S. Hintz is a historian with the Lemelson Center for the Study of Invention and Innovation at the Smithsonian's National Museum of American History. He specializes in the history of invention and is currently working on a book that considers the changing fortunes of American independent inventors from 1900 to 1950, an era of expanding corporate R&D. He holds a BS in aerospace engineering from the University of Notre Dame and an MA and a PhD in the history and sociology of science from the University of Pennsylvania.

Marie Stettler Kleine is a doctoral candidate in Virginia Tech's Department of Science, Technology, and Society. Her research interests include the intersection of the social study of engineering, international development, and religion. She received a graduate certificate in the Interdisciplinary Graduate Education Program in Human Centered Design. She holds a BS in mechanical engineering and international studies from Rose-Hulman Institute of Technology and an MS in science and technology studies from Virginia Tech.

Dutch MacDonald is the director of BCG Platinion|MAYA Design, a digital innovation and engineering studio. Formerly, he was the president and CEO of MAYA, which was acquired by the Boston Consulting Group in 2017. He is a past president of the Pittsburgh chapter of the American Institute of Architects and serves on the board of the Carnegie Museum of Art. He holds a BArch from Carnegie Mellon University and has also studied at the École Polytechnique Fédérale de Lausanne and the University of Pittsburgh.

Mickey McManus is a senior advisor at BCG and is a visiting research fellow at Autodesk. A pioneer in pervasive computing, collaborative human/machine innovation, human-centered design, and education, he holds ten patents in the area of connected products, vehicles, and services. While chairman of MAYA he coauthored the award-winning book on the internet of things and beyond called *Trillions* (Wiley, 2012) and is currently working on his next book, code-named *Primordial: When Things Wake Up*.

Sebastian Pfotenhauer is an assistant professor for innovation research in the Munich Center for Technology in Society at the Technische Universität München. His research interests include innovation theory; national, regional, and institutional innovation strategies; and the governance of complex sociotechnical systems. He holds an MS in technology and policy from MIT and a PhD in physics from the Friedrich-Schiller-Universität in Jena, Germany.

Natalie Rusk researches and develops new technologies for learning at the MIT Media Lab. She is one of the creators of the Scratch programming language, which young people around the world use to code and share interactive animations, stories, and games. She initiated the Computer Clubhouse, which provides creative learning opportunities for youth in more than ninety community centers in twenty countries. She earned a PhD in child development at Tufts and an EdM from the Harvard Graduate School of Education.

Andrew L. Russell is a professor of history and dean of the College of Arts and Sciences at SUNY Polytechnic Institute in Utica, New York. He is the author of *Open Standards and the Digital Age: History, Ideology, and Networks* (Cambridge University Press, 2014) and coeditor (with Robin Hammerman) of *Ada's Legacy: Cultures of Computing from the Victorian to the Digital Age* (ACM/Morgan & Claypool, 2015). With Lee Vinsel, he is a founder of the Maintainers, a global, interdisciplinary network of scholars and practitioners with interests in maintenance, infrastructure, repair, and the myriad forms of labor and expertise that sustain our human-built world.

Lucinda M. Sanders is cofounder and CEO of the National Center for Women and Information Technology (NCWIT), housed at the University of Colorado, Boulder. She held executive positions at AT&T/Bell Labs for over twenty years and was awarded the Bell Labs Fellow Award. She is a recipient of the 2013 U.S. News STEM Leadership Hall of Fame Award and holds six patents related to communications technology.

Brenda Trinidad is a doctoral candidate in human and social dimensions of science and technology at Arizona State University and a research assistant at the Center for Nanotechnology in Society. She has written on space tourism for Arizona State's Consortium for Science, Policy, and Outcomes and for *Slate.com.*

Lee Vinsel is an assistant professor of science, technology, and society at Virginia Tech. His work centers on the relationship between governance and technological change. His book manuscript *Taming the American Idol: Cars, Risks, and Regulations*, examines the history of auto regulation from the birth of the industry to the present. With Andrew L. Russell, he is a founder of the Maintainers, a global, interdisciplinary network of scholars and practitioners with interests in maintenance, infrastructure, repair, and the myriad forms of labor and expertise that sustain our human-built world. He earned a PhD in history from Carnegie Mellon University and held a postdoctoral fellowship at the Harvard Kennedy School.

Matthew Wisnioski is an associate professor of science, technology, and society and a senior fellow of the Institute for Creativity, Arts, and Technology at Virginia Tech. He is the author of *Engineers for Change: Competing Visions of Technology in 1960s America* (MIT Press, 2012). He is at work on a book titled *Every American an Innovator* that documents the rise of today's ubiquitous culture of innovation. He is also a critical participant in the NSF's Revolutionizing Engineering and Computer Science Department initiative at Virginia Tech. He received an MS in materials science and engineering from Johns Hopkins University and a PhD in history from Princeton University.

Index

Note: Figures and tables are indicated by "f" and "t" respectively, following page numbers.